T0206043

By measuring the direction and intensity of magnetism in rocks of different ages, a record of the Earth's magnetic field in the past can be obtained. This book deals with the particular case of reversals of the Earth's magnetic field. These have played a major role in the development of plate tectonics and in establishing a geological timescale. The magnetism of rocks is discussed in some detail, with a warning of possible misinterpretations of the record. The latest observational results and theories are reviewed, with special attention to the structure and geometry of the transition field.

Changing conditions at the core–mantle boundary, their effect on reversals, the generation of plumes and the possible correlation of reversals with tectonic changes, ice ages or mass extinctions are thoroughly discussed, including suggested periodicities in the reversal record and in other geophysical data.

Reversals of the Earth's Magnetic Field

Reversals of the Earth's Magnetic Field

Second Edition

J. A. Jacobs
Institute of Earth Studies,
University of Wales, Aberystwyth

CAMBRIDGE
UNIVERSITY PRESS

CAMBRIDGE UNIVERSITY PRESS
Cambridge, New York, Melbourne, Madrid, Cape Town, Singapore, São Paulo

Cambridge University Press
The Edinburgh Building, Cambridge CB2 2RU, UK

Published in the United States of America by Cambridge University Press, New York

www.cambridge.org
Information on this title: www.cambridge.org/9780521450720

First published by Adam Hilger Ltd 1984
Second edition published by Cambridge University Press 1994
This digitally printed first paperback version 2005

A catalogue record for this publication is available from the British Library

Library of Congress Cataloguing in Publication data
Jacobs, J. A. (John Arthur), 1916–
Reversals of earth's magnetic field / J. A. Jacobs – 2nd ed.
   p.   cm.
Includes bibliographical references and index.
ISBN 0 521 45072 1
1. Geomagnetism.   I. Title.
QC815.2.J33   1994
538′.72 – dc20   93-50683 CIP

ISBN-13 978-0-521-45072-0 hardback
ISBN-10 0-521-45072-1 hardback

ISBN-13 978-0-521-67556-7 paperback
ISBN-10 0-521-67556-1 paperback

Contents

Preface

This book was finished in June 1993, almost 11 years since the first edition of *Reversals of the Earth's Magnetic Field* was finished (published by Adam Hilger Ltd 1984). It has often been said that the typical doubling period for the accumulation of scientific knowledge during the last two centuries is about 15 years. This is certainly true for reversals of the Earth's magnetic field if we substitute data and theoretical modelling for scientific knowledge. We have accumulated a vastly increased amount of data, most of it with much greater precision. Our knowledge of the physics of the geodynamo has also been greatly expanded, but we still do not know the detailed mechanism of the generation of the field, and even less about the reversal process.

This book is an attempt to summarize the most important advances that have been made in the last decade. The general layout of the first edition has been preserved. The first chapter is a brief overview of the Earth's magnetic field and the second chapter discusses the magnetization of rocks. This chapter has been expanded and discusses some of the problems that have now been highlighted in the acquisition of natural remanent magnetization. Chapter 3 discusses in general terms the morphology of geomagnetic reversals and gives some of the early development of the subject. These three chapters lay the background for the next three chapters, which form the heart of the book. Chapter 4 deals with excursions of the magnetic field and chapter 5 with models and possible reversal mechanisms. Chapter 6 is a new chapter on transition fields, a topic that has attracted much attention in recent years and is still highly controversial. Chapter 7 (the old Chapter 6) deals with magnetostratigraphy and gives a much more detailed account of how a magnetic polarity timescale is constructed. Chapter 8 (the old Chapter 7) discusses the controversial question of possible correlations of the Earth's magnetic field with near-surface phenomena – climate, mass extinctions, tectonics, mantle plumes. As noted before, this is a highly speculative area, but nevertheless one of increasing popular appeal.

J. A. Jacobs

1
The Earth's magnetic field

1.1 Introduction

In geomagnetism we are measuring extremely small magnetic fields – at its strong-est near the poles, the Earth's magnetic field is several hundred times weaker than that between the poles of a toy horseshoe magnet. In a magnetic compass, the needle is weighted so that it will swing in a horizontal plane, its deviation from geographical north being called the declination, D. A non-magnetic needle which is balanced horizontally on a pivot becomes inclined to the vertical when magnetized. Over most of the northern hemisphere the north-seeking end of the needle will dip downwards, the angle it makes with the horizontal being called the magnetic dip or inclination, I. The total intensity F, the declination D and the inclination I completely define the magnetic field at any point. The horizontal and vertical components of F are denoted by H and Z. H may be further resolved into two components X and Y, X being the component along the geographical meridian (northward) and Y the orthogonal component (eastward). Figure 1.1 illustrates these different magnetic elements. They are simply related to one another by the following equations:

$$H = F \cos I \qquad Z = F \sin I \qquad \tan I = Z/H \qquad (1.1)$$

$$X = H \cos D \qquad Y = H \sin D \qquad \tan D = Y/X \qquad (1.2)$$

$$F^2 = H^2 + Z^2 = X^2 + Y^2 + Z^2 \qquad (1.3)$$

The variation of the magnetic field over the Earth's surface is best illustrated by isomagnetic charts, i.e. maps on which lines are drawn through points at which a given magnetic element has the same value. Contours of equal intensity in any of the elements X, Y, Z, H or F are called isodynamics. Figures 1.2–1.4 are world maps showing contours of equal declination (isogonics), equal inclination (isoclinics) and total intensity for the year 1990. Palaeomagnetists have tradition-ally used the oersted as the unit of magnetic field strength and the gauss (Γ) as the unit of magnetic induction. The distinction is somewhat pedantic in geophysical applications, since the permeability of air is virtually unity in cgs units. In SI units, which will be used throughout this book,

$$1 \ \Gamma = 10^{-4} \ \text{Wb m}^{-2} \ (\text{weber/m}^2) = 10^{-4} \ \text{T (tesla)}$$

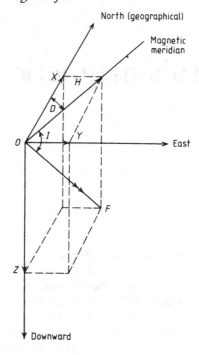

Figure 1.1

IGRF Declination 1990.0

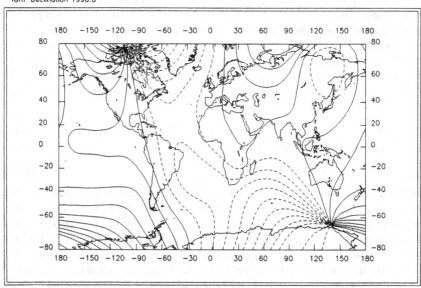

Figure 1.2 World map showing contours of equal declination *D* for 1990. The zero and positive contours are shown as solid lines and the negative contours as dashed lines. Contours are shown for the following values: −150°, −120°, −90° to +90° with a contour interval of 10°, +120° and +150°. Provided by D. R. Barraclough. Courtesy of the Geomagnetism Unit, British Geological Survey.

IGRF Inclination 1990.0

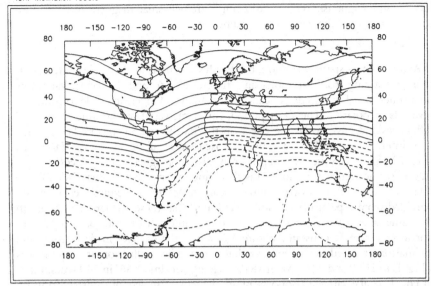

Figure 1.3 World map showing contours of equal inclination *I* for 1990. The zero and positive contours are shown as solid lines and the negative contours as dashed lines. The contours are from −80° to +80° with a contour interval of 10°. Provided by D. R. Barraclough. Courtesy of the Geomagnetism Unit, British Geological Survey.

IGRF Total intensity 1990.0

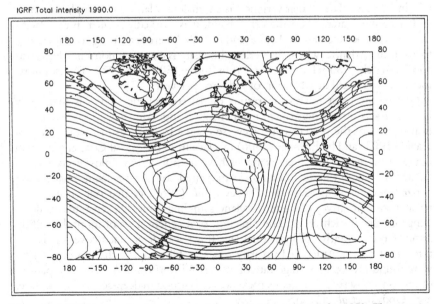

Figure 1.4 World map showing contours of equal total intensity *F* for 1990. The contour interval for *F* is 200 nT and the contours range from 24 000 nT, centred over southern Brazil, to 66 000 nT, centred approximately on the southern dip-pole south of Australia. Provided by D. R. Barraclough. Courtesy of the Geomagnetism Unit, British Geological Survey.

Since in geomagnetism we are measuring extremely small magnetic fields, a more convenient unit is the gamma (γ), defined as

$$1 \gamma = 10^{-9} \, T = 1 \, nT$$

Apart from its spatial variation, the Earth's magnetic field also shows temporal changes ranging from variations on a timescale of seconds to secular variations on a timescale of hundreds of years and on an even longer timescale to complete reversals of polarity. The short-period, transient, variations are due to external influences and have no lasting effect on the Earth's main magnetic field, which is of internal origin. They will not be discussed at all in this book. Variations over $10-10^4$ a may be determined from archaeomagnetic and palaeomagnetic studies of the secular variation. This time range is probably characteristic of core fluid motions. If successive annual mean values of a magnetic element are obtained from a particular station, it is found that these secular changes are in the same sense over a long period of time, although the rate of change does not usually remain constant. Figure 1.5 shows the changes in declination and inclination at London, Boston and Baltimore. The declination at London was 11½° E in 1580 and 24½° W in 1819, a change of almost 36° in 240 years. Lines of equal secular change (isopors) in an element form sets of ovals centring on points of local maximum change (isoporic foci). Figures 1.6 and 1.7 show the secular change in Z for the years 1922.5 and 1942.5. It can be seen that the secular variation is a regional rather than a planetary phenomenon and that considerable changes can take place in the general distribution of isopors even within 20 years. The secular variation is anomalously large and complicated over and around Antarctica; on the other hand, it is markedly smaller in the Pacific hemisphere (between about 120° E and 80° W). There is a strong secular change focus in the Atlantic, where the vertical intensity is changing non-linearly. Z was approximately $-50 \, nT \, a^{-1}$ in the 1960s and $-150 \, nT$ in 1978. The isophoric foci drift westward at a fraction of a degree per year.

One of the most interesting results of palaeomagnetic studies is that many igneous rocks show a permanent magnetization approximately opposite in direction to that of the present field. Reverse magnetization was first discovered by David (1904) and by Brunhes (1906) in a lava from the Massif Central mountain range in France – since then examples have been found in almost every part of the world. About one-half of all rocks measured are found to be normally magnetized and one-half reversely. Dagley et al. (1967) carried out an extensive palaeomagnetic survey of Eastern Iceland sampling some 900 separate lava flows lying on top of each other. The direction of magnetization of more than 2000 samples representative of individual lava flows was determined covering a time interval of 20 Ma. At least 61 polarity zones, or 60 complete changes of polarity, were found giving an average rate of at least 3 inversions/Ma. The same pattern of reversals observed in igneous rocks was also found in deep-sea sediments (see, e.g., Opdyke et al., 1966).

In 1839 Gauss showed that the field of a uniformly magnetized sphere, which is the same as that of a dipole at its centre, is an excellent first approximation to the Earth's magnetic field. Gauss further analysed the irregular part of the Earth's

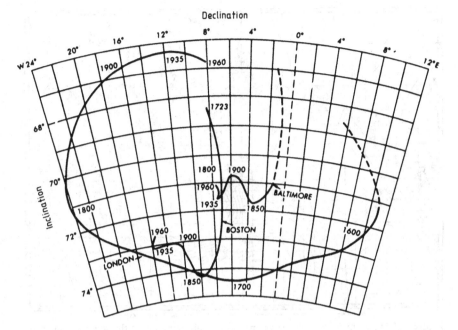

Figure 1.5 Secular change of declination and inclination at London, Boston and Baltimore. After Nelson *et al.* (1962).

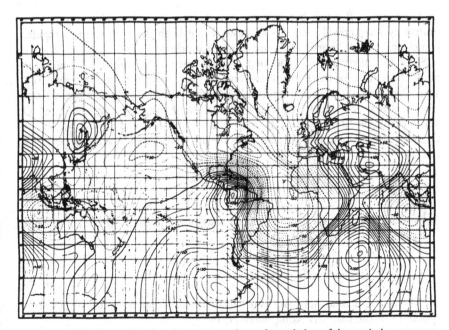

Figure 1.6 World map showing the geomagnetic secular variation of the vertical component Z. Epoch 1922.5. After Vestine *et al.* (1947a).

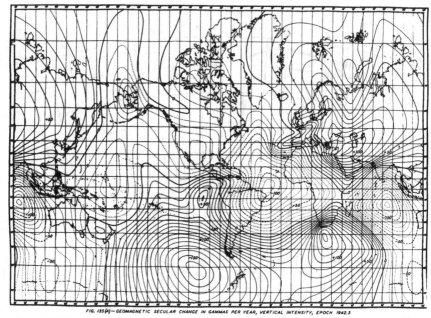

FIG. 135(A)—GEOMAGNETIC SECULAR CHANGE IN GAMMAS PER YEAR, VERTICAL INTENSITY, EPOCH 1942.5

Figure 1.7 World map showing the geomagnetic secular variation of the vertical component *Z*. Epoch 1942.5. After Vestine *et al.* (1947a).

field, i.e. the difference between the actual observed field and that due to a uniformly magnetized sphere, and showed that both the regular and irregular components of the Earth's field are of internal origin.

Since the north-seeking end of a compass needle is attracted towards the northern regions of the Earth, those regions must have opposite polarity. Consider therefore the field of a uniformly magnetized sphere whose magnetic axis runs north–south, and let *P* be any external point distant *r* from the centre *O* and *θ* the angle *NOP*, i.e. *θ* is the magnetic co-latitude (see Figure 1.8). If *m* is the magnetic moment of a geocentric dipole directed along the axis, the potential at *P* is

$$V = \frac{m}{4\pi} \frac{\cos \theta}{r^2} \qquad (1.4)$$

The inward radial component of force corresponding to the magnetic component *Z* is given by

$$Z = \mu_0 \frac{\partial V}{\partial r} = \frac{-\mu_0 m}{2\pi} \frac{\cos \theta}{r^3} \qquad (1.5)$$

and the component at right angles to *OP* in the direction of decreasing *θ*, corresponding to the magnetic component *H*, by

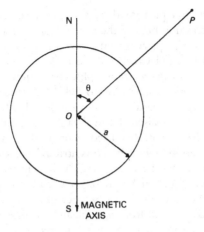

Figure 1.8

$$H = \mu_0 \frac{1}{r} \frac{\partial V}{\partial \theta} = \frac{-\mu_0 m}{4\pi} \frac{\sin \theta}{r^3} \tag{1.6}$$

where μ_0 is the permeability of free space.

The inclination I is then given by

$$\tan I = Z/H = 2 \cot \theta \tag{1.7}$$

and the magnitude of the total force F by

$$F = (H^2 + Z^2)^{1/2} = \frac{\mu_0 m}{4\pi r^3} (1 + 3 \cos^2 \theta)^{1/2} \tag{1.8}$$

Thus intensity measurements are a function of latitude.

The geomagnetic poles, i.e. the points where the axis of the geocentric dipole which best approximates the Earth's field meets the surface of the Earth, are situated at approximately 79° N, 70° W and 79° S, 110° E. The geomagnetic axis is thus inclined at about 11° to the Earth's geographical axis.

Studies of the secular variation in historic times (HSV) show that the dipole intensity has continued to drop over the last 150 a, the rate of decrease increasing since about 1965. This appears to be accompanied by an increase in the non-dipole field near the core–mantle boundary (CMB), implying that there is no observable energy-density loss in the core field. The possible significance of this will be discussed later (Section 5.8) when we consider reversal models. The westward drift of the non-dipole field also appears to have been decreasing over the last 100 a. Newitt and Dawson (1984) find no evidence for it over North America for the last 100–200 a and Kalinin and Rozanova (1983) find significant regional differences in the westward drift rate.

Data on palaeomagnetic secular variation (PSV) can be obtained from archaeological kilns and fireplaces, from lava flows and from sequences of lake sediments. The longest records are from sedimentary sequences, mainly lacustrine, but some from marine cores – see Figure 1.9, which shows type curves of the secular variation obtained from lake sediments from east-central North America. This type of record can in principle provide continuous detailed data, but suffers from low resolution due to smoothing of the signal. Hanna and Verosub (1989) have given a review of lacustrine palaeomagnetic records from western North America covering the past 40 ka and Creer and Tucholka (1983) an overview of type PSV curves suitable for magnetostratigraphic correlation and dating. Particularly interesting is the PSV record from Mono Lake, California (Lund et al., 1988), which shows a distinctive periodic vector waveform that follows an excursion of the magnetic field – this is discussed in detail in Section 4.4. Levi and Karlin (1989) obtained a 60 ka record from sediments in the Gulf of California which show recurring fluctuations about the geocentric axial dipole value. They suggest that these fluctuations may be related to a generally reduced dipole moment between about 20 and 5 ka which may be connected with possible geomagnetic excursions at 51–49 ka and 29–26 ka (see Sections 4.2–4.4).

Comparison of PSV records from Europe and North America show that PSV waveforms with periods longer than a few hundred years can be correlated over ~ 3000 km. Sproul and Banerjee (1989) noted that the Elk Lake Minnesota PSV record shows striking similarity to European archaeomagnetic records (Kovacheva, 1980) if the Elk Lake record is offset by 520 a corresponding to a westward drift rate of 0.23°/a. Distinctive PSV waveforms which reappear about every 2400–3000 a have been seen at some sites (e.g. Lake St Croix, Mono Lake). Lund and Banerjee (1975) have suggested that this is evidence of westward (or eastward) drift of a complex non-dipole waveform which changes very slowly with time compared with the time it takes for the waveform to drift entirely round the Earth (2400–3000 a).

That the secular variation of the present-day geomagnetic field is lower in the Pacific Ocean region than over the rest of the world has been known for some time (see, e.g., Vestine and Kahle, 1966; Doell and Cox, 1971; Bingham and Stone, 1972). Early palaeomagnetic studies of Brunhes age Hawaiian lava flows ($\leq$ 0.7 Ma) suggested that a low non-dipole field in the Pacific region was not just a recent feature (Doell and Cox, 1971; Doell, 1972). Studies of other areas of the Pacific (the Galapagos Islands – Doell and Cox, 1972; Easter Island – Isaacson and Heinrichs, 1976; and the Society Islands – Duncan, 1975) revealed no anomalously low secular variation thus constraining the region of a low non-dipole field. Because the eruption of lava flows is episodic and dating is not very precise, some authors have suggested that the low non-dipole field in Hawaii is either a very recent phenomenon or the result of incomplete sampling (McElhinny and Merrill, 1975; Coe et al., 1978). More recently McWilliams et al. (1982), using ^{14}C-dated lava flows in Hawaii, concluded that lower secular variation is a real feature of the Central Pacific, although their data come mainly from younger samples (the oldest is 31 ka). Their conclusion is reinforced by Peng and King (1992), who obtained a palaeomagnetic record for the past 13 ka from sediments in Lake Waiau near the summit of Mauna Kea, Hawaii. Their data are similar to those obtained from lava flows (Holcomb et al., 1986), but are of much higher resolution. It seems that anomalously low secular variation has occurred beneath the central Pacific area on timescales of 10^4 a.

The determination of the palaeointensity of the geomagnetic field is more

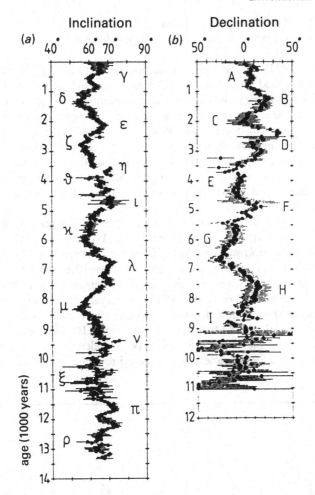

Figure 1.9 Type curves of the secular variation showing (a) inclination and (b) declination for east-central North America (Great Lakes – Minnesota region) obtained by stacking individual records at 40-year increments. The bars represent standard errors stacked at each level. The Greek and English labels identify the principal maxima and minima which were used for between lake correlation. After Creer and Tucholka (1982).

difficult to establish than its palaeodirection, since the intensity of the magnetic remanence is not only dependent on the intensity of the field but is also strongly related to the material. Chauvin *et al.* (1991) have obtained palaeointensity measurements using the Thellier method for two late Quaternary volcanic sequences on Réunion Island in the southern hemisphere (21° S). Values of the strength of the palaeofield varied from 19 to 55 μT. For significantly older rocks (0.6–2 Ma) from Réunion Island, Senanayake *et al.* (1982) had obtained field strengths in the range 19.4–40.2 μT, which is not substantially different from the late Quaternary values. Chauvin *et al.* found that for the youngest sequence

(5–12 ka) VDMs were mostly in the range 7.5×10^{22} to 9.9×10^{22} A m², but for the oldest sequence (82–98 ka) the range was from 4.1×10^{22} to 8.8×10^{22} A m².

Tric *et al.* (1992) obtained high-resolution records of the palaeointensity of the geomagnetic field for the past 80 ka from five marine cores – three from the Tyrrhenian Sea, one from the eastern Mediterranean and one from the southern Indian Ocean. The results correlate well with archaeomagnetic data before 10 ka and with volcanic data for the last 40 ka (see Figure 1.10). The dipole field moment shows large-scale changes – it fell to 22 and 28 % of its present value 39 and 60 ka ago, respectively. These lows alternated with periods of higher intensity. There is some indication of a periodic nature in these intensity variations, but the record is not long enough to allow of any conclusion being reached.

Meynadier *et al.* (1992) obtained relative magnetic field intensities for the last 140 ka from three marine cores in the Somali basin, Western Indian Ocean (a time versus depth correlation was established from the $\delta^{18}O$ record). The remanence intensity was normalized with respect to the anhysteretic remanent magnetization. The quasi-cyclic pattern for the past 80 ka confirms the results obtained in the Mediterranean by Tric *et al.* (1992) – see Figure 1.11. A power spectrum analysis indicated two dominant peaks centred at 100 ka and 22–25 ka and two smaller peaks at 19 and 43 ka. The longer period (100 ka) reflects the large parabolic aspect of the curve and cannot be considered for the limited time interval studied – a longer time series is needed to come to any conclusions on

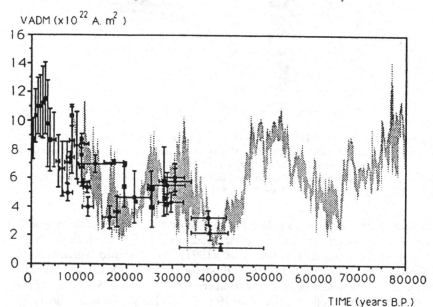

Figure 1.10 Comparison of the intensity of the geomagnetic field between the sedimentary records and archaeomagnetic and volcanic data for the period 0–40 ka. Data from the sedimentary records are shown hachured and extend to 80 ka. After Tric *et al.* (1992).

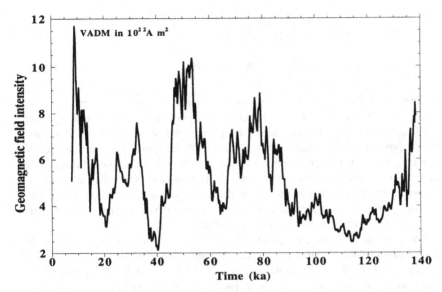

Figure 1.11 Geomagnetic intensity calculated from the stack of results in the Mediterranean and Indian Ocean. After Meynadier *et al.* (1992).

the significance of these spectral peaks. It is interesting, however, that these values correspond to the frequencies of the Earth's orbital parameters – see Section 8.2. Figure 1.11 also shows a broad change in the field intensity to lower values between 140 and 80 ka, reaching a minimum value of one-third of that of the present field, 115 ka ago. It is interesting that this coincides with an excursion of the Earth's magnetic field – the Blake event (see Section 4.6). Low field intensity was also observed during the Laschamp event, about 40 ka ago (Roperch *et al.*, 1988) – see Section 4.2.

Small-scale marine anomalies (25–100 nT amplitude, 8–25 km wavelength) have been observed on the more generally recognized seafloor spreading pattern. There has been much debate as to whether these 'tiny wiggles' are short polarity intervals or palaeointensity fluctuations. By normalizing and stacking multiple profiles, Cande and Kent (1992) have shown that this pattern of tiny wiggles is a high-resolution recording of the palaeodipole field. They concluded that the tiny wiggles are most likely caused by palaeointensity fluctuations of the dipole field and are a ubiquitous background signal to most fast-spreading magnetic profiles. This question is discussed further in Chapter 6.

1.2 Spherical harmonic analysis of the Earth's magnetic field

Assuming that there is no magnetic material near the ground, the Earth's magnetic field can be derived from a potential function V which satisfies Laplace's equation and can thus be represented as a series of spherical harmonics

$$V = \frac{a}{\mu_0} \sum_{n=1}^{\infty} \sum_{m=0}^{n} P_n^m(\cos\theta) \left\{ \left[c_n^m \left(\frac{r}{a}\right)^n + (1 - c_n^m) \left(\frac{a}{r}\right)^{n+1} \right] A_n^m \cos m\phi \right.$$

$$\left. + \left[s_n^m \left(\frac{r}{a}\right)^n + (1 - s_n^m) \left(\frac{a}{r}\right)^{n+1} \right] B_n^m \sin m\phi \right\} \tag{1.9}$$

Written in this form the coefficients A_n^m and B_n^m have the dimensions of magnetic field, and c_n^m and s_n^m are numbers lying between 0 and 1, and represent the fractions of the harmonic terms $P_n^m(\cos\theta) \cos m\phi$ and $P_n^m(\cos\theta) \sin m\phi$ in the expansion of V which, on the surface of the sphere ($r = a$), are due to matter outside the sphere. There is no term with $n = 0$, which would correspond to a magnetic monopole within the Earth. It is also assumed that there are no electric currents flowing across the surface of the Earth – if there were, they would set up a non-potential field and thus contribute a part of the Earth's magnetic field which could not be represented by Equation (1.9).

The potential V cannot be measured directly; what can be determined are the three components of force $X = (\mu_0/r)/(\partial V/\partial\theta)$ (horizontal, northward), $Y = (-\mu_0/r \sin\theta)(\partial V/\partial\phi)$ (horizontal, eastward) and $Z = \mu_0(\partial V/\partial r)$ (vertical, downward) at the Earth's surface, $r = a$. Z (at $r = a$) may be expanded as a series of spherical harmonics

$$Z = \mu_0 \frac{\partial V}{\partial r} = \sum_{n=1}^{\infty} \sum_{m=0}^{n} P_n^m(\cos\theta)(\alpha_n^m \cos m\phi + \beta_n^m \sin m\phi) \tag{1.10}$$

and the coefficients α_n^m, β_n^m determined from the observed values of Z.

By differentiating Equation (1.9) with respect to r and then writing $r = a$, we have

$$Z = \mu_0 \frac{\partial V}{\partial r} = \sum_{n=1}^{\infty} \sum_{m=0}^{n} P_n^m(\cos\theta)\{[nc_n^m - (n+1)(1 - c_n^m)]A_n^m \cos m\phi$$

$$+ [ns_n^m - (n+1)(1 - s_n^m)]B_n^m \sin m\phi\} \tag{1.11}$$

The coefficients of each separate harmonic term for each n and m must be equal in the two expansions of Z given by Equations (1.10) and (1.11). Hence,

$$\alpha_n^m = [nc_n^m - (n+1)(1 - c_n^m)]A_n^m$$

$$\beta_n^m = [ns_n^m - (n+1)(1 - s_n^m)]B_n^m \tag{1.12}$$

Again from an analysis of the observed values of X and Y, the coefficients in the following two expansions derived from Equation (1.9) may be obtained:

$$Y_{r=a} = \left(\frac{-\mu_0}{r \sin\theta} \frac{\partial V}{\partial\phi} \right)_{r=a}$$

$$= \frac{1}{\sin\theta} \sum_{n=1}^{\infty} \sum_{m=0}^{n} P_n^m(\cos\theta)(mA_n^m \sin m\phi - mB_n^m \cos m\phi) \tag{1.13}$$

$$X_{r=a} = \left(\frac{\mu_0}{r} \frac{\partial V}{\partial \theta} \right)_{r=a}$$

$$= \sum_{n=1}^{\infty} \sum_{m=0}^{n} \frac{d}{d\theta} P_n^m (\cos \theta)(A_n^m \cos m\phi + B_n^m \sin m\phi) \qquad (1.14)$$

Both these equations contain A_n^m and B_n^m, and if values of X are known all over the world, values of Y can be deduced. If there is disagreement between observed and calculated values of Y, it would imply that the field was not completely derivable from a potential V, and, hence, that Earth–air currents do exist. When Gauss first carried out such calculations in 1839, he found no discrepancy. From a knowledge of the coefficients A_n^m, B_n^m, α_n^m and β_n^m, Equations (1.12) determine c_n^m and s_n^m. Gauss found from data available at that time that $c_n^m = s_n^m = 0$, i.e. the source of the Earth's magnetic field is entirely internal. The coefficients of the field of internal origin are

$$g_n^m = (1-c_n^m) A_n^m \qquad h_n^m = (1 - s_n^m)B_n^m \qquad (1.15)$$

and are known as Gauss coefficients. If the external field is negligible, Equations (1.15) reduce to $g_n^m = A_n^m$ and $h_n^m = B_n^m$. Barraclough (1978) has listed the coefficients of all published models known to him at that time together with a description of the data and methods used in their derivation. Langel (1987) has given the coefficients for a number of selected models. Table 1.1 gives the values up to $m = n = 4$ for the models listed in Table 1.2.

It is clear that by far the most important contribution to V comes from the term containing g_1^0 which is proportional to $P_1(\cos \theta)/r^2$, i.e. $\cos \theta/r^2$, and corresponds to the field of a geocentric dipole oriented along the z axis. Harmonic analysis of most force fields yields a first-degree term, but it must be emphasized that the formal calculation of the term provides little insight into whether it describes a distinct physical entity or process. The g_1^1 term is proportional to $\sin \theta \cos \phi/r^2$. If γ is the angle between the x and r axes,

$$\cos \gamma = \sin \theta \cos \phi \qquad (1.16)$$

so that the g_1^1 term corresponds to a geocentric dipole oriented in the x direction. Similarly, the h_1^1 term is proportional to $\sin\theta \sin\phi/r^2$ and corresponds to a geocentric dipole oriented in the y direction. The terms g_1^0, g_1^1, h_1^1 can be combined to give the total geocentric dipole, which at present is inclined at about $11\frac{1}{2}°$ to the rotation axis. The terms involving $n = 2$ (r^{-3} in the potential) represent a geocentric quadrupole, the terms $n = 3$ (r^{-4} in the potential) a geocentric octupole, etc. It must be emphasized that a spherical harmonic analysis is nothing more than a mathematical procedure for representing the field by hypothetical sources at the Earth's centre. The actual sources of the Earth's magnetic field must almost certainly lie in the fluid outer core and not in the solid inner core.

It is sometimes convenient to represent the Earth's field as a dipole displaced from the centre a distance d along the z axis (Wilson, 1970, 1971). The potential of such an offset dipole is proportional to

Table 1.1. Gauss coefficients of selected main-field models listed in Table 1.2

g/h	n	m	(1)	(2)	(3)	(4)	(5)	(6)	(7)	(8)	(9)	(10)	(11)
g	1	0	−32347.7	−32271.0	−31733.8	−30950.0	−30570.0	−30550.0	−30401.2	−30375.0	−30205.9	−30055.7	−29989.6
g	1	1	−3110.6	−2751.0	−2356.3	−2260.0	−2100.0	−2270.0	−2163.8	−2087.0	−2066.4	−2017.0	−1958.6
h	1	1	6245.5	5931.0	5984.9	5920.0	5800.0	5900.0	5778.2	5769.0	5744.6	5670.5	5608.1
g	2	0	513.8	19.0	−522.8	−890.0	−1260.0	−1520.0	−1540.1	−1648.0	−1791.7	−1932.0	−1994.8
g	2	1	2923.4	2740.0	2826.4	2990.0	2960.0	3030.0	2997.9	2954.0	2997.1	3001.3	3027.2
h	2	1	121.6	−141.0	−717.1	−1240.0	−1660.0	−1900.0	−1932.0	−1995.0	−2058.2	−2044.4	−2127.3
g	2	2	−19.9	76.0	677.1	1440.0	1630.0	1580.0	1590.3	1579.0	1608.6	1619.7	1661.6
h	2	2	1573.9	1326.0	1494.8	840.0	540.0	240.0	202.9	116.0	43.0	−69.2	−196.1
g	3	0	263.7	859.0	937.1	1020.0	1150.0	1180.0	1307.1	1164.0	1289.9	1267.1	1279.9
g	3	1	−1402.9	−1098.0	−1232.7	−1570.0	−1730.0	−1910.0	−1988.9	−2033.0	−2070.8	−2127.2	−2179.8
h	3	1	−545.4	−412.0	−296.3	−440.0	−520.0	−450.0	−425.4	−389.0	−369.9	−343.5	−334.4
g	3	2	1320.7	1374.0	1428.7	1180.0	1210.0	1260.0	1276.8	1299.0	1276.0	1259.4	1251.4
h	3	2	410.8	207.0	26.2	120.0	180.0	290.0	227.8	230.0	245.6	263.2	270.7
g	3	3	−61.7	239.0	399.0	760.0	880.0	910.0	881.2	880.0	833.4	818.0	833.0
h	3	3	828.7	884.0	683.4	240.0	30.0	−90.0	−133.8	−141.0	−188.0	−208.5	−251.1
g	4	0	869.4	673.0	794.7	880.0	920.0	950.0	949.3	930.0	947.5	953.8	938.3
g	4	1	963.4	619.0	513.3	680.0	780.0	800.0	803.5	811.0	800.9	786.1	782.5

h	4	1	−404.8	98.0	283.5	250.0	140.0	150.0	160.3	142.0	161.7	196.7	211.6
g	4	2	408.9	421.0	517.2	650.0	580.0	580.0	502.9	490.0	457.9	437.8	398.4
h	4	2	−380.1	−244.0	−171.9	−120.0	−280.0	−310.0	−274.3	−276.0	−275.8	−257.0	−256.7
g	4	3	−330.3	−145.0	−309.3	−400.0	−380.0	−380.0	−397.7	−402.0	−396.0	−412.8	−419.2
h	4	3	3.0	−205.0	−225.6	−130.0	−80.0	−40.0	2.3	5.0	18.5	20.1	52.0
g	4	4	−195.0	0.0	158.4	180.0	300.0	310.0	266.5	262.0	243.6	232.3	199.3
h	4	4	−150.0	−139.0	−73.5	−30.0	−130.0	−170.0	−246.6	−264.0	−278.8	−287.5	−297.6
g	5	0	0.0	−23.0	−167.8	−220.0	−240.0	−270.0	−233.5	−179.0	−214.5	−214.2	−217.4
g	5	1	0.0	356.0	395.3	260.0	280.0	320.0	355.7	357.0	359.5	357.4	357.6
h	5	1	0.0	−227.0	−269.3	−150.0	30.0	20.0	5.1	30.0	15.7	31.4	45.2
g	5	2	0.0	383.0	285.9	180.0	200.0	200.0	228.4	248.0	249.0	256.1	261.0
h	5	2	0.0	38.0	−34.5	10.0	70.0	100.0	117.8	135.0	142.0	150.7	149.4
g	5	3	0.0	8.0	15.3	−40.0	−60.0	−40.0	−28.8	−20.0	−29.0	−42.8	−73.9
h	5	3	0.0	−41.0	21.6	0.0	−20.0	−50.0	−114.8	−123.0	−131.0	−137.4	−150.3
g	5	4	0.0	−49.0	−7.0	−90.0	−140.0	−150.0	−157.9	−171.0	−166.9	−166.7	−162.0
h	5	4	0.0	−64.0	−76.0	−130.0	−140.0	−140.0	−108.9	−100.0	−91.1	−81.9	−78.1
g	5	5	0.0	108.0	0.0	0.0	−70.0	−70.0	−62.2	−64.0	−58.2	−58.9	−48.3
h	5	5	0.0	60.0	0.0	0.0	80.0	90.0	82.4	84.0	80.8	86.2	91.8

After Langel (1987).

Table 1.2. Models tabulated in Table 1.1

Model number	Reference	Epoch	Degree/Order
(1)	Gauss (1839)	1835	4/4
(2)	Fritsche (1899)	1842	6/5
(3)	Schmidt (1895)	1885	6/4
(4)	Dyson and Furner (1923)	1922	6/4
(5)	Vestine *et al.* (1947b)	1945	6/6
(6)	Finch and Leaton (1957)	1955	6/6
(7)	Cain *et al.* (1967)	1960	10/10
(8)	Leaton *et al.* (1965)	1965	8/8
(9)	Hurwitz *et al.* (1974)	1970	12/12
(10)	Peddie and Fabiano (1976)	1975	12/12
(11)	Langel *et al.* (1980)	1979.85	13/13

After Langel (1987).

$$\cos\theta(r^2 + d^2 - 2rd\cos\theta)^{-1} = \frac{\cos\theta}{r^2}\left(1 - \frac{2d}{r}\cos\theta + \frac{d^2}{r^2}\right)^{-1} \qquad (1.17)$$

Expanding in a Taylor series for $d < r$, it can be seen that the offset dipole is equivalent to a dipole plus a series of higher-degree multipole fields at the Earth's centre (James and Winch, 1967). The larger the value of d, the larger the number of zonal harmonics required to approximate the offset dipole. Bartels (1936) showed that

$$g_2^0/2g_1^0 = d/a \quad \text{and} \quad g_3^0/g_2^0 = 3d/2a \qquad (1.18)$$

The above example illustrates the non-uniqueness of a spherical harmonic analysis and emphasizes that separate terms in such an analysis do not represent separate real magnetic sources. An outstanding feature of Table 1.1 is the secular variation of the individual coefficients, which is clearly apparent in spite of individual scatter. There appears to have been an overall decrease in the dipole moment of the Earth's field of about 5% per century, while the quadrupole moment has increased substantially since AD 1800. Over the same time interval, the inclination of the dipole axis appears to have remained sensibly constant.

The non-dipole components of the Earth's field, though much weaker than the dipole component, show more rapid changes. The timescale of the non-dipole changes is measured in decades and that of the dipole in centuries. The isoporic foci also drift westward at a fraction of a degree per year. Both eastward and westward drifts of the non-dipole field have been inferred from palaeomagnetic studies. Observatory records from Sitka, Alaska, indicate an eastward drift during the past 60 years, in contrast to the pre-dominance of a westward direction of drift observed over most areas of the world for the past several hundred years (Skiles, 1970; Yukutake, 1962). Yukutake and Tachinaka (1969) have shown that the non-dipole field can be decomposed into two parts, one standing and the other drifting westward at about the same rate as the secular variation. The standing and drifting parts are approximately of the same size and intensity. The drifting field consists mainly of low harmonics ($n \leqslant 3$), whereas the standing field has a more complicated distribution. Although there is still a poor distribution and uneven quality of the raw data from ground studies, the situation has improved considerably through the use of aircraft and more recently with satellites. It has become apparent that secular changes cannot be extrapolated accurately over intervals longer than about four or five years.

In practice the series expansion of the magnetic potential V is truncated at a maximum value N of m and n, usually in the range 8–15. The International Association of Geomagnetism and Aeronomy (IAGA) adopted in 1968 an International Geomagnetic Reference Field (IGRF) describing the main geomagnetic field at 1965 by means of 80 spherical harmonic coefficients ($N = 8$). An additional set of 80 coefficients describing the secular variation was included for use in extending the main-field model in time, both backward (not earlier than 1955) and forward (not later than 1975). By the early 1970s it was becoming obvious that inaccuracies in the secular variation coefficients were causing unacceptably large errors in field values computed for current epochs from the model and the IGRF was revised by IAGA in 1975. This revision was limited to the provision of a revised set of 80 secular variation coefficients to be used for deriving field values for dates after 1975, the original and revised versions of the IGRF being continuous at 1975. By the late 1970s the cumulative effects of uncertainties in the secular variation models led to unacceptable inaccuracies in the IGRF and a second revision was made by IAGA in 1981. By that time the results of NASA's *Magsat* satellite mission had been analysed. *Magsat* was launched on 30 October 1979 into a sun-synchronous (twilight) orbit with an apogee of 561 km and a perigee of 352 km. Re-entry occurred on 11 June 1980. The instrumentation included a caesium vapour magnetometer to measure the magnitude of the field, a fluxgate magnetometer to measure the field components and an optical system to measure the orientation of the fluxgate magnetometer. The scalar magnetometer was used to calibrate the vector magnetometer in flight. The data are estimated to be accurate to about 2 nT in magnitude and 6 nT in each component. Preliminary results of the *Magsat* mission were given in *Geophysical Research Letters* (Vol. 9, No. 4, 1982).

The new version of the IGRF consisted of five models: four models of the main geomagnetic field for 1965, 1970, 1975 and 1980 and a model of the secular variation valid for the interval 1980–5. The four main-field models each consist of 120 coefficients ($N = 10$), while the secular variation model has, as before, 80 coefficients. The main-field models for 1965, 1970 and 1975 were designated Definitive International Geomagnetic Reference Fields (DGRF 1965, DGRF 1970 and DGRF 1975, respectively) since further revision is not envisaged. Linear interpolation between neighbouring models is used for dates that lie between the epochs of the models. The main-field model for 1980 is not continuous with the earlier series of IGRF models. For dates between 1975 and 1980 a Provisional International Geomagnetic Reference Field (PGRF 1975) is defined by linear interpolation between DGRF 1975 and the main-field coefficients of IGRF (1980). The models were derived from three sets of proposed models by taking weighted means, the weights being assigned according to the apparent accuracy of the proposed models. A special issue of the *Journal of Geomagnetism and Geoelectricity* (Vol. 34, No. 6, 1982) is devoted to papers on this revision of the IGRF (Peddie 1982).

At the IAGA General Assembly in 1985, a third revision was made. DGRF 1980 replaced IGRF 1980, IGRF 1985 including a secular variation model for 1985–90 was adopted and main-field IGRF models were adopted for 1945, 1950, 1955 and 1960. The fourth revision replaced the models for 1945, 1950, 1955 and 1960 with DGRFs. Details of the derivations and characteristics of these models have been given in a special issue of *Physics of the Earth and Planetary Interiors* (Vol. 48, Nos 3–4, 1987). At the twentieth General Assembly of the International Union of Geodesy and Geophysics (IUGG) in 1991, it was recommended that IGRF 1985 be replaced by a newly derived DGRF 1985 and that the IGRF be extended to 1995 by adoption of IGRF 1990, comprising a model of the main field at 1990 and a predictive model of the secular variation. The spherical harmonic coefficients for the nine DGRF models spanning the interval 1945–85 and for IGRF 1990 are given in Table 1.3. Details of this latest model are given in a special issue of the *Journal of Geomagnetism and Geoelectricity* (Vol. 44, No. 9, 1992). This issue also contains papers on the evaluation of candidate models over Australia, South Africa, New Zealand and Antarctica, Canada and India, and marine magnetic anomalies in the North Atlantic.

A number of people have attempted to carry out spherical harmonic analyses of the palaeomagnetic field of the Earth back through the Mesozoic and into the Palaeozoic. It is not surprising that different conclusions have been reached, since the past distribution of the continents plays a critical role in any such spherical harmonic analysis, and the distribution of continents through the late Palaeozoic is not known for certain. A meaningful global analysis of the palaeomagnetic field is impossible when the observation points cover only about one-third of the Earth's surface and their relative positions are not precisely known. McElhinny and Merrill (1975) noted that nearly half of the palaeomagnetic poles derived from rocks during the last 5 Ma have 95% circles of confidence about them that do not include the present geographic poles. Wilson and Ade-Hall (1970) first noted that late Tertiary and Quaternary palaeomagnetic poles from Europe and Asia all tended to lie too far away from the observation site along the great circle joining the site to the geographic pole. Successive analyses of worldwide data by Wilson (1970, 1971, 1972) and Wilson and McElhinny (1974) confirmed that this far-sided effect occurs globally. It can be produced by an axial dipole that is not geocentric but is displaced a distance d northward along the axis of rotation. For the time-averaged late Tertiary field, Wilson and McElhinny obtained a value for d of 325 ± 57 km. As shown earlier (Equation 1.17), if the potential due to an offset axial dipole is expanded in terms of spherical harmonics, only zonal terms occur. For $d = 325$ km, g_3^0 is less than 8% of the magnitude of g_2^0, so that the offset dipole is approximately equivalent to a spherical harmonic model including only the g_1^0 and g_2^0 terms.

Table 1.3.

						DGRF					IGRF	
n	m	1945	1950	1955	1960	1965	1970	1975	1980	1985	1990	1990–5
g 1 0		−30594	−30554	−30500	−30421	−30334	−30220	−30100	−29992	−29873	−29775	18.0
g 1 1		−2285	−2250	−2215	−2169	−2119	−2068	−2013	−1956	−1905	−1851	10.6
h 1 1		5810	5815	5820	5791	5776	5737	5675	5604	5500	5411	−16.1
g 2 0		−1244	−1341	−1440	−1555	−1662	−1781	−1902	−1997	−2072	−2136	−12.9
g 2 1		2990	2998	3003	3002	2997	3000	3010	3027	3044	3058	2.4
h 2 1		−1702	−1810	−1898	−1967	−2016	−2047	−2067	−2129	−2197	−2278	−15.8
g 2 2		1578	1576	1581	1590	1594	1611	1632	1663	1687	1693	0.0
h 2 2		477	381	291	206	114	25	−68	−200	−306	−380	−13.8
g 3 0		1282	1297	1302	1302	1297	1287	1276	1281	1296	1315	3.3
g 3 1		−1834	−1889	−1944	−1992	−2038	−2091	−2144	−2180	−2208	−2240	−6.7
h 3 1		−499	−476	−462	−414	−404	−366	−333	−336	−310	−287	4.4
g 3 2		1255	1274	1288	1289	1292	1278	1260	1251	1247	1246	0.1
h 3 2		186	206	216	224	240	251	262	271	284	293	1.6
g 3 3		913	896	882	878	856	838	830	833	829	807	−5.9
h 3 3		−11	−46	−83	−130	−165	−196	−223	−252	−297	−348	−10.6
g 4 0		944	954	958	957	957	952	946	938	936	939	0.5
g 4 1		776	792	796	800	804	800	791	782	780	782	0.6
h 4 1		144	136	133	135	148	167	191	212	232	248	2.6
g 4 2		544	528	510	504	479	461	438	398	361	324	−7.0
h 4 2		−276	−278	−274	−278	−269	−266	−265	−257	−249	−240	1.8
g 4 3		−421	−408	−397	−394	−390	−395	−405	−419	−424	−423	0.5
h 4 3		−55	−37	−23	3	13	26	39	53	69	87	3.1
g 4 4		304	303	290	269	252	234	216	199	170	142	−5.5
h 4 4		−178	−210	−230	−255	−269	−279	−288	−297	−297	−299	−1.4

Table 1.3. (cont.)

						DGRF					IGRF	
	n	m	1945	1950	1955	1960	1965	1970	1975	1980	1985	1990	1990–5
g	5	0	-253	-240	-229	-222	-219	-216	-218	-218	-214	-211	0.6
g	5	1	346	349	360	362	358	359	356	357	355	353	-0.1
h	5	1	-12	3	15	16	19	26	31	46	47	47	-0.1
g	5	2	194	211	230	242	254	262	264	261	253	244	-1.6
h	5	2	95	103	110	125	128	139	148	150	150	153	0.5
g	5	3	-20	-20	-23	-26	-31	-42	-59	-74	-93	-111	-3.1
h	5	3	-67	-87	-98	-117	-126	-139	-152	-151	-154	-154	0.4
g	5	4	-142	-147	-152	-156	-157	-160	-159	-162	-164	-166	-0.1
h	5	4	-119	-122	-121	-114	-97	-91	-83	-78	-75	-69	1.7
g	5	5	-82	-76	-69	-63	-62	-56	-49	-48	-46	-37	2.3
h	5	5	82	80	78	81	81	83	88	92	95	98	0.4
g	6	0	59	54	47	46	45	43	45	48	53	61	1.3
g	6	1	57	57	57	58	61	64	66	66	65	64	-0.2
h	6	1	6	-1	-9	-10	-11	-12	-13	-15	-16	-16	0.2
g	6	2	6	4	3	1	8	15	28	42	51	60	1.8
h	6	2	100	99	96	99	100	100	99	93	88	83	-1.3
g	6	3	-246	-247	-247	-237	-228	-212	-198	-192	-185	-178	1.3
h	6	3	16	33	48	60	68	72	75	71	69	68	0.0
g	6	4	-25	-16	-8	-1	4	2	1	4	4	2	-0.2
h	6	4	-9	-12	-16	-20	-32	-37	-41	-43	-48	-52	-0.9
g	6	5	21	12	7	-2	1	3	6	14	16	17	0.1
h	6	5	-16	-12	-12	-11	-8	-6	-4	-2	-1	2	0.5
g	6	6	-104	-105	-107	-113	-111	-112	-111	-108	-102	-96	1.2
h	6	6	-39	-30	-24	-17	-7	1	11	17	21	27	1.2

g	7	0	70	65	65	67	75	72	71	72	74	77	0.6
g	7	1	−40	−55	−56	−56	−57	−57	−56	−59	−62	−64	−0.5
b	7	1	−45	−35	−50	−55	−61	−70	−77	−82	−83	−81	0.6
g	7	2	0	2	2	5	4	1	1	2	3	4	−0.3
b	7	2	−18	−17	−24	−28	−27	−27	−26	−27	−27	−27	0.2
g	7	3	0	1	10	15	13	14	16	21	24	28	0.6
b	7	3	2	0	−4	−6	−2	−4	−5	−5	−2	1	0.8
g	7	4	−29	−40	−32	−32	−26	−22	−14	−12	−6	1	1.6
b	7	4	6	10	8	7	6	8	10	16	20	20	−0.5
g	7	5	−10	−7	−11	−7	−6	−2	0	1	4	6	0.2
b	7	5	28	36	28	23	26	23	22	18	17	16	−0.2
g	7	6	15	5	9	17	13	13	12	11	10	10	0.2
b	7	6	−17	−18	−20	−18	−23	−23	−23	−23	−23	−23	0.0
g	7	7	29	19	18	8	1	−2	−5	−2	0	0	0.3
b	7	7	−22	−16	−18	−17	−12	−11	−12	−10	−7	−5	0.0
g	8	0	13	22	11	15	13	14	14	18	21	22	0.2
b	8	1	7	15	9	6	5	6	6	6	6	5	−0.7
g	8	1	12	5	10	11	7	7	6	7	8	10	0.5
b	8	2	−8	−4	−6	−4	−4	−2	−1	0	0	−1	−0.2
g	8	2	−21	−22	−15	−14	−12	−15	−16	−18	−19	−20	−0.2
b	8	3	−5	−1	−14	−11	−14	−13	−12	−11	−11	−11	0.1
g	8	3	−12	0	5	7	9	6	4	4	5	7	0.3
b	8	4	9	11	6	2	0	−3	−8	−7	−9	−12	−1.1
g	8	4	−7	−21	−23	−18	−16	−17	−19	−22	−23	−22	0.3
b	8	5	7	15	10	10	8	5	4	4	4	4	0.0
g	8	5	2	−8	3	4	4	6	6	9	11	12	0.4
b	8	6	−10	−13	−7	−5	−1	0	0	3	4	4	−0.1
g	8	6	18	17	23	23	24	21	18	16	14	11	−0.5
g	8	7	7	5	6	10	11	11	10	6	4	3	−0.5

Table 1.3. (cont.)

					DGRF						IGRF	
n	m	1945	1950	1955	1960	1965	1970	1975	1980	1985	1990	1990–5
h 8	7	3	−4	−4	1	−3	−6	−10	−13	−15	−16	−0.3
g 8	8	2	−1	9	8	4	3	1	−1	−4	−6	−0.6
h 8	8	−11	−17	−13	−20	−17	−16	−17	−15	−11	−11	0.6
g 9	0	5	3	4	4	8	8	7	5	5	4	0.0
g 9	1	−21	−7	9	6	10	10	10	10	10	10	0.0
h 9	1	−27	−24	−11	−18	−22	−21	−21	−21	−21	−21	0.0
g 9	2	1	−1	−4	0	2	2	2	1	1	1	0.0
h 9	2	17	19	12	12	15	16	16	16	15	15	0.0
g 9	3	−11	−25	−5	−9	−13	−12	−12	−12	−12	−12	0.0
h 9	3	29	12	7	2	7	6	7	9	9	10	0.0
g 9	4	3	10	2	1	10	10	10	9	9	9	0.0
h 9	4	−9	2	6	0	−4	−4	−4	−5	−6	−6	0.0
g 9	5	16	5	4	4	−1	−1	−1	−3	−3	−4	0.0
h 9	5	4	2	−2	−3	−5	−5	−5	−6	−6	−6	0.0
g 9	6	−3	−5	1	−1	−1	0	−1	−1	−1	−1	0.0
h 9	6	9	8	10	9	10	10	10	9	9	9	0.0
g 9	7	−4	−2	2	−2	5	3	4	7	7	7	0.0
h 9	7	6	8	7	8	10	11	11	10	9	9	0.0
g 9	8	−3	3	2	3	1	1	1	2	1	2	0.0
h 9	8	1	−11	−6	0	−4	−2	−3	−6	−7	−7	0.0
g 9	9	−4	8	5	−1	−2	−1	−2	−5	−5	−6	0.0
h 9	9	8	−7	5	5	1	1	1	2	2	2	0.0
g 10	0	−3	−8	−3	1	−2	−3	−3	−4	−4	−4	0.0
g 10	1	11	4	−5	−3	−3	−3	−3	−4	−4	−4	0.0

h	g	10	1	5	13	−4	4	2	1	1	1	1	0.0
g	h	10	2	1	−1	−1	4	2	2	2	2	3	0.0
h	g	10	2	1	−2	0	1	1	1	1	0	0	0.0
g	h	10	3	−2	13	2	0	−5	−5	−5	−5	−5	0.0
h	g	10	3	−20	−10	−8	0	2	3	3	3	3	0.0
g	h	10	4	−5	−4	−3	−1	−2	−1	−2	−2	−2	0.0
h	g	10	4	−1	2	−2	2	6	4	4	6	6	0.0
g	h	10	5	−1	4	7	4	4	6	5	5	5	0.0
h	g	10	5	−6	−3	−4	−5	−4	−4	−4	−4	−4	0.0
g	h	10	6	8	12	4	6	4	4	4	3	3	0.0
h	g	10	6	6	6	1	1	0	0	−1	0	0	0.0
g	h	10	7	−1	3	−2	1	0	−1	−1	−1	−1	0.0
h	g	10	7	−4	−3	−3	−1	−2	−1	−1	−1	−1	0.0
g	h	10	8	−3	2	6	−1	2	0	0	−2	2	0.0
h	g	10	8	−2	6	7	6	3	3	3	4	4	0.0
g	h	10	9	5	10	−2	2	2	3	3	3	3	0.0
h	g	10	9	0	11	−1	0	0	1	1	0	0	0.0
g	h	10	10	−2	3	0	0	0	−1	−1	0	0	0.0
h	g	10	10	−2	8	−3	−7	−6	−4	−5	−6	−6	0.0

Using different techniques and different data sets, Creer *et al.* (1973), Wells (1973) and Merrill and McElhinny (1977) all confirmed the existence of a significant g_2^0 component in the Quaternary and Late Tertiary palaeofield. Hence, the existence of at least one long-term non-dipole component appears to be well established. However, general agreement on the significance of any other long-term non-dipole components is lacking. Deviations from a geocentric axial dipole field can be expressed only in terms of ratios of Gauss coefficients because of the lack of corresponding palaeointensity data (see Table 1.4). Merrill and McElhinny (1977) were unable to give an accurate estimate of the errors involved in estimating the coefficients in Table 1.4: however, they were able to show that the north–south asymmetry, as reflected by the g_3^0/g_1^0 ratio, is significant at the 90% confidence level. They suggested that this north–south asymmetry may reflect differences in boundary conditions at the CMB, such as temperature, topography and electrical conductivity. These questions are discussed further in Section 8.5.

Table 1.5 (after Coupland and Van der Voo, 1980) gives the results of a spherical harmonic analysis of the Earth's magnetic field for a number of time periods, using all available data for the last 130 Ma (meeting certain criteria), returned to pre-continental drift site locations and orientations using spreading poles based on sea-floor magnetic anomalies. From about 95 to 50 Ma ago, the long-term non-dipole field consisted of a large g_2^0 component, while g_3^0 was near zero. Between 50 and 40 Ma ago, $g_2^0/|g_1^0|$ apparently changed from about -0.10 to $+0.05$, while the $g_3^0/|g_1^0|$ component changed from near zero to -0.10. Changes since then have been much more gradual. The large values of g_0^0 are incompatible with an offset dipole as a physical model for the generation of the long-term non-dipole field.

Between 100 and 75 Ma ago, $g_2^0/|g_1^0|$ apparently increased from near zero to about -0.20. Coupland and Van der Voo also found that over the period 10–50 Ma BP the value of g_2^0 determined for dominantly reversed data is about

Table 1.4. Gaussian coefficients normalized to the
geocentric axial dipole term g_1^{0a}

| Term | Present field | Palaeomagnetic field (0–5Ma) | |
		Normal	Reversed
g_1^1/g_1^0	$+0.067$	-0.017	$-$?
h_1^1/g_1^0	-0.190	$+0.030$	$+$?
g_2^0/g_1^0	$+0.063$	$+0.050$	$+0.083$
g_3^0/g_1^0	-0.043	$+0.017^b$	$+0.034^b$

^{a}The term g_1^0 is negative for the present (normal) field
and positive for the reversed field. The table therefore
indicates that the sign of all the coefficients changes
with that of g_1^0.
^{b}Sign corrected from original paper.
After Merrill and McElhinny (1977).

Table 1.5. Results of spherical harmonic analyses of the Earth's magnetic field

Time period	Extent (Ma)	Data set	No. of sites	g_2^0	g_3^0
Upper Cenozoic	0–26	extended	251	−0.0549	−0.0381
		restricted	102	−0.0488	−0.0426
Quaternary	0–2	extended	81	−0.0633	−0.0469
		restricted	45	−0.0581	−0.0522
Pliocene	2–7	extended	67	−0.0916	−0.0201
		restricted	18	−0.0767	−0.0661
Miocene	7–26	extended	58	0.0034	−0.0793
		restricted	23	−0.0081	−0.0463
Lower Tertiary	26–65	extended	51	−0.1343	
		restricted	22	−0.1552	
Upper Cretaceous	65–100	extended	42	−0.1279	
		restricted	18	−0.1446	
Lower Cretaceous	100–136	extended	34	(0.1952)	
		restricted	11	(−0.0213)	

After Coupland and VanVoo (1980).

twice as large (in magnitude) as it is for dominantly normal data. This is in agreement with the results of Wilson (1972), who found that the dipole offset calculated for 24 reversed Quaternary and Upper Tertiary directions from the USSR was much greater than the dipole offset calculated for 36 normal directions. However, if g_3^0 is included in the analysis, the magnitude of g_2^0 is no longer greater in the dominantly reversed data, nor is g_3^0 greater.

Livermore *et al.* (1983), using data from palaeomagnetic poles determined from records from continents, islands and sea mounts and magnetic inclinations from deep-sea cores, found that $g_2^0 \sim 0.05$ g_1^0 throughout the last 35 Ma. For the last 5 Ma they modelled the palaeomagnetic field by fitting all terms up to degree 3. They found that $g_3^0 \sim 0.02$ g_1^0 and that most non-zonal terms were small with the exception of the equatorial quadrupole term h_2^1, which was ≈ 0.03 g_1^0.

Livermore *et al.* (1984) later modelled the palaeomagnetic field for the last 200 Ma, using as data palaeomagnetic poles estimated from the major plates returned to their locations at the estimated dates of magnetization in fixed hotspot co-ordinates. They neglected any contributions from an axial octupole (g_3^0) and found that a small axial quadrupole of the same sign as the axial dipole may have persisted throughout the Cenozoic. During the late Cretaceous, g_2^0 appears to have changed sign with respect to g_1^0. Negative values seem to have obtained throughout the long Cretaceous normal superchron.

Most applications of palaeomagnetism assume that the Earth's magnetic field has, on the average, the form of a geocentric axial dipole (GAD), although it has not been possible to prove this rigorously. Palaeomagnetic observations generally support the GAD hypothesis, although small second-order effects have been known for some time. A convenient way of examining the effects of a non-dipole field is the inclination anomaly (Cox, 1975), which is defined as the difference between the observed inclination and the geocentric dipole inclination,

$$\Delta I = I \text{ (observed)} - I \text{ (dipole)}$$

The inclination anomaly for an axial quadrupole contribution is largely symmetric about the equator, with its greatest effect at the equator and diminishing towards the poles. For an axial octupole the contribution is mostly antisymmetric – it has no effect at the equator and poles and is greatest at mid-latitudes with opposite sign in opposing hemispheres (see Figure 1.12).

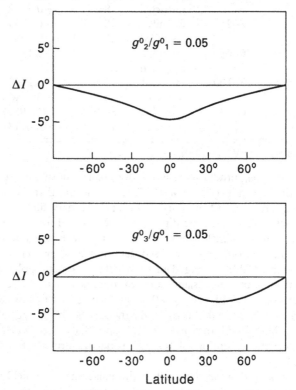

Figure 1.12 Comparison of expected inclination anomaly with latitude for low-degree axial fields. Upper panel: anomaly for axial quadrupole non-dipole field having a g^0_2/g^0_1, ratio of 0.05. Lower panel: anomaly for an axial octupole non-dipole field having a g^0_3/g^0_1 ratio of 0.05. After Schneider and Kent (1988).

Schneider and Kent (1988) have estimated ΔI values recorded in Plio–Pleistocene (0–5 Ma) deep-sea sediments from the equatorial Indian Ocean. The anomalies are small but significant and consistently negative. There is also a statistically significant difference in the magnitude of the anomaly, being larger (~5°) during times of reversed polarity and smaller (~2°) during times of normal polarity. These results are consistent with previous spherical harmonic analyses of global palaeomagnetic data (Merrill and McElhinny, 1977; Coupland and

Van der Voo, 1980; Livermore *et al.*, 1983). The presence of this asymmetry supports earlier suggestions (e.g. Merrill and McElhinny, 1977) of the existence of a standing component of the non-dipole field which does not invert during reversals of the main field.

Schneider and Kent (1990a) later analysed palaeomagnetic inclination records for the last 2.5 Ma from 186 deep-sea cores from low to middle latitudes ($\pm$ 45°). Although palaeomagnetic declinations cannot be obtained from piston cores, they do permit of better geographical distribution of sampling sites compared with land-based data, particularly in the Pacific and southern hemisphere. They can also be more reliably dated (using biostratigraphic events and magnetic reversals), enabling plate tectonic corrections to be readily made. Schneider and Kent (1990a) found that the amplitude of the axial quadrupole varied with the polarity of the main field (0.026 g_1^0, for normal and 0.046 g_1^0, for reverse). The axial octupole, on the other hand, showed no appreciable difference (-0.029 g_1^0 for normal and -0.021 g_1^0 for reverse). Their estimates of the quadrupole component agree well with earlier determinations for the Plio–Pleistocene. However, the octupole component is of opposite sign to the previous determinations. Schneider and Kent suggest that spurious inclination shallowing may account for the earlier positive octupole estimates using land-based data. Depositional remanent magnetization in deep-sea sediments may be largely free from inclination shallowing.

Schneider and Kent reflect on the result that the octupole changes sign during a transition of the main dipole field, while the quadrupole does not. They suggest that this may be because the octupole and main dipole belong to the dipole family, whereas the quadrupole field belongs to the quadrupole family.* A small quadrupole field may thus persist and be completely unaffected by polarity reversals of the main field. Such a standing quadrupole field would have opposite effects on the two dipole polarity states and could thus account for the observed polarity dependence of this term.

Schneider and Kent (1990b) later investigated the non-dipole field during the Tertiary, using data from three widely separated sites on the African plate (the western Indian Ocean basin, the area of the Walris ridge and Angola Abyssal plain, and Gubbio in Italy). They found that the axial octupole contribution was negligible (0–1% of the axial dipole) in the Paleocene (60–65 Ma) and no greater than 5–6% in the Oligocene (30–40 Ma). Since a quadrupole component would not give rise to a detectable anomalous variation in palaeomagnetic direction over a single plate, Schneider and Kent used additional data from North America. The additional data do not change their estimates of the octupole contribution obtained from the African plate data alone. Their range of values for the quadrupole contribution is rather large, with best-fitting values for the Paleocene being $\sim$ 10% and for the Oligocene 5–6%. It is interesting to compare the results of Schneider and Kent with the earlier work of Coupland and Van

---

* Roberts and Stix (1972) have shown that the dynamo field is made up of two families and that under certain symmetry conditions in the core these two families are non-interactive. This is discussed in more detail in Section 5.7.

der Voo (1980) and Livermore *et al.* (1984). For the interval between 30 and 40 Ma, Coupland and Van der Voo found a quadrupole of ~ −3% and an octupole of ~ 12% (see Figure 1.13).* These estimates are outside the limits obtained by Schneider and Kent. For the period 60–65 Ma, Coupland and Van der Voo obtained a smaller value for the octupole (~ 4% and a large quadrupole of ~ 16%. Both these estimates are within the acceptable limits found by Schneider and Kent.

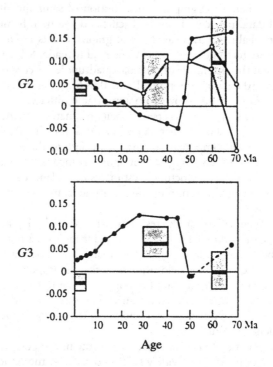

Figure 1.13 Comparison of allowable ranges for quadrupole ($G2 = g_2^0/g_1^0$) and octupole ($G3 = g_3^0/g_1^0$) non-dipole components. Solid circles show quadrupole and octupole model of Coupland and Van der Voo (1980). Open circles show quadrupole model of Livermore *et al.* (1984). Shaded areas indicate 95% errors on non-dipole estimates. After Schneider and Kent (1990b).

1.3 Origin of the Earth's magnetic field

There has been much speculation on the origin of the Earth's magnetic field and its secular variation, and many possible sources have been suggested, most of which have proved to be inadequate. The only possible means seems to be some form of electromagnetic induction, electric currents flowing in the Earth's fluid,

---

* Coupland and Van der Voo normalized their non-dipole field estimates by the absolute magnitude of the dipole field, so the sign of their estimates must be inverted to make a comparison.

electrically conducting core. Palaeomagnetic measurements have shown that the Earth's main field has existed throughout geological time and that its strength has never differed significantly from its present value. In a bounded, stationary, electrically conducting body, any system of electric currents will decay. The field or current may be analysed into normal modes each of which decays exponentially with its own time constant. The time constant is proportional to σl^2, where σ is the electrical conductivity and l a characteristic length representing the distance in which the field changes by an appreciable amount. For a sphere the size of the Earth, the most slowly decaying mode is reduced to $1/e$ of its initial strength in a time of the order of 100 000 a. Since the age of the Earth is more than 4000 Ma, the geomagnetic field cannot be a relic of the past, and a mechanism must be found for generating and maintaining electric currents to sustain the field. The most likely source of the electromotive force needed to maintain these currents is the motion of core material across the geomagnetic lines of force. The study of this process, in which the currents generated reinforce the magnetic field which gives rise to the driving electromotive force, is known as the homogeneous dynamo problem. A number of detailed accounts of dynamo theory have been given (see, e.g., Busse, 1978; Parkinson, 1983; Merrill and McElhinny, 1983; Roberts and Gubbins, 1987; Roberts, 1987) and no attempt will be made to review the subject in this book. Those aspects of the different models that have been proposed which have a bearing on the question of reversals will be discussed later. More general accounts of magnetohydrodynamics have been given by Moffatt (1978), Roberts and Soward (1978) and Gubbins and Roberts (1987).

The dynamo problem involves the solution of a highly complicated system of coupled partial differential equations – electrodynamic, hydrodynamic and thermodynamic. Elsasser (1954) showed by a dimensional analysis that in geophysical and astrophysical problems the displacement current and all purely electrostatic effects are negligible, as are all relativistic effects of order higher than U/c, where U is the fluid velocity. Thus the electromagnetic field equations are the usual Maxwell equations,

$$\nabla \times \boldsymbol{E} = - \partial \boldsymbol{B}/\partial t \tag{1.19}$$

$$\nabla \times \boldsymbol{B} = \mu_0 \boldsymbol{j} \tag{1.20}$$

$$\nabla \cdot \boldsymbol{B} = 0 \tag{1.21}$$

where $\boldsymbol{B}$ and $\boldsymbol{E}$ are the magnetic and electric fields, respectively, and $\boldsymbol{j}$ the electric current density. The electromotive forces which give rise to $\boldsymbol{j}$ are due both to electric charges and to motional induction, so that the total current $\boldsymbol{j}$ is given by

$$\boldsymbol{j} = \sigma(\boldsymbol{E} + \boldsymbol{U} \times \boldsymbol{B}) \tag{1.22}$$

Assuming the electrical conductivity σ to be constant, taking the curl of Equation (1.20), and using Equations (1.22) and (1.19), $\boldsymbol{E}$ can be eliminated, leading to the equation

$$\nabla \times (\nabla \times \boldsymbol{B}) = \mu_0 \sigma \left(-\frac{\partial \boldsymbol{B}}{\partial t} + \nabla \times (\boldsymbol{U} \times \boldsymbol{B}) \right) \tag{1.23}$$

Since $\nabla \times (\nabla \times B) = \nabla(\nabla \cdot B) - \nabla^2 B = -\nabla^2 B$, on using Equation (1.21), we finally obtain

$$\partial B/\partial t = \nabla \times (U \times B) + v_m \nabla^2 B \qquad (1.24)$$

where

$$v_m = 1/\mu_0 \sigma \qquad (1.25)$$

is the magnetic diffusivity. Equations (1.21) and (1.24) give the relationships between B and U which have to be satisfied from electromagnetic considerations. The term $\nabla \times (U \times B)$ in Equation (1.24) is the source term, and represents the physical process by which magnetic induction is 'created' through the flow of fluid across lines of force. The term $v_m \nabla^2 B$ represents the tendency for the field to decay through ohmic dissipation by the electric currents supporting the field. The balance between these two terms, at a particular point, determines how the magnetic field changes with time at that point.

If the material is at rest, Equation (1.24) reduces to

$$\partial B/\partial t = v_m \nabla^2 B \qquad (1.26)$$

This has the form of a diffusion equation, and indicates that the field leaks through the material from point to point. Dimensional arguments indicate a decay time of the order L^2/v_m where L is a length representative of the dimensions of the region in which current flows. For conductors in the laboratory this decay time is very small – even for a copper sphere of radius 1 m it is less than 10 s. For cosmic conductors, on the other hand, because of their enormous size, it can be very large. As an alternative limiting case, suppose that the material is in motion but has negligible electrical resistance. Equation (1.24) then becomes

$$\partial B/\partial t = \nabla \times (U \times B) \qquad (1.27)$$

This equation is identical with that satisfied by the vorticity in the hydrodynamic theory of the flow of a non-viscous fluid, where it is shown that vortex lines move with the fluid. Thus, Equation (1.27) implies that the field changes are the same as if the magnetic lines of force were 'frozen' into the material.

When neither term on the right-hand side of Equation (1.24) is negligible, both the above effects are observed, i.e. the lines of force tend to be carried about with the moving fluid and at the same time leak through it.

If L, T, V represent the order of magnitude of a length, time and velocity, respectively, transport dominates leak if $LV \gg v_m$. The condition for the onset of turbulence in a fluid is that the non-dimensional Reynolds number $R_e = LV/v$ be numerically large. By analogy, a magnetic Reynolds number R_m may be defined as

$$R_m = LV/v_m \qquad (1.28)$$

Thus, the condition for transport to dominate leak is that $R_m \gg 1$. This condition is only rarely satisfied in the laboratory – in cosmic masses, however, it is easily satisfied because of the enormous size of L. Thus, under laboratory conditions, lines of force slip readily through the material in cosmic masses, on the other hand, the leak is very slow and the lines of force can be regarded as very nearly frozen into the material.

The generation of magnetic fields in astrophysical bodies relies on fluid motion having a large R_m – in the case of a dynamo operation in the Earth's core, it has been estimated that $R_m \gtrsim 10$. This is necessary so that the magnetic field can be distorted enough by the fluid motion to reinforce the large-scale field. Additionally, however, it is necessary that magnetic field lines diffuse sufficiently through the fluid and dissipate rapidly enough to keep the field topology simple. Otherwise the main effect of the fluid motion would simply be to tangle the field lines without producing efficient regeneration.

To the electromagnetic equations must be added the hydrodynamical equation of fluid motion in the Earth's core (the Navier–Stokes equation) together with the equation of continuity, which, for an incompressible fluid (the speed of flow is much less than the speed of sound in the Earth's core) reduces to

$$\nabla \cdot U = 0 \tag{1.29}$$

The Navier–Stokes equation is

$$\rho\left(\frac{\partial U}{\partial t} + (U \cdot \nabla)U + 2\Omega \times U - \nu\nabla^2 U\right) - \frac{1}{\mu_0}(\nabla \times B) \times B = -\nabla p + \rho\nabla W \tag{1.30}$$

where U is the velocity relative to a system rotating with angular velocity Ω, p the pressure, W the gravitational potential (in which is absorbed the centrifugal force), and ρ and ν the density and kinematic viscosity, respectively. Equations (1.24) and (1.30) contain only the vectors U and B and are the basic equations of field motion.

There is no *a priori* reason why the Earth's magnetic field should have a particular polarity and there is no fundamental reason why its polarity should not change. It is easy to see that dynamos can produce a field in either direction. The induction equation (1.24) is linear and homogeneous in the field and the Navier–Stokes equation (1.30) inhomogeneous and quadratic. Thus, if a given velocity field will support either a steady or a varying magnetic field, then it will also support the reversed field and the same forces will drive it. This, however, merely shows that the reversed field satisfies the equations – it does not prove that reversal will take place.

Because of the complexity of the equations describing the hydromagnetic conditions in the Earth's core, most effort has been directed to seeking solutions of Maxwell's equations for a given velocity distribution. This approach, known as the kinematic dynamo problem, is linear and has been the subject of much investigation (see, e.g., Roberts and Gubbins, 1987). Although it is now known that kinematic dynamos exist, solutions in which Maxwell's equations are solved for specified velocities are of limited geophysical interest, since there is no guarantee that there exist forces in the Earth's core that can sustain them. Without a satisfactory theory to account for the driving force, the problem is not realistic and has only been 'pushed one stage further back'. In a dynamical theory the velocities would be calculated from assumed forces – very little work has been done on this more difficult non-linear problem. A number of possible driving mechanisms for the Earth's dynamo have been proposed (for a comprehensive review, see Gubbins and Masters, 1979) – a general discussion of this problem is beyond the scope of this book.

The power source must replace the energy lost by electric currents and be efficient enough to do so without generating too much heat (the observed surface heat flux is about 4×10^{13} W, most of which comes from radioactive decay in the crust, with at most about 25% or 10^{13} W coming from the core). The power source must also be long-lasting, since palaeomagnetic measurements have shown that the Earth's magnetic field has existed for at least 3500 Ma. Gubbins and Masters (1979) discuss in detail five possible power sources – radioactive heating, primordial heat and freezing of the inner core, gravitational energy release, chemical effects such as the heat of reaction between components, and precessional and tidal forcing. They favour stirring of the core by differentiation associated with the growth of the inner core. Shock-wave data indicate that the outer core is predominantly iron, with some 8–15% light alloying elements, and the inner core pure iron, with perhaps a little nickel. An advantage of this mechanism is its greater efficiency (see, e.g. Gubbins, 1977; Loper, 1978; Gubbins *et al.*, 1979). All the gravitational energy released by rearrangement of core material is available to generate magnetic field. This process stirs the core directly, the magnetic field being the medium through which mechanical work is converted into heat. It must be stressed that the concentration of light material in the core is a different *component* in the mixture. Busse (1972) and Malkus (1973) have considered slurries in which the concentration is that of the solid phase. The essential difference is that the concentration of a phase can change rapidly because of pressure changes, since it only requires melting or freezing rather than actual molecular diffusion of one component.

Schloessin and Jacobs (1980) have suggested a modification to the above model which might also explain reversals of the Earth's magnetic field. In their model the light material rises all the way to the CMB and does not recombine at some higher level in the core. Thus, the inner core and lower mantle have been growing from some time soon after the Earth formed, at the expense of an initially more extended and probably entirely liquid core. On this assumption motions in the outer core are caused and sustained by currents which offset density inhomogeneities due both to changes in concentration of certain constituent components and to changes in temperature generated at the advancing CMB and inner core (ICB) boundary. In this model the energy available for fluid core motions and, hence, for maintenance of the magnetic field is directly related to the time rate of change of the growth of the solid components at the ICB and CMB. Magnetic polarity reversals might be explained as due to epochs during which the solid growth rate which dominates the fluid motion shifts from the ICB to the CMB and vice versa. Further possible effects of changing conditions at the CMB or ICB on reversals of the Earth's magnetic field are discussed in Sections 5.8 and 8.5.

1.4 Mean-field electrodynamics

Most dynamo models use large-scale, highly ordered fluid motions, i.e. motions in which the characteristic length of the velocity field is not much less than the radius of the Earth. In the early 1950s several attempts were made to produce models in which turbulent (i.e. random and small-scale) velocities might act as dynamos. This theory, which has been called mean-field electrodynamics, has been developed independently by Moffatt (1970) in Britain and by Krause, Rädler and Steenbeck in Germany. An account of the German work has been given by Krause and Rädler (1980).

In mean-field dynamo models the velocity U and magnetic field B are each represented as the sum of a statistical average and a fluctuating part. We thus write

$$U = U_0 + u \quad \langle u \rangle = 0 \tag{1.31}$$

$$B = B_0 + b \quad \langle b \rangle = 0 \tag{1.32}$$

The average fields U_0 and B_0 are assumed to vary on a length scale L, while the fluctuating fields u and b (with zero statistical average) are assumed to vary on a length scale l ($l \ll L$). This separates the velocity and magnetic fields into mean, slowly varying and fluctuating parts. The induction equation may then be divided into its mean and fluctuating parts,

$$\partial B_0 / \partial t = \nabla \times (U_0 \times B_0) + \nabla \times \varepsilon + \nu_m \nabla^2 B_0 \tag{1.33}$$

$$\partial b / \partial t = \nabla \times (U_0 \times b) + \nabla \times (u \times B_0) + \nabla \times G + \nu_m \nabla^2 b \tag{1.34}$$

where

$$\varepsilon = \langle u \times b \rangle \quad \text{and} \quad G = u \times b - \langle u \times b \rangle \tag{1.35}$$

ε can be regarded as an extra mean electric force arising from the interaction of the turbulent motion and magnetic field. If the velocity field is isotropic, it can be shown that

$$\varepsilon = \alpha B_0 - \beta \nabla \times B_0 \tag{1.36}$$

where α and β depend on the local structure of the velocity field.

The induction equation (1.33) satisfied by the mean field is

$$\partial B_0 / \partial t = \nabla \times (\alpha B_0 + U_0 \times B_0) + (\nu_m + \beta) \nabla^2 B_0 \tag{1.37}$$

The term αB_0 represents an electric field parallel to B_0. The quantity β is an eddy diffusivity similar in its effects to the ohmic diffusivity ν_m. It operates by mixing magnetic fields transported from neighbouring regions – its effect is to replace ν_m by a total diffusivity $\theta_T = \nu_m + \beta$. Parker (1955) first drew attention to the possibility that $\varepsilon = \alpha B_0$, and Steenbeck and Krause (1966) christened this the α effect. A key concept in the theory is the helicity, defined as $u \cdot (\nabla \times u)$. Parker (1955, 1970) showed that convective fluid motions having non-zero helicity could distort lines of magnetic force in such a way as to produce a regeneration mechanism. A non-vanishing helicity indicates that the vorticity $\nabla \times u$ tends to turn predominantly either in a clockwise or anticlockwise sense about the direction of the velocity (for further details see Moffatt, 1978; Krause and Rädler, 1980). The helicity of fluid motions is a consequence of the action of the Coriolis force on convection in rotating bodies. This is the reason for the importance of rotation in the dynamo mechanism. It must be remembered, however, that the Coriolis force can only change the direction of flow – it cannot change the speed, since it acts at right angles to the direction of flow, and so cannot drive it against the retarding effects of other forces. Since the magnetic field B is divergence free ($\nabla \cdot B = 0$), it can be represented by a toroidal and a poloidal magnetic field. The toroidal field has no radial component and is thus confined to the Earth's core. The poloidal field has a radial component in general, and joins continuously with the external observed field. There are two possible types of dynamo using the α effect – $\alpha\omega$ dynamos and α^2 dynamos. In an $\alpha\omega$ dynamo, the α effect is used in conjunction with a large-scale shear flow (the ω effect). Parker (1955, 1971, 1979) is primarily responsible for the development of $\alpha\omega$ dynamos, in which an α effect from cyclonic turbulence in the core generates poloidal field from toroidal field and differential rotation creates toroidal field from poloidal field, thereby completing the cycle. In α^2 dynamos, both the toroidal and poloidal magnetic fields are generated by α effects.

A dynamo in which both an α effect and an ω effect help to generate a toroidal field from a poloidal field is called an $\alpha^2\omega$ dynamo. In order for dynamo action to occur in α^2 dynamos, the helicity must exceed a critical value. The helicity may be positive or negative – the only requirement for the dynamo to operate is that the helicity be of sufficient magnitude, regardless of sign. A good account of mean-field electrodynamic dynamos has been given by Roberts and Gubbins (1987). Olson (1983) has investigated reversals of the magnetic field based on the α^2 dynamo. He derived the response of α to fluctuations in core turbulance by considering fluctuations in the core's net helicity. He showed that the α effect is inversely proportional to the internal magnetic field energy through a 'source time' function – the magnitude of this function determines the magnitude of the equilibrium magnetic fields, and the sign their polarity. An account of Olson's work is given in Section 5.5. Olson and Hagee (1990) later investigated the structure of the transition field during a reversal for both α^2 and $\alpha^2\omega$ dynamos.

2
The magnetization of rocks

2.1 Introduction

Although most rock-forming minerals are non-magnetic, all rocks show some magnetic properties due to the presence of various iron oxide and sulphide minerals making up only a few per cent of the rock. These minerals occur as small grains dispersed through the magnetically inert matrix provided by the more common silicate minerals that make up most rocks. After the initial formation of the rock, subsequent weathering, diagenetic or metamorphic processes may produce iron oxides from these minerals, and such oxides may themselves carry a component of remanent magnetism which dates from the time of alteration.

When a rock forms, it usually acquires a magnetization parallel to the ambient magnetic field referred to as a *primary magnetization*. This can give information about the direction and intensity of the magnetic field in which the rock formed. However, subsequent to formation, the primary magnetization may decay either partly or wholly and further components may be added by a number of processes. These subsequent magnetizations are called *secondary magnetization*. The magnetization first measured in the laboratory is called the *natural remanent magnetization* (NRM). A major problem in palaeomagnetic investigations is to recognize and eliminate secondary components.

The mineralogy of rock magnetism is complicated by a multiplicity of phases and solid solutions of iron oxide, particularly with titanium dioxide. Most magnetic minerals are within the ternary systems $FeO-Fe_2O_3 - TiO_2$. There are essentially two types of mineral: the strongly magnetic cubic oxides magnetite (Fe_3O_4), maghemite (γFe_2O_3) and the solid solutions of magnetite with ulvospinel (Fe_2TiO_4) known as titanomagnetite; and the more weakly magnetic, rhombohedral minerals based on hematite (αFe_2O_3) and its solid solutions with ilmenite ($FeTiO_3$). The spontaneous magnetization of magnetite is considerably greater than that of hematite, so that magnetite-bearing rocks (such as basic igneous rocks) normally have much stronger remanent magnetizations than rocks whose principal magnetic carriers are hematite (e.g. many red bed formations). The physical properties of rock magnetism have been discussed in detail by Fuller (1970), Stacey and Banerjee (1974) and Creer *et al.* (1975).

There are several different types of magnetism. In a typical substance there is no overall magnetism arising from the orbital motion of the electrons round the central nucleus. However, if it is placed in a magnetic field, a force is exerted on each of the orbital electrons tending to modify its orbit slightly. Since the resulting effect tends to oppose the applied field, the magnetization acquired in this way is negative (i.e. the susceptibility is negative) and is called *diamagnetism*. All substances possess diamagnetism, which is independent of temperature, although it is often obscured by other superimposed effects. Many common minerals such as quartz and feldspar are predominantly diamagnetic. The electrons also have a spin motion about their axes in addition to their orbital motion around the central nucleus. They are usually randomly oriented, but in the presence of a magnetic field they tend to line up in the direction of the field, giving an increase in the magnetization. This effect is called *paramagnetism*. In natural minerals, only a few important ions show significant paramagnetic properties, the commonest being Mn^{2+}, Fe^{3+} and Fe^{2+}. Since an applied magnetic field tends to orient the spins while thermal fluctuations tend to randomize them, paramagnetic effects are strongly dependent on temperature.

Some materials exhibit a permanent (spontaneous) magnetization even in the absence of an external field. This phenomenon of *ferromagnetism* is due to the 'exchange' interaction between atoms. Some substances (e.g. iron, cobalt, nickel) contain unpaired electrons which are magnetically coupled between neighbouring atoms. This interaction results in a strong spontaneous magnetization (i.e. without the application of an external field) and in the property of being able to retain the alignment imparted by an applied field after it has been removed. These properties are several orders of magnitude greater than those of diamagnetic atoms in the same substance. Below some critical temperature (the Curie temperature) the interaction dominates, but above this temperature thermal disordering takes over and the behaviour is that of a simple paramagnetic material.

Some substances are characterized by subdivision into two sublattices, the atomic moments of which are each aligned, but antiparallel to one another. The ferromagnetic effects cancel one another out when the moments of the two sublattices are equal and there is no net magnetic moment. This phenomenon is known as *antiferromagnetism*. Such substances do not have a Curie temperature, because there is no net ferromagnetism. In this case the ordering of the atomic moments is destroyed at a critical temperature called the Néel temperature, above which the substances exhibit paramagnetism. If the atomic moments of the two sublattices are unequal, there is a net spontaneous magnetization – a weak ferromagnetism known as *ferrimagnetism*. Most naturally occurring magnetic minerals are either antiferromagnetic or ferrimagnetic. Ferrimagnetic minerals include magnetite, maghemite and some members of the ilmenite – hematite solid solution series; antiferromagnetic minerals include hematite, ilmenite and ulvospinel. These last two minerals, however, have Curie or Néel temperatures well below room temperature. Some minerals possess a feeble spontaneous magnetization which is superposed on an antiferromagnetic structure and which disappears along with the antiferromagnetism at the Néel temperature. This is called imperfect antiferromagnetism and may be due to imperfect antiparallel alignment (canting), lattice imperfections or a small parasitic component (see Figure 2.1(a), (b)).

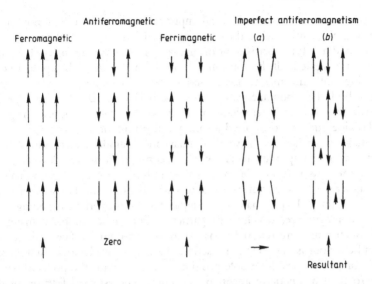

Figure 2.1 Schematic representation of spontaneous magnetization in crystals. The arrows represent the elementary moments, the resultants being given at the bottom. After Irving (1964).

The best-known canted antiferromagnetic is hematite, in which equal sublattice magnetizations are not quite antiparallel so that there is a small net magnetization normal to the average spin axis.

A certain amount of energy (magnetostatic energy) is required for a body to exhibit remanence. While the exchange energy tends to line up all the spins, the magnetostatic energy attempts to prevent this in order to minimize the total energy. The result is a balance of energy in which small zones, a few microns in size, are uniformly magnetized while adjacent zones may have their magnetization in some other direction. These zones are called *magnetic domains*, domain walls separating adjacent regions in which the spontaneous magnetization is in different directions. A sample of ferromagnetic material may thus have only a weak overall spontaneous magnetization, although in individual domains it may be quite large.

Within any crystal there are certain 'easy' directions of minimum energy along which the magnetic dipoles prefer to be aligned. However, in domain walls separating adjacent regions of uniform magnetization, some spins cannot be aligned in the preferred magnetocrystalline direction; these particular spins require extra energy to overcome the magnetocrystalline effect. In the absence of any external influence, domains form in such a manner as to reduce the total magnetostatic energy and wall energy to a minimum. When a body is placed in a magnetic field, the domain walls can move fairly easily, allowing more of the grain to become magnetized in the direction of the applied field. Such magnetization is reversible when the applied field is small, and the domain walls move back again when the field is removed. If the field is increased, the domain walls are forced over small imperfections and impurities in the grain and cannot return to their original position when the field is removed. The process is then no longer reversible and a definite permanent magnetization is left in the body. If higher fields still are applied, all the spins line up in the direction of the applied field, overcoming both the magnetostatic and magnetocrystalline energies. At this point the body has spontaneous or saturation magnetization. For small grains no domain walls can occur (single-domain grains). At some critical size, the grain will subdivide into two or more domains and form multidomain grains.

The magnetic behaviour of single-domain and multidomain grains is quite different and it is important to know which configurations are relevant to the magnetic minerals in rocks. Hematite commonly behaves as single-domained while magnetite grains are generally multidomained. Some small multi-domained grains (diameter $\lesssim 15$ μm) have some single-domain properties, including high stability of remanence, which is of particular importance in palaeomagnetic studies. Such grains are referred to as pseudo-single-domain. The basic reason for such behaviour is due to the interaction of a domain wall with crystal defects so that none of its possible stable positions coincides with the position which gives zero total magnetic moment for the grain. The essential features of the behaviour of single-domain grains in rocks over geological time can be adequately described in terms of Néel's (1949, 1955) simple theory.

2.2 Processes of magnetization in rocks

The mechanism by which NRM is acquired depends upon the mode of formation and subsequent history of the rocks as well as the characteristics of the magnetic minerals they contain. Rocks may become magnetized by a number of different natural processes.

Thermoremanent magnetization (TRM)

Igneous rocks are formed by the cooling of magma from temperatures well over 1000 °C accompanied by the crystallization of various mineral phases (including iron oxides). As the iron oxides continue to cool, an exchange interaction between atoms begins to dominate thermal disorder at a particular temperature (the Curie point). A strong spontaneous magnetization is set up within these crystals parallel to the ambient geomagnetic field. The Curie temperatures for magnetite and hematite are ~ 575 °C and 675 °C respectively. The response time of the magnetic moments of the grains to changes in the magnetic field (the relaxation time) is very short (a few seconds) just below the Curie temperature. However, the relaxation time increases exponentially with increasing grain volume and with decreasing temperature. On cooling through the 'blocking temperature', the relaxation time rapidly exceeds 100 s and the magnetism of the grain becomes effectively locked in the direction of the ambient magnetic field. As the grain cools further, the relaxation time may exceed 10^9 a, so that a record of the geomagnetic field at the time of cooling is effectively 'frozen' into the rock for all time. The total magnetic moment is the sum of all the magnetic moments of the individual grains. The magnetic moment is not stabilized until the grain is cooled below the blocking temperature. A piece of clay will normally contain minerals with blocking temperatures from room temperature up to 680 °C. This type of magnetization is called thermoremanent magnetization (TRM).

If a rock is cooled through various temperature intervals in the presence of a magnetic field, the TRM acquired in each interval is found to be independent of that acquired in each of the other intervals, a result known as the principle of partial thermoremanent magnetism (PTRM) superposition. The sum of all the PTRM values is equal to the total TRM acquired by cooling from above the Curie temperature to room temperature. Thus, if a rock is reheated at a later stage in its history (e.g. through deep burial, intrusion of a nearby igneous body or some tectonic event), a new component of TRM will be acquired, its direction being parallel to the magnetic field at the time of reheating. It will be carried by those grains that have blocking temperatures up to the maximum reheating temperature. A given rock unit may therefore carry several components of TRM corresponding to different heating episodes in its history.

Depositional and post-depositional remanent magnetization (DRM and PDRM)

This type of magnetization is acquired by some sediments which contain small magnetic particles. The magnetic particles will have been eroded from pre-existing rock formations, most of the grains having been derived from igneous or meta-morphic rocks. These grains will carry a magnetization that dates from their time of cooling and they will therefore act as small magnets. When they fall through the water column, they will become aligned in the direction of the ambient magnetic field. This orientation may be preserved during the depositional process, so that the resultant sediment acquires a remanent magnetization which is parallel to the field at the time of deposition. This is called depositional remanent mag-netization (DRM). A post-depositional remanent magnetization (PDRM) may be acquired if the carrier grains are free to rotate in the voids of the sediment matrix as they attempt to follow the secular variation of the geomagnetic field which occurred subsequent to deposition. This PDRM is locked into the sedi-ment by consolidation, which may take from a few days or weeks up to tens of thousands of years. The PDRM process has been studied experimentally in several laboratories during recent years and is now generally accepted as being the more important. In particular, Tucker (1980) and Hamano (1980) have shown that the efficiency of the magnetization process depends on the relative diameters of the carrier and bulk matrix grains, being most effective when the carrier grains are smaller than the matrix grains. The size of the voids progressively decreases as a consequence of compaction which occurs as a result of the deposition of more and more sediment as time goes by. Thus, the ability of the carrier grains to rotate towards the geomagnetic field direction is progressively impeded and decreases sharply at a critical value of the porosity. There is therefore a critical depth (below the water–sediment interface) above which the magnetic particles can realign with the magnetic field. This depth is called the 'locking-in depth'. PDRM is thought to be primarily responsible for the NRM of deep-sea sediments, and either DRM or PDRM for that of lake sediments.

The processes which contribute to PDRM acquisition and 'lock-in' depth are not well understood. Laboratory experiments have shown that there is often a considerable lag between the field change and the actual depth at which it is recorded (Løvlie, 1974; Hamano, 1980). In high-porosity surface sediments, fine magnetic grains are free to rotate in the pore spaces. As the sediment is buried below the bioturbation zone, pore spaces are reduced and the ambient magnetization direction becomes locked in as the consolidation process overcomes the torque of the Earth's magnetic field. Thus, the depth of PDRM 'lock-in' is related to factors influencing early consolidation – such as grain size and shape, sediment composition and sedimentation rate. Estimates for the PDRM 'lock-in' zone in marine sediments range from 0.1 to 1.0 m below the sediment surface. de Menocal *et al.* (1990) analysed eight deep-sea sediment cores and found the average 'lock-in' depth for higher sedimentation rate cores to be ~ 16 cm. They also confirmed the observation of Burns (1989) that low-accumulation-rate sediments have relatively deep PDRM acquisition. Burns attributed this to relatively higher clay content and, hence, higher porosity. de Menocal *et al.*, however, suggest that the deeper PDRM acquisition of low-accumulation-rate sediments is more likely to be due to time-dependent accumulation processes in clay-rich sediments than to the clay content itself. Low-accumulation-rate sediments have sufficient time to develop increased inter-particle rigidity (structural strength) which inhibits early compaction and, hence, PDRM acquisition. de Menocal *et al.* also investigated the relative stratigraphic positions of the Brunhes–Matuyama boundary, oxygen isotope interglacial stage 19.1 and the widespread Australasian microtektite strewn field (see Section 8.4).

Niitsuma (1977) proposed a model for the acquisition of remanent magnetization in sediments in which the magnetization is gradually acquired in a zone ~ 30–40 cm in thickness and at a depth of ~ 40 cm below the depositional surface. Okada and Niitsuma (1989) tested this model, using data from three high-deposition-rate cores within a distance of 14 km of one another in the Boso peninsula, Japan. They estimated a depth lag of magnetization of 42 cm and the average width of the magnetization zone to be a few tens of centimetres.

Isothermal remanent magnetization (IRM)

Magnetic materials may acquire a remanent magnetization without heating if exposed to a magnetic field. In many natural materials, the IRM acquired in the Earth's magnetic field is less than the TRM, although this may not be true for magnetically soft materials.

Viscous remanent magnetization (VRM)

Viscous remanent magnetism is essentially time-dependent IRM. If a magnetic material is exposed to a magnetic field, it may slowly acquire a magnetization in the direction of that field. The rate at which VRM is acquired (or decays) depends on the temperature and particle size. For any particle the relaxation time $\tau \propto \exp$ (v/T), where v is the volume of the grain and T the temperature. Very fine ferromagnetic particles (less than domain size) have very short relaxation times even at normal temperatures, and are said to be *superparamagnetic*. A small increase in volume can change a particle from superparamagnetic to single-domain; a small decrease in temperature leads to the same effect. A material which contains many small particles will have some which are superparamagnetic and some which are single-domain. For each grain there is a critical blocking temperature T_B at which τ becomes small but which might also be below the Curie temperature. Similarly, at any given temperature T there is a critical blocking diameter d_B (corresponding to a sphere of volume v_B) at which τ becomes small.

Chemical remanent magnetization (CRM)

Physicochemical changes may take place in rocks subsequent to their initial formation and may involve crystallization of new magnetic mineral phases. After nucleation, when the grains are very small, their relaxation times will be very short. As they grow through a critical volume (the blocking volume), their magnetization becomes aligned with the ambient geomagnetic field and the rock acquires a component of magnetization dating from this time. This is called chemical remanent magnetization (CRM), and may be important in sedimentary rocks as a result of lithification or chemical weathering.

In the majority of igneous rocks the NRM consists mainly of TRM, even though CRM and VRM may often contribute significantly. The NRM is carried primarily by titanomagnetites – and sometimes by titanohematites. The magnetic mineralogy of deep-sea sediments is not well known; the NRM appears to be carried mainly by titanomagnetites, although contributions from pyrrhotite (in certain anoxic environments) and ferromanganese oxides and oxyhydroxides also contribute. The magnetic mineralogy of terrestrial sediments is highly variable, since the sediments originate in part from pre-existing igneous, sedimentary and metamorphic rocks, and also because magnetic minerals sometimes form in sediments by authigenesis and diagenesis. The origin of the magnetization of the so-called 'red beds' (principally red sandstones and shales) is discussed briefly later.

The simplest magnetization to interpret is that carried by a lava. The lavas cool through the range of blocking temperatures, in which the primary NRM is acquired as a TRM in a time which is short compared with the time constants of the secular variation. Hence, each lava gives a spot reading of the local field direction. Determinations of the ancient field intensity are particularly reliable from such rocks. However, it is impossible to determine the absolute ages with significant resolution to establish the time of the spot readings provided by the flows – a sequence of lava flows may have been erupted over a few years or thousands of years or more. Lavas can thus give excellent palaeointensity data and accurate spot readings of the field direction with no knowledge of the direction between successive flows.

Sedimentary records, on the other hand, have the advantage that sedimentation is, in general, a more continuous process than is the extrusion of lavas. However, the type of sedimentary environment which gives the most continuous sedimentation is also one in which absolute sedimentation rates are low, so that resolution of the record tends to be poor. Another problem is that it is often difficult to know whether a hiatus in sedimentation has occurred. However, if the sedimentation rate is rapid and constant, sedimentary records (oceanic or on land) may give a detailed description of field changes. Although the data are filtered, one can reasonably expect that there are no gaps in the record and that the higher-frequency components of the transition field have been removed by smoothing.

Hoffman and Slade (1986) stress that progressive time-averaging of palaeomagnetic reversal records can reduce resolution to such an extent as to significantly alter features and waveforms that are intrinsic to actual core processes. They illustrate their point by comparing actual data from the reverse to normal Miocene record obtained from thick sequences of basaltic lavas at Steens Mountain, Oregon (see also Section 6.7) with records obtained by increased smoothing of the data. Features in the declination, inclination and intensity records become obscured by even a small amount of artificial smoothing, and are completely unrecognizable as the degree of smoothing is increased.

Fujiwara and Ohtake (1975) carried out a palaeomagnetic study of late Cretaceous alkaline rocks from Hokkaido, Japan. Both normal and reversed polarities were found in one complex of picritic dolerites. They suggested that the existence of two different palaeomagnetic polarities in one rock complex might be explained on the assumption that the geomagnetic field had been normal at the time when the rock at the margins cooled through the Curie temperature of the ferromagnetic minerals in the rock, but had reversed its polarity when the rock inside the intrusive body became magnetized during cooling. This would imply that the time of the polarity transition was quite short, since it occurred during the cooling of a relatively thin intrusive body.

Denham (1981) has examined the effect of field reversals on the length of time a rock can preserve a particular polarity. The longevity of a magnetic polarity message depends on the decay time constants of the constituent magnetic grains, and Denham estimated the minimum time constants necessary for the preservation of an ancient polarity message against the alternating sequence of polarity reversals. He showed that the viscous decay of a magnetic polarity message in rocks is slowed by a factor of 3–6, owing to reversals of the field compared with the effect in a field that is constantly polarized opposite to the original remanence. It should be noted, however, that increases in temperature have a far greater effect than reversals of polarity, e.g. for original time constants of 1–100 Ma, halving and quartering the relaxation time has the same effect as raising the ambient temperature from 273 K to nominal values of only 276.5 K and 280 K, respectively. Raising the temperature to 300 K would lower the time constants by more than two orders of magnitude.

Channell (1978) has examined the magnetic properties in a bed of pelagic limestones of Cretaceous age from Sicily, which vary in colour from red to white. Samples taken from the red portion of the bed showed normal polarity, and those from the white reversed polarity, the directions being almost antiparallel after partial demagnetization. The remanence in the case of the white variety is due to detrital magnetite, and that in the red to hematite. The hematite magnetization significantly postdates the detrital magnetization.

Red beds have been extensively sampled for palaeomagnetic investigations because of their relatively strong magnetization. In most cases the remanence is carried by hematite, but there is much debate as to whether it is acquired at an early stage of the rock history or much later through diagenetic processes. Many examples have been found in which the magnetization is DRM – in other cases the magnetization is CRM and may have been acquired over several million years.

Many believe that red bed remanence faithfully mirrors the geomagnetic field present at (or soon after) the time of deposition. According to this view, remanence data from red beds can be used to study both large-scale as well as small-scale features of the ancient geomagnetic field. Elston and Purucker (1979) believe that the rapid acquisition of remanence is due to depositional remanence resulting from physical alignment of detrital hematite grains. On the other hand, Helsley and Steiner (1974) suggest that it is due to chemical remanence, as a by-product of hematite authigenesis occurring within several thousand years after deposition. Purucker *et al.* (1980) suggest that both processes may have occurred. Larson *et al.* (1982), on the other hand, believe that none of the above explanations is viable and that red bed remanence primarily consists of multiple, commonly near-antiparallel, components acquired through chemical alteration discontinuously over long time intervals. If this view is correct, the remanence in red beds rarely reflects the geomagnetic field at the time of deposition.

In 1975 a detailed study of the palaeomagnetic, rock magnetic and petrographic properties of the Triassic Moenkopi Formation was carried out. The results of the petrographic study were published by Walker *et al.* (1981), and the rock magnetic and palaeomagnetic study by Larson *et al.* (1982). Larson *et al.* found both normal and reversed Triassic components in many of the same samples at many localities, and that there was a close correlation between the magnetic polarity and intensity of magnetic remanence and the lithologic characteristics in the Sinbad Valley section. Moreover, the intermediate remanence directions throughout the formation were of weak intensity, suggesting the presence of two or more nearly balanced antiparallel components. They concluded that the principal remanence carried by the formation is CRM acquired diagenetically over a geologically long time interval, and they presented a model to explain how it was acquired through natural processes acting intermittently at geologically reasonable rates.

There is still some controversy over the modes and timing of the acquisition of the components of NRM in red sediments. Larson and Walker (1982) interpreted the results of further work they carried out in the Moenkopi Formation which indicate that the characteristic NRM components carried by hematite are acquired as long as 10 Ma after deposition. Liebes and Shive (1982) also studied the characteristic NRM in the Moenkopi and Chugwater Formations, and concluded that it was formed after deposition, but prior to burial by about 1 m. They interpret the characteristic NRM as an early-acquired CRM.

Herrero-Bervera and Helsley (1983) claimed to have seen a polarity transition in fine-grained red beds of the Chugwater Formation. Successive sample directions display a tendency to backtrack by tens of degrees along the transition path. Tauxe and Badgley (1984) suggested that this backtracking could be the result of the varying prominence of components acquired at slightly different times in a rapidly evolving transitional field. Larson and Walker (1985) argued that the data could just as well be explained by a mixture of nearly antiparallel normal and reversed polarity CRM components. However, Helsley and Herrero-Bervera (1985) pointed out that the transitional record did not conform to a great circle path, as would be the case for a mixture of polarities. Further work on the Chugwater Formation has been carried out by Shive *et al.* (1984). They concluded that the characteristic NRM was acquired within a few hundred years of deposition, basing their conclusion on the strong correlation of polarity patterns over distances of 200 km or more and the lack of any evidence for normal and reversed components in individual samples.

Tauxe and Badgley (1984) obtained further evidence for the early acquisition of characteristic NRM in red beds from an analysis of a polarity transition in Miocene Siwalik, Pakistan, red beds in which they were able to decipher the order of acquisition of the components of the NRM. Tauxe *et al.* (1980) had shown earlier that the highest unblocking temperature component is a DRM or PDRM. Tauxe and Badgley (1984) therefore deduced that the components of CRM with lower unblocking temperatures were formed within 100–500 a after deposition while the geomagnetic field was undergoing rapid directional change.

Maillot and Evans (1992) have investigated small-scale variations in the magnetization of a red bed sequence in the Permian Lodève Basin in southern France. They found that magnetic intensity variations (as much as an order of magnitude) can occur over stratigraphic distances of only a few centimetres, with peaks near the top of individual strata. Rock magnetic experiments and petrographic observations suggest that no variation of intrinsic properties of the magnetic grains can explain the magnetic data, and they concluded that the intensity changes must be the result of changes in the acquisition process. It is difficult to envisage a chemical process producing such large intensity variations, and Maillot and Evans suggested that they arise from depositional processes. Tauxe and Kent (1984) and Ellwood (1984) have shown that the intensity of magnetization can be enhanced by perturbations of the still unconsolidated sediments. Moreover, in a continental, semi-arid environment like that in which the rocks of Lodève were formed, the deposition is essentially discontinuous – periods of rapid sedimentation alternating with stages of no deposition. Maillot and Evans therefore conclude that the most likely cause of the variations in intensity is grain mobility which depends on the magnitude of mechanical perturbations, and that the remanence carried by the Lodève strata was acquired during, and very shortly after, deposition. It is not possible to generalize their model to other red bed formations, but, as the authors point out, it would be interesting to know if stratigraphically distributed intensity variations are a common feature of red beds known to carry a stable primary remanence.

Hoffman (1984) has raised another problem concerning the time of acquisition of magnetization in basaltic rocks containing titanomagnetite rich in titanium ($Fe_{3-x} Ti_x O_4$, $x > 0.75$). If not oxidized during cooling, such a magnetic carrier will have a Curie temperature too low to insure the acquisition of a stable TRM. A stable CRM may ultimately be acquired through room temperature oxidation of titanomagnetite to titanomaghemite. Such a process could occur over a time period much later than that of the initial cooling.

van Hoof and Langereis (1991) found 14 reversals in marine marls in the southern coast of Sicily. They carried out a detailed study of two R–N reversals – the lower Nunivak and the lower Cochiti. Thermal demagnetization diagrams showed a viscous and/or secondary component which is removed below 200 °C, a low-temperature (LT) component removed between 200 and 480 °C and a high-temperature (HT) component carried by single-domain magnetite which is removed at 580 °C. The demagnetization diagrams show that the LT and HT components were acquired at different times. A stable HT component in magnetite is generally considered to carry the primary remanence. In a record where the HT component is reversed and the LT component normal, the latter is interpreted to represent a (partial) overprint by secondary remanences. However, if the LT component is reversed and the HT component normal, as is the case in the lower Nunivak reversal record, the LT component is not a secondary overprint. Just after the R→N transition, the HT component records, at a certain depth below the sediment–water interface, the existing N geomagnetic field, whereas at that depth the LT component has already locked into the previously existing R magnetic field. There is thus a lag between the two components. In the lower Nunivak record, this lag is ~ 35 cm (see Figure 2.2). van Hoof and Langereis estimate that this corresponds to a time lag of ~ 5 ka. For the lower Conchiti record, they estimate the time lag to be ~ 14–16 ka. Thus, in these two cases the HT component does not reflect geomagnetic changes during the reversal. The origin of the LT and HT components is probably related to physical and chemical processes below the sediment–water interface, resulting in diagenetic alteration and authigenic formation of magnetic minerals (Lund and Karlin, 1990). The HT component is carried by fine-grained single-domain magnetite (van Velzen and Zijderveld, 1990) possibly of biogenic origin. The origin of the LT component is uncertain. Because of the strong decay of this component between 200 and 330–360 °C, van Hoof and Langereis (1991) suggest that either pyrrhotite or greigite may be the carrier of the remanence.

Quidelleur *et al.* (1992) have also looked into the question of multicomponent magnetization, using records of reversals from three continental sections of the altiplano in Bolivia. In one section (Viscachani) the magnetization is carried by magnetite. In the other two sections (Achocalla and Irpavi) the magnetization

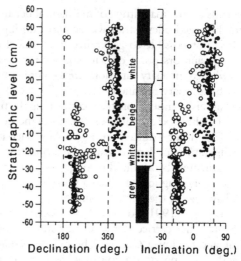

Figure 2.2 Declinations and inclinations during the lower Nunivak reversal. Open (closed) symbols denote LT (HT) directions. The LT component reverses polarity 35 cm higher than the HT component and shows more scatter. The weathering colours (carbonate contents) of each sedimentary cycle are grey (70%), white (80%), beige (60%) and white (80%), whereas fresh colours show more gradual changes from dark blue to light blue. Dots in the lower white layer refer to brown oxidation spots. The 0 cm level was (arbitrarily) put at a clearly recognizable (sedimentary) level. After van Hoof and Langereis (1991).

is characterized by a complex magnetic mineralogy dominated by magnetic sulphides (mainly pyrrhotite) and magnetite. At Achocalla the low-temperature (LT) and high-temperature (HT) components show the same polarity pattern. However, the transitional directions of the LT component are located ~ 50 cm below the abrupt polarity change recorded by magnetite. These observations are very similar to those obtained by van Hoof and Langereis (1991) from a marine sequence in Sicily. In both cases the succession of polarity intervals recorded by the LT and HT components is identical. In the last section (Irpavi) the LT and HT components have opposite polarities. Although the magnetic mineralogy indicates the presence of sulphides, Quidelleur *et al.* do not rule out the possibility that the LT component has a secondary origin due to coarse magnetic grains.

An excellent introduction to the fundamental principles of palaeomagnetism has been given in a recent book by Butler (1992). The book also contains a practical discussion of the techniques of measurement and analysis, topics not covered in this book.

2.3 Self-reversal

There are two major classes of self-reversal mechanisms – TRM, which will be reproducible in laboratory experiments, and CRM, which will not be reproducible. Natural non-reproducible self-reversals are obviously much more difficult, if not impossible, to confirm. Merrill and Grommé (1969) argued that the coexistence of normally magnetized magnetite and reversely magnetized hemoilmenite in the dioritic outer rim of the Bucks batholith, California, implies that the hemoilmenite has self-reversed by some non-reversible mechanism. However, McClelland and Goss (1993) have pointed out that their data could equally well be explained by the two magnetic minerals not being coeval and mineralogical changes having taken place significantly later than the original acquisition of remanence. McClelland (1987) also invoked non-reproducible self-reversal to explain anomalous magnetic behaviour in a section of a profile through a contact aureole of a Tertiary dike intruding Siluro-Devonian andesites near Mull, Scotland.

McClelland and Goss (1993) have continued the work of Hedley (1968), who showed that self-reversal occurs on the transformation of maghemite to hematite after its formation by dehydration of synthetic acicular lepidocrocite. Their experiments suggest that self-reversal of hematite remanence only occurs when the parent maghemite is still blocked at the temperature of its transformation to hematite. When the maghemite is unblocked, the resulting hematite remanence is normally magnetized. They further conjecture that the self-reversal may be a feature of the maghemite to hematite transition only – it may not be necessary for the maghemite to have been formed from lepidocrocite, its contribution being to produce maghemite of the appropriate grain size to remain blocked at the temperature of its transformation to hematite. Transformation of maghemite to hematite in nature is likely to happen at much lower temperatures, greatly extending the grain-size range in which this type of self-reversal might occur.

Although evidence (to be discussed later) is overwhelmingly in favour of reversals of the Earth's field as the cause of reversed magnetization, before such an explanation is accepted, it must be asked whether any physical or chemical processes exist whereby a material could acquire a magnetization opposite in direction to that of the ambient field. Graham (1949) found some sedimentary rocks of Silurian age which were reversely magnetized. He was able to identify the precise geological horizon over a distance of several hundred kilometres by the presence of a rare fossil which existed only during a short geological period. He found that some parts of the horizon were normally magnetized and some reversely, and argued that this could not be accounted for by a reversal of the Earth's field, which would affect all contemporaneous strata alike. (However, if the timescale of reversals in Silurian times was as short as has been found in some later rocks, Graham's fossil might well have survived at least one reversal.) Graham therefore wrote to Professor Néel at Grenoble and asked him whether he could think of any process by which a rock could become magnetized in a direction opposite to that of the ambient field. Néel (1951, 1955) suggested, on theoretical grounds, four possible mechanisms: by means of magnetostatic or superexchange reactions between either different minerals or different compositional volumes within the same mineral grain. Magnetostatic interactions are now thought to be too weak (Uyeda, 1955; Stacey, 1963).

The first and third of Néel's mechanisms involve only reversible physical and/or chemical changes. In his first mechanism Néel imagined a crystalline substance with two sublattices A and B, with the magnetic moments of all the magnetic atoms in lattice B oppositely directed to those of lattice A. If the spontaneous magnetization of the two sets of atoms J_A and J_B vary differently with temperature, Néel suggested that the resultant magnetization of the whole, $J_A + J_B$, could reverse with change in temperature (Figure 2.3). Gorter and Schulkes (1953) later synthesized a range of substances with the properties predicted by Néel, although no naturally occurring rock has been found which behaves in this manner. In ferrites the two constituents A and B are the two interpenetrating cubic sublattices which make up the crystal structure. Because the exchange interaction between the cation sites in the two sublattices is negative, the B sublattice acquires a spontaneous magnetization exactly antiparallel to that of the A sublattice when the ferrite cools through its Curie temperature. Given appropriate temperature coefficients for the two spontaneous magnetizations, reproducible self-reversal will occur on further cooling.

Néel's second mechanism is a modification of the first, in which $J_A > J_B$ at all temperatures, so that no reversal would take place. However, Néel suggested that subsequent to the formation of such a substance, chemical or physical changes might occur which would lead to the demagnetization of lattice A, leaving the reverse magnetization of lattice B predominant. No evidence of such a possibility actually occurring in nature has yet been found.

In his third mechanism Néel considered a substance containing a mixture of two different types of grains, A and B, one with a high Curie point T_A and a low intensity of magnetization J_A, and the other with a low Curie point T_B and a high magnetization J_B (Figure 2.4). When such a substance cools from a high temperature, substance A, because of its higher Curie point, becomes magnetized

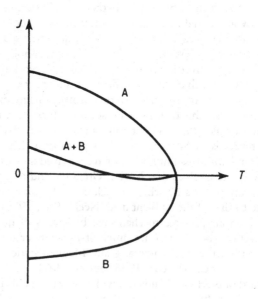

Figure 2.3

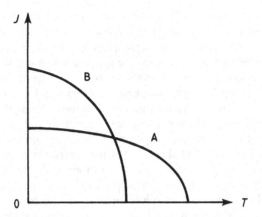

Figure 2.4

first in the direction of the ambient field. When the temperature falls below T_B, substance B becomes magnetized but will be subject to the dual influence of the ambient field and of the field due to the grains of substance A. Néel suggested that under suitable geometrical conditions, the resultant direction of magnetization of B could on the average be opposite to that of the ambient field. At room temperature the greater value of J_B causes the resultant magnetization of the whole to be in the opposite direction to the ambient field.

Néel's fourth mechanism, like his second, involves the possibility of subsequent demagnetization by physical or chemical changes. The reverse magnetization might be possible later in time, even though initially the intensity of the B component was not large enough. Since discrete magnetized grains free to rotate must align themselves along and not against the field, only the second and fourth of Néel's mechanisms could apply to sedimentary rocks; igneous rocks could, in theory, become reversely magnetized by any of them.

In 1952, Nagata *et al.* found a dacite pumice from Mount Haruna, Japan, which was shown in the laboratory to be self-reversing and to contain an intimate mixture of two magnetic minerals, an ilmenohematite and a titanomagnetite, indicating reversal by magnetostatic interaction (Nagata *et al.*, 1952; Uyeda, 1955). However, subsequent separation of the minerals revealed that the self-reversal was an intrinsic property of one of them alone and that the presence of titanomagnetite was irrelevant (Uyeda, 1958; Nagata and Uyeda, 1959). Uyeda (1955) showed the magnetic mineral responsible to be a member of the ilmenite–hematite solid solution series ($x\mathrm{FeTiO_3} - (1-x)\mathrm{Fe_2O_3}$), where x is the mole fraction ilmenite. Uyeda later (1958) showed the reverse TRM to be intrinsic to series members having a range of compositions near the centre of the series ($0.60 > x > 0.45$), and intimately related to the occurrence of a cation order–disorder transformation observed for these compositions. Ordering in this system involves the magnetism of titanium ions to alternate cation layers, a process which develops a strong ferrimagnetic moment. Uyeda suggested that self-reversal resulted from some form of antiparallel spin coupling between coexisting cation ordered and disordered phases, with the Curie point of the disordered phase being higher than that of the ordered phase. Much work then followed on the magnetic properties of the strongly ferrimagnetic, cation-ordered phase. However, progress on the magnetic properties of the cation-disordered phase was slower. This is because this phase (referred to as the 'x phase') is neither sufficiently magnetic nor abundant enough in size or volume to be detected through thermomagnetic experiments or X-ray diffraction.

Ishikawa and Syono (1963) showed that the reversal is connected with the ordering and disordering of iron and titanium ions in the lattice. Self-reversal occurs only when ordered and disordered phases are both present in metastable equilibrium, and seems to be the result of an antiparallel superexchange coupling between the phases. Specimens which are completely ordered or disordered did not show reversed TRM. The mechanism is essentially different from any of Néel's mechanisms. Moreover, the self-reversing properties were a function of cooling rate. Since the rate of cooling in laboratory experiments is many orders of magnitude greater than that at which rocks cool in nature, the results may have no geophysical significance. Westcott-Lewis and Parry (1971) later found that the self-reversing property of ilmenohematites disappeared for grains smaller than 4 μm, which acquire only normal TRM. They suggested that this is because chemical inhomogeneities become important in small grains which do not contain sufficient of both the ordered and metastable phases. Carmichael (1959, 1961) has also produced a self-reversal in the laboratory in an ilmenite–hematite solid solution containing much less ilmenite. He suggested that the self-reversal may be related to the exchange of electrons between Fe^{2+} and Fe^{3+} ions in the oppositely directed magnetic sublattices of the solid solution.

Verhoogen (1956, 1962) suggested that a possible self-reversal mechanism could occur as a result of the ionic migration between different lattices, thus changing the relative magnetic strengths of lattices of opposite polarity in antiferromagnetic minerals. However, O'Reilly and Banerjee (1967) pointed out that the composition range over which such mechanisms might occur is small and corresponds to a very high degree of oxidation that is metastable under most geological conditions. Moreover, it is difficult to recognize such a process, since it is irreversible and cannot be repeated by reheating and cooling in the laboratory.

Hoffman (1992) suggested a new model of self-reversal in the ilmenite–hematite system based on the results of his earlier (1975) thermomagnetic experiments on two synthesized $Il_{60}Hem_{40}$ samples. He assumes that the x phase is an Fe-rich material formed along the boundary between transformation-induced cation domains antiphase in composition. The cation-disordered boundary material is treated magnetically as if it were the end-member hematite with a canted, or tilted spin, arrangement. He then suggests the reverse TERM arises within the bulk, ferrimagnetic material through both interphase superexchange (with the boundary material) and intraphase superexchange which ultimately produces a non-collinear spin arrangement.

Heller and Egloff (1974) have examined the NRM of a granite-aplite dike of the Bergell Massif, Switzerland, which showed stripes of normal and reversed polarity. Thermal experiments indicated 'imperfect' self-reversal due to the presence of two interacting phases with different Curie points within the exsolved ilmenohematite. Heller later (1980) carried out thermomagnetic experiments on the Olby–Laschamp lavas from the Chaîne des Puys, France, and showed that some of the samples of the Olby flow undergo complete self-reversal during thermal demagnetization – other samples showed partial self-reversal of the NRM at room temperature. He suggested that the thermomagnetic behaviour is due to magnetic phases of titanomagnetite with different Curie temperatures. Heller and Petersen (1982a,b) later carried out magnetic and optical examination of some of the samples and suggested that the self-reversal is caused by a negative magnetostatic coupling between titanomagnetite phases of widely varying degrees of oxidation. This has been contested by Roperch *et al.* (1988), who showed that the Laschamp flow samples exhibited magnetic stability at high temperatures (400–650 °C). (The magnetization of the Olby–Laschamp lava flows is discussed further in Section 4.2.)

The duration of the formation of the magnetic phases is highly important. If the interacting phases were present prior to the acquisition of remanence, then in a cooling lava a high-Curie-point phase would be magnetized first and the magnetization of this phase would control the interactions leading to a reversal during further cooling. If the secondary interacting phases were formed after the acquisition of remanence, by low-temperature physical or chemical processes, then a primary phase having a low Curie point might be responsible for the reversal of a later phase having a higher Curie temperature.

Heller and Petersen (1982b) showed that the titanomagnetites of the Olby and Laschamp flows were oxidized under different conditions and to a varying extent. They suggested that the observed self-reversal is caused by interaction of two magnetic phases: little-oxidized or unoxidized primary titanomagnetite having T_c around 180 °C, and secondary titanomaghemite having T_c up to 400 °C.

Kennedy (1981) has also found self-reversal of TRM in a dacite from Mount Natib in the Philippines which may be explained by Ishikawa and Syono's (1963) model. The Fe-rich metastable phase produced during ordering in the minerals of the ilmenite–hematite solid solution represents the initial stages of the exsolution process. Exsolution-unmixing of an Fe-enriched phase aids ordering in the remainder of the mineral by increasing its Ti content.

Lawson *et al.* (1981) have examined synthetic ilmenite–hematite samples by transmission electron microscopy and found well-defined antiphase domains (APD) and antiphase domain boundaries (APB) (APD are chemical domains and, as such, are distinct from magnetic domains). Samples synthesized or annealed at 900 °C, which is below the order–disorder transition for $Ilm_{70}Hem_{30}$, have much larger domains or no domains at all and consequently have a much smaller volume of APB than samples synthesized or annealed at 1300 °C. Only the high-temperature samples acquire reverse TRM when cooled in an applied magnetic field. Lawson *et al.* hypothesized that some critical volume of APB is necessary for the acquisition of reverse TRM. In samples containing less than the critical volume, the APB are unable to couple magnetically with the main body of the domains effectively, and a normal TRM results.

Although examples of Néel's first mechanism (Figure 2.3) have been found in ternary ferrites (Gorter and Schulkes, 1953), the compositions are unlike those found in normal rocks. Two examples of this type of ferrimagnetism have, however, been reported. Schult (1968) found self-reversal of NRM in basalt specimens from Germany when cooled below 200 K; at the temperature of inversion the saturation magnetization showed a minimum. No specimens were found to give self-reversals above about 200 K. The only example of this type of behaviour reproducible in the laboratory with an inversion temperature above 300 K was reported by Kropaček (1968) for a tin-substituted hematite embedded in a cassiterite rock.

Apart from the Haruna dacite, a few other examples have been found of self-reversal of the NRM in volcanic rocks – in certain oceanic basalts (Ozima and Ozima, 1967; Sasajima and Nishida, 1974) and in a basalt from Germany (Schult, 1976). A number of cases of partial reversal have also been discovered.

Oceanic basalts may become magnetized as they cool and initially have the polarity of the ambient field, but chemical changes (such as serpentinization, maghematization, . . .) may occur at a later time when the field is of a different polarity. The chemical remanent magnetization associated with these later processes will thus have an opposite polarity to that of the unoxidized magnetic grains and, on heating, the polarity could reverse.

A dacite pitchstone from Mount Asio, Japan (Nagata et al., 1953), and an iron sand from Sokoto, Japan (Uyeda, 1958), acquire a reversed TRM between 200 and 300 °C, but because it is small compared with that acquired parallel to the field in other temperature intervals, the total TRM is directed along the field. Everitt (1962) first demonstrated the self-reversal behaviour of pyrrhotite in the laboratory. Later Bhimasankaram (1964) showed that reversal in natural and synthetic pyrrhotites takes place between two components with Curie temperatures 560 °C and 310 °C coupled antiparallel to each other. Robertson (1963) has also discovered self-reversal in pyrrhotite in a monozonite from Australia.

Zapletal (1992) measured the remanent magnetization induced at room temperature (IRM) in massive specimens separated from a pyrrhotite Fe_7S_8 crystal. An anomalous dependence of IRM on the applied field was observed, including self-reversal of the IRM at fields of the order of 10–100 mT. Both normal and reversed IRM vectors always lay in the maximum susceptibility plane perpendicular to the crystal c' axis. The reverse IRM was stable for several hours, but was spontaneously changed to normal IRM after several months in some cases. The crystal was composed of thin monocrystalline layers (twins), the orientation of the axes of easy magnetization differing between neighbouring layers. Zapletal suggests that crystal twinning and low coercivity in each twin contribute to the self-reversal of IRM in the ferrimagnetic pyrrhotite crystal. The observed self-reversal cannot be due to magnetic transitions, phase changes or chemical alterations, since there were no temperature changes.

Other partial self-reversals have been detected in pyrrhotite-bearing rocks and inferred in a few other cases. However, pyrrhotite is not common in rocks and is absent from most reversely magnetized rocks. Partial self-reversal leading only to a reduction of the NRM intensity at room temperature, but not to a completely antiparallel alignment of the NRM direction with respect to the ambient magnetic field, has been observed in a historical lava flow from Mount Etna (Heller *et al.*, 1979). It has also been shown that the TRM produced in continental basalts under moderate- or high-temperature oxidation conditions in the laboratory can acquire self-reversal or at least partial self-reversal characteristics at room temperature (Havard and Lewis, 1965; Creer and Petersen, 1969; Petersen and Bleil, 1973; Ryall and Ade-Hall, 1975; Petherbridge, 1977; Tucker and O'Reilly, 1980).

Hoffman (1982) has analysed some Oligocene basalt samples, cored from Yarraman Creek in the Liverpool Range, New South Wales, Australia, and found that they contained two remanence components with opposing directions. The self-reversed component was clearly identified in several samples subjected to detailed thermal demagnetization and was most evident in samples showing the highest degree of low-temperature oxidation. Several flows under investigation were extruded during what appears to have been an excursion of the geomagnetic field, each possessing a magnetization direction far from the full polarity state. The possibility that the self-reversed moments are simply secondary components, acquired in an opposing field direction, is, therefore, ruled out. Optical examinations and thermomagnetic curves indicated that the samples contain from nearly unoxidized to moderately low-temperature oxidized titanomagnetite. The self-reversed component is most clearly removed during thermal demagnetization between 230 and 290 °C above the observed Curie point of the titanomagnetite, and is believed to be associated with titanomaghemite.

Heller *et al.* (1986) collected dacitic andesite pumice fragments which had been hurled several kilometres during the 1985 eruption of the Nevado del Ruiz volcano (Columbia) and showed that they carried a stable but reversed NRM. Heating experiments showed that the magnetization is due to a self-reversal mechanism which also induces a reversed TRM in the laboratory field. The pumice was magnetized in the present (normal) geomagnetic field, and Heller *et al.*'s studies have shown for the first time that self-reversal has actually controlled the process of NRM acquisition. They suggested that possibly two intimately mixed, chemically very similar, phases within the lower-Curie-point phase are interacting. If a high enough temperature is reached, the normally magnetized phase, with slightly higher Curie temperature, becomes magnetized, so that on cooling in zero field no interaction takes place with the other phase of slightly lower Curie temperature. Such a mechanism is similar to the exchange interaction model of Uyeda (1958).

In a later paper Haag *et al.* (1993) investigated the temperature dependence of the magnetic domain pattern in titanomagnetite and titanohemite in the self-reversing andesites. During heating – cooling cycles in the laboratory, the natural remanent magnetization loses its self-reversal properties between 110 and 180 °C – the titanohematite domains also disappear in the same temperature interval. Haag *et al.* also observed along the phase boundary between a titanomagnetite and a titanohematite grain, but within the titanohematite, a 2 μm wide chemically different domain, which has a higher iron content than the main titanohematite. The two titanohematite phases represent two phases, one with a high Curie temperature and low spontaneous magnetization and the other with a low Curie temperature and a high spontaneous magnetization. This is just the case of Néel's third mechanism for self-reversal discussed earlier. Haag *et al.* had proposed in 1990 that two interacting titanohematite phases were responsible for the self-reversal.

Another case of self-reversal of TRM has been found by Ozima *et al.* (1992) in pyroclastics from the 1991 eruption of Mount Pinatubo, Phillippines. At one site the directions of the NRM of dacite pumice fragments were dispersed, indicating that these fragments settled after their temperature reached their Curie points. At another site almost all the dacite pumice fragments showed reversed NRM, implying that these settled before their temperature reached the Curie point, thereby acquiring a self-reversed TRM in the ambient geomagnetic field. Almost all the dacite pumice samples from both sites acquired self-reversed TRM in the laboratory. As a result of detailed laboratory studies, Ozima *et al.* concluded that the mechanism of the self-reversal of the TRM in the Pinatubo pumice is similar to that of the Haruna dacite pumice.

2.4 Field reversal or self-reversal?

To prove that a reversed rock sample has become magnetized by a reversal of the Earth's field, it is necessary to show that it cannot have been reversed by any physicochemical process. This is almost impossible to do, since physical changes may have occurred since the initial magnetization or may occur during laboratory tests. More positive results can only come from the correlation of data from rocks of various types at different sites and by statistical analyses of the relation between the polarity and other chemical and physical properties of the rock sample.

There have been many cases where reversely magnetized lava flows cross sedimentary layers. Where the sediments have been baked by the heat of the cooling lava flow, they are also found to be strongly magnetized in the same reverse direction as the flow. In fact, in about 95% of all cases the direction of magnetization of the baked sediment is the same as that of the dike or lava which heated it, whether normal or reversed. It seems improbable that the adjacent rocks as well as the lavas themselves should possess a self-reversal property, and such results seem difficult to interpret in any other way than by a reversal of the Earth's field.

The same pattern of reversals observed in igneous rocks has also been found in deep-sea sediments (see, e.g., Figure 3.1). No two substances could be more different or have more different histories than the lavas of California and the pelagic sediments of the Pacific. The lavas were poured out, hot and molten, by volcanoes and magnetized by cooling in the Earth's field: the ocean sediments, on the other hand, accumulated grain by grain by slow sedimentation and by chemical deposition in the cold depths of the ocean. If these two materials show the same pattern of reversals, then it must be the result of an external influence working on both and not due to a recurrent synchronous change in the two materials. The evidence seems compelling that reversals of the Earth's field are the cause of the reversals of magnetization.

Reversals have been found during the Precambrian and have been observed in all subsequent periods. There is no evidence that periods of either polarity are systematically of longer or shorter duration. However, during the Kiaman – a period of about 6 Ma within the upper Carboniferous and Permian (about 235– 290 Ma ago) – the polarity of the Earth's field appears to have been almost always reversed: until quite recently no normal intervals at all were known within this period. McElhinny (1969) and Burek (1970) have now both reported a normal event at about 280 Ma and Creer *et al.* (1971) another at about 263 Ma. If the field reversal hypothesis is incorrect, it follows that mineral assemblages necessary for self-reversal were abundant in Carboniferous and Triassic rocks (both these periods have many reversals), but were all but missing in all Permian rocks. Such a conclusion is very difficult to believe; it is far more plausible to assume that the field itself alternated very rarely during the Permian.

If the dipole field of the Earth has reversed, it is most probably a result of physical processes occurring in the core of the Earth, and should thus be quite uncorrelated with physical processes associated with the crust and upper mantle or the atmosphere, such as orogenic and volcanic activity or climatic changes. It should be pointed out, however, that the lack of correlation between a rock's polarity and other properties does not necessarily rule out self-reversal, because the self-reversal mechanism may be intrinsically undetectable.

Self-reversals can only occur in minerals that have (or have had) specific compositions. One would therefore expect that there might be a correlation between the observed polarity and the mineralogy. A number of workers have reported chemical differences between normally and reversely magnetized lava sequences from various parts of the world. The ferromagnetic minerals in reversely magnetized lavas appear to be more highly oxidized than those which are normally magnetized. This was first noted by Balsey and Buddington (1954, 1958) in the Precambrian rocks of the Adirondacks; it has since been confirmed by several workers in rocks from Mull, Siberia, Iceland, Japan and elsewhere. The minerals concerned are mixed oxides of iron and titanium. The chemical evidence is that iron is in the high-valency (Fe^{3+}) state more frequently in reversed rocks; the petrological evidence is that the iron–titanium oxide crystals are visibly in different, more highly oxidized, crystal form, while some of the silicate minerals are noticeably more reddened.

Larson and Strangway (1966) have argued that the correlation between magnetic polarity and petrology is probably fortuitous, and cite extensive investigations of basaltic samples from Oregon, New Mexico, California and Japan in which there was no such correlation. Wilson (1966a), however, continued to find distinct petrological differences between normal and reversed Tertiary basalts from Japan and Iceland and Carboniferous lavas from Scotland.

The Columbia Plateau basalts, which are of Miocene age, have been studied in more detail than any other sequence (Wilson and Watkins, 1967). Here there are five series each of up to 20 lavas with reversed magnetization which are predominantly but not exclusively highly oxidized alternating with five sequences with normal magnetization and a predominantly low degree of oxidization. Not all lava sequences show such a clear correlation: in fact, Watkins and Haggerty (1968) have shown that the state of oxidation often varies greatly in a single flow, being higher in the centre than near the upper or lower margins. The variations in oxidation appear to be an original feature of the flows and to have been produced during the original solidification and cooling, probably by oxygen derived from the decomposition of water. Larson and Strangway (1968) do not accept the correlation found by Wilson and Watkins (1967), and in their reanalysis of the same data find no significant correlation when field relations and silicate petrography are taken into account.

Watkins and Haggerty (1968) later examined the magnetic polarity and oxidation state of over 550 specimens in single lavas and dikes in Eastern Iceland. They found no correlation between polarity and the oxidation state in the dikes, although there was a strong correlation between the percentage of reversed polarity and higher oxidation in the lavas. In a later paper Ade-Hall and Watkins (1970) found no correlation between opaque petrological parameters and polarity in specimens of Canary Island lava flows of Miocene to Pliocene age. The proposed correlation, therefore, does not seem to hold on a worldwide scale and a satisfactory explanation of those positive correlations that have been found has not yet been given. Domen (1969) has reported consistent differences between the demagnetization curves and Curie points of normally and reversely magnetized basalts from Kawajiri, Japan, and shown the existence of self-reversals in the reversed samples by comparison of thermal demagnetization curves (the explanation of these effects is not clear).

Kristjansson and McDougall (1982) have carried out a detailed study of the late Tertiary magnetic field in Iceland, using more than 2400 lava flows, and found that magnetic moment versus latitude curves were similar for normal and reverse polarities. From this, it would seem that both the actual mean strength of the dipole field and the rock magnetic properties of the basalts are independent of polarity. The supposed positive correlation between oxidation state and reverse magnetic polarity in rocks cannot have applied in this case (as highly oxidized lavas are generally more intensely magnetized and more stable than others), unless the reverse primary magnetic fields were weaker than normal fields by a factor precisely cancelling the effects of the oxidation. Kristjansson and McDougall suggest that it is likely that much of the original sample material on which this correlation was based simply had not received adequate demagnetization treatment to rid the less oxidized samples of Brunhes age viscous magnetization.

There is no known causal relationship between the oxygen fugacity and polarity, and the general consensus is that the observed correlations are fortuitous. Merrill (1985) has reviewed all the evidence and come to the conclusion that the reported correlation between magnetic polarity and oxidation state is due to a data artefact.

Moberly and Campbell (1984) claimed that volcanism along the Hawaiian hot-spot chain occurs preferentially during times when the magnetic field is of normal polarity. Support for such a correlation was found by Merrill (1985) by computer simulation of magnetic anomalies. However, in spite of this Merrill believes that the correlation may be due to a data artefact and suggests that a VRM overprint mechanism is the most likely explanation for the observed bias. Verhoef *et al.* (1985) claim that the majority of sea mounts in the North Atlantic as well as in the Pacific are normally magnetized, and came to the same conclusion as Merrill, viz. the predominance of normal polarity may be the result of a general overprint of TRM by a strong VRM.

Hildebrand and Staudigel later (1986) investigated the magnetic polarity of sea mounts in the Pacific Basin, excluding those on the Hawaiian Ridge. They found that on crust older than 65 Ma the majority of sea mounts have normal polarity. On the other hand, on crust younger than 65 Ma sea mounts show normal, reversed and mixed polarity. They suggest that a large number of sea mounts were formed during a major magmatic event lasting from about 110 to 70 Ma ago. This period overlaps the long Cretaceous superchron (118–84 Ma ago) when the Earth's field was normal most of the time and therefore could explain the preponderance of normal polarity observed on sea mounts older than 65 Ma. The question of possible correlations between events observed at the Earth's surface, mantle plumes and reversals of the Earth's magnetic field is discussed in Section 8.5.

If the origin of reversals is one of the instantaneous self-reversal mechanisms, then normal and reversely magnetized rocks should be randomly distributed throughout a group of rocks of different ages. If reversals are due to one of the time-dependent self-reversal mechanisms, reversals should be increasingly abundant in older rocks. If, on the other hand, reversals are due to geomagnetic field reversals, normal and reversely magnetized groups of rocks should be exactly the same age over the entire Earth; and unless it so happened that the Earth's field suffered more reversals in the past, the proportion of reversed magnetizations should not be greater among older rocks. Almost all rocks baked by igneous bodies agree in polarity with that of the baking rock, irrespective of their compositions (Wilson, 1962, 1966b). Self-reversal would produce baking and baked rock pairs of opposite polarity because the supposed self-reversal property is unlikely to be common to each member of a given pair. Furthermore, in cases where a wide variety of rock types in a single locality all possess reversed polarity, it is extremely unlikely that they all contain the self-reversing property. Again all rocks of a given age, at least within the past 4 Ma, have the same polarity irrespective of rock type, composition or location. Self-reversal would have produced a random distribution of reversely magnetized rocks throughout this period. Moreover, the fact that about half of all the rocks examined are reversely magnetized is just what would be expected from field reversal – self-reversal would produce this proportion only by coincidence.

Examples have been found of time-ordered sequences of lavas that apparently record the change of the ancient magnetic field from one polarity to the other, with directionally intermediate steps. Self-reversal cannot explain a gradual swing of direction of magnetization from one lava flow to another. A more reasonable explanation is that these examples record field reversal actually taking place.

Finally, it is found that the range and mean value of the ancient field intensity that is deduced from normally magnetized rocks is the same as that deduced from reversely magnetized rocks. It is unlikely that a self-reversal mechanism would produce the same values as occur in normally magnetized rocks and this suggests, again, that the Earth's field has reversed. Thus, although it has been established that self-reversal does occur in some rocks, the total evidence is overwhelmingly in favour of field reversal in the great majority of cases.

The discovery of laboratory self-reversal in the Haruna dacite (Nagata *et al.*, 1952) established the reasonableness of the self-reversal hypothesis, but delayed by many years the general acceptance of field reversal.

3
The morphology of geomagnetic reversals

3.1 Introduction

Reversals of the geomagnetic field were originally found in lava flows. Later Opdyke *et al.* (1966) found a polarity record in deep-sea sediments going back 3.6 Ma in which the pattern of reversals was remarkably similar to that found in igneous rocks on land (see Figure 3.1).

During a polarity change, the direction of the geomagnetic field swings

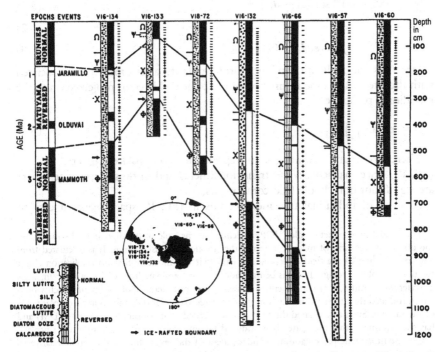

Figure 3.1 Correlation of magnetic stratigraphy in seven cores from the Antarctic. Minus signs indicate normally magnetized specimens; plus signs, reversely magnetized. Greek letters denote faunal zones. Inset: source of cores. After Opdyke *et al.* (1966).

through about 180°, the virtual geomagnetic poles* following widely different paths for different transitions. This could be interpreted as the result of a decrease in the main dipole field so that the observed field is dominated by the non-dipole component. Alternatively, the field could just tip over, being still mainly controlled by the dipole term with independently varying axial and equatorial components. In the first model the pole paths for transitions would not be expected to be the same whether recorded at different places at the same time or the same place at different times, because of variations in the non-dipole field. The second model suggests that there could be strong similarities between pole paths for the same transition observed anywhere on the Earth.

Hoffman (1984) has given an alternative description of the palaeomagnetic field which involves a simple co-ordinate transformation of the record in (D,I) space to one in (D′, I′) space. This is achieved by a single rotation of (D,I) space about the east–west axis such that the normal polarity axial dipole field direction is given a +90° inclination i.e. the +z axis in the local Cartesian system is made to coincide with the normal polarity axial dipole field direction. When plotted stereographically in (D′,I′) space, palaeomagnetic directional movements appear as if one were literally looking down the normal polarity axial dipole direction. Hoffman showed that the relationship between the two co-ordinate systems (x^1, y^1, z^1) and (x^1, y^1, z^1) is given by

$$x^1 = (x^2 + z^2)^{1/2} \sin (I_d - \theta) \tag{3.1}$$

$$y^1 = y \tag{3.2}$$

$$z^1 = (x^2 + z^2)^{1/2} \cos (I_d - \theta) \tag{3.3}$$

where I_d is the inclination of the normal polarity axial dipole field associated with the site and θ is the inclination of the palaeofield vector projected on to the meridional plane, i.e.

$$\theta = \arctan \left(\frac{z}{x}\right) \tag{3.4}$$

There are many advantages in representing the palaeomagnetic field in (D′, I′) space, particularly for transitional fields and in those cases when there are significant non-dipole field components.

Analysis of individual transition records in (D′, I′) space makes it easy to

---

* The virtual geomagnetic pole (VGP) is defined as the pole of the dipolar field which gives the observed direction of magnetization at the site under consideration. It is calculated from any spot reading of the field direction, the word 'virtual' meaning that no implication about the position of an average dipole is being made. To compare data from different sampling sites at different latitudes, it is convenient to calculate the equivalent dipole moment which would have produced the measured intensity at the calculated palaeolatitude (assuming a dipolar field) of the sample. Such a calculated dipole moment is called a virtual dipole moment (VDM) and has the advantage that no scatter is introduced by any wobble of the main dipole, since the determined magnetic palaeolatitude is independent of the orientation of the dipole relative to the Earth's rotational axis. However, it is not possible to determine that portion of the observed intensity (and therefore of the VDM) which arises from non-dipole components. Hence, a true dipole moment (TDM) cannot be obtained directly from palaeointensity studies.

determine the angular distance of a given palaeodirection from that of the axial dipole field and the number of strictly transitional directions contained within a given record. The degree of near-sidedness or far-sidedness (see Section 5.4) of each palaeodirection is also made clearer. Hoffman (1984) also showed how these characteristics revealed in (D', I') space can be used to infer the harmonic content of the transitional field. Figure 3.2 shows the palaeodirections corresponding to the Olduvai–Matuyama (N→R) transition recorded on Molokai, Hawaii, plotted in (D', I') space.

Four major normal and reversed sequences have been found during the past 3.6 Ma. These major groupings were originally called geomagnetic polarity epochs and named by Cox *et al.* (1964) after people who have made significant

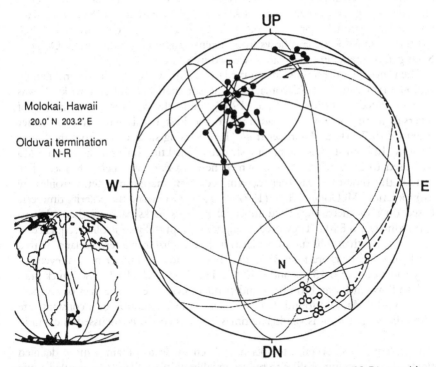

Figure 3.2 Palaeodirections corresponding to the Olduvai–Matuyama (N–R) transition recorded on Molokai, Hawaii, plotted in (D', I') space. The vector data are projected sequentially from the centre of a sphere and are geometrically displayed as though observed in their actual geographic surroundings. The 'pseudo-poles' correspond to the normal (N) and reverse (R) axial dipole field directions for the site. The 'pseudo-equator' is the great circle containing all possible directions 90° (midway) from the axial dipole directions and the small circles (shown for scale) correspond to directions making a 30° angle from that of the nearest axial dipole. Inset: path of the corresponding VGP (that is, the sequence of geomagnetic poles that would occur on the Earth if the recorded directions were associated with purely dipole palaeomagnetic fields). After Hoffman (1991).

contributions to geomagnetism. Superimposed on these polarity epochs are brief fluctuations in magnetic polarity with a duration that is an order of magnitude shorter. These have been called polarity events and have been named after the localities where they were first recognized (see Figure 3.3, which shows one of the first timescales for geomagnetic reversals). The terms 'epoch' and 'event' have now been replaced by 'chron' and 'subchron'. The question of terminology is discussed in more detail in Chapter 7 on magnetostratigraphy. Cases have also been found of more irregular movements of the poles with even shorter directions and sometimes not even establishing opposite polarity. These have been called 'excursions' and are discussed in Chapter 4. It has been suggested that excursions may be aborted reversals. It is not clear whether some short polarity intervals should be classified as subchrons or excursions. Clement and Kent (1986/7) suggest that there is no real distinction and that the geomagnetic field exhibits a complete spectrum of polarity behaviour. Even more rapid changes have been reported – e.g. Mankinen et al. (1985) and Prévot et al. (1985), who have reported directional changes of 58 ± 21°/a and intensity changes of 6700 ± 2700 nT/a in Miocene lava flows from Steens Mountain, south-central Oregon. Such geomagnetic 'impulses' are discussed in Section 6.7.

The timescale of reversals during the last 3.6 Ma was originally determined by radiometric dating and palaeomagnetic measurements on volcanic rocks. It was later extended to 11 Ma by Vine (1966), using the palaeomagnetic record of reversals in the oceanic crust adjacent to the East Pacific Rise. It was further extended to 76 Ma by Heirtzler et al. (1968), using similar data from the South Atlantic. In both studies the previously determined timescale for the past 3.5 Ma was used to calibrate the rate at which new sea-floor was being formed. The rest of the timescale was then determined from marine magnetic profiles by extrapolation. McDougall et al. (1977) were able to extend the polarity timescale based on radiometrically dated subaerial igneous rocks back another 2 Ma to approximately 6.5 Ma. They used a long stratigraphically controlled sequence of rocks – more than 400 successive lavas in Borarfyördur, western Iceland, with a total thickness of more than 3500 m. They found two further normal events – very short ones at 6.43 Ma and 6.12 Ma. Harrison et al. (1979) used thick piles of lava flows in Iceland to obtain independent evidence for the ages of reversals back to 13.0 Ma, and found that, in general, the assumption of a constant rate of sea-floor spreading is valid. The timescale of reversals is discussed in detail in Chapter 7.

In recent years several attempts have been made to obtain a more detailed timescale by stacking profiles to reduce incoherent noise. This has revealed a fine structure of previously undetected short polarity intervals (Blakely and Cox, 1972; Blakely, 1974). For the time interval from 22.7 to 7.3 Ma Blakely found 18 new reversals, increasing the average frequency of reversals by 50%. It is very difficult to resolve polarity intervals shorter than about 20 000 a. New oceanic crust does not form at the rises by a smooth, steady state process – rather it forms by a sequence of discrete volcanic eruptions separated in space by several kilometres and in time by about 14 000 a. As a result, even when a magnetometer

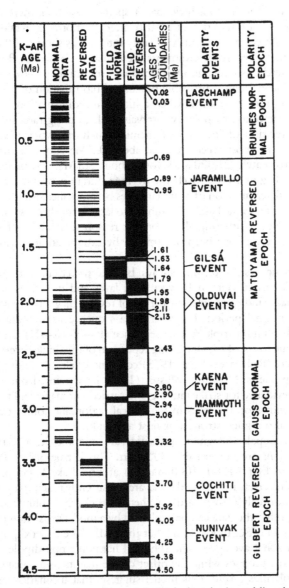

Figure 3.3 Timescale for geomagnetic reversals. Each short horizontal line shows the age as determined by K–Ar dating and the magnetic polarity (normal or reversed) of one volcanic cooling unit. Normal polarity intervals are shown by the solid portions of the 'field normal' column, and reversed polarity intervals by the solid portions of the 'field reversed' column. The duration of events is based in part on palaeomagnetic data from sediments and profiles. After Cox (1969).

is towed near the bottom to achieve better spatial resolution, there is too much noise of geological origin to permit polarity intervals shorter than about 20 000 a to be resolved. Attempts to identify polarity intervals as short as 20 000 a using the palaeomagnetic record of sediments accumulated on the sea floor has also, in general, proved unsuccessful. Data obtained from different cores of the same age are often inconsistent – probably because of bioturbation of the sediments. It is also often difficult to know when a hiatus in sedimentation has occurred. Finally, in the detection of polarity intervals on land using radiometric dating, the analytic precision of the various chemical analyses involved is about 3%. Thus, even for the most recent reversal about 0.7 Ma ago, this introduces an error of approximately 21 000 a. Beyond about 3.5 Ma, the precision of the K–Ar method begins to approach the average duration of individual polarity events and the determination of boundary ages becomes increasingly ambiguous. Also K–Ar ages of oceanic basalts are subject to large errors because of hydrothermal alteration and weathering. It is for this reason that the older portions of the polarity timescale have been determined by extrapolating marine magnetic anomaly data.

Palaeomagnetic results cannot always be interpreted unambiguously. For example, work carried out on Keweenawan lavas and intrusives around Lake Superior, Canada, indicates an older period of reverse magnetization giving way to a younger normal period. However, in a sequence from Mamainse Point, Palmer (1970) found a triple reverse sequence (R→N→R→N), and suggested that this was due to a strike fault causing repetition of the single reversal sequence found elsewhere. Robertson later (1973) confirmed Palmer's findings and supported his hypothesis. However, Massey (1979) later pointed out that there is no geological evidence for strike faulting and carried out major and trace element geochemical studies on the lavas in the critical region and found that the geochemical data also argued strongly against any faulting.

In some cases palaeomagnetic polarities are asymmetric, i.e. normal and reversed directions are not exactly 180° apart. This is true of late Precambrian Keweenawan rocks (~ 1200–1000 Ma old) around Lake Superior. Here the inclinations of reversely magnetised rocks are consistently much steeper (upward) than those of normally (downward) magnetized rocks. Nevanlinna and Pesonen (1983) showed that the data can be explained by a model consisting of a geocentric dipole and an offset dipole located at the CMB. The offset dipole simulates the long-term zonal average of the spherical harmonic non-dipole field. If the geocentric dipole reverses while the offset dipole maintains a constant polarity, asymmetric polarities result. This model suggests that inclination asymmetries are due to a different non-dipole ratio during normal and reversed polarity and also satisfactorily explains the three asymmetric reversals at Mamainse Point referred to earlier.

Pesonen and Halls (1983) also examined the palaeomagnetic field in late Precambrian Keweenawan rocks and found that on the average the palaeointensities of reversely magnetized rocks are 40% higher than those with normal polarity. They showed that this result cannot be explained by any difference in remanence characteristics, grain size or cooling rate (dike width) between the rocks of the two polarity groups. The differences, however, disappear when the data are reduced to the palaeoequator in accordance with a geocentric axial dipole field. The palaeointensity results are thus consistent with the suggestion that the reversal asymmetry is caused by the North American plate occupying a lower palaeolatitude when the field had normal polarity. However, such an explanation becomes invalid if three successive asymmetric reversals occurred at Mamainse Point. In this case, the two-dipole-field configuration of Nevanlinna and Pesonen (1983) best explains the data.

Another feature of the geomagnetic field is polarity 'bias'. This is illustrated in Figure 3.4, which shows the percentage of normal polarity plotted as a function of time. The even balance between normal and reversed polarity during the Cenozoic reoccurs in the lower Triassic and in the late Silurian to lower Devonian, but over the entire Phanerozoic it is comparatively rare. For long intervals of geological time (as in the late Palaeozoic) it was reversed most of the time. This confirms earlier work of McElhinny (1971). Irving and Pullaiah (1976) showed that the change in polarity bias is not due to the occurrence of quiet intervals superimposed on a general background of disturbed intervals in which the polarity is evenly balanced, because bias in the disturbed intervals themselves shows the same cyclic change.

Cox (1981) has shown how stochastic models of the geomagnetic field can be generalized to take account of polarity bias. Defining polarity bias B as the fraction of time when the polarity is normal, during the Cenozoic $B \simeq 0.5$. During the last part of the Mesozoic, approximately 110–80 Ma ago, the field had a strong normal bias ($B > 0.9$) while during the late Palaeozoic, approximately 310–230 Ma ago, the field had a strong reversed bias ($B < 0.1$) (see Figure 3.4). Intervals during which the bias remains constant are variable in length, ranging from 30 to more than 100 Ma. During times of strong polarity bias the frequency of reversals is generally low and during times of low bias ($B \simeq 0.5$) the frequency is high and sometimes variable, as in the Cenozoic. During times of extreme polarity bias, as in the late Mesozoic and late Palaeozoic, the dynamo is not locked into one polarity state. This is demonstrated by the presence in the late Mesozoic record of a few short, irregularly spaced reversed intervals

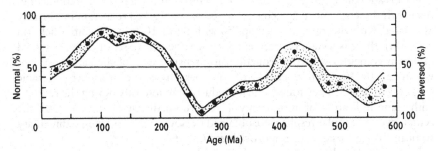

Figure 3.4 Polarity bias of the geomagnetic field during the Phanerozoic. Overlapping 50 Ma averages of polarity ratios as observed in palaeomagnetic results are shown together with the limits of the standard errors. After Irving and Pullaiah (1976).

and by the presence during the late Palaeozoic of a few short irregularly spaced normal intervals (Irving and Pullaiah, 1976). The question of bias and the frequency of magnetic reversals is discussed further in Section 3.4.

Cox (1981) suggested that symmetry is probably the key to understanding the question of bias. In the differential equations that describe the dynamo, there is no asymmetry with respect to the rotation axis. The total energy is also the same for both normal and reversed fields, and both are equally probable. However, the non-dipole component of the field is highly irregular and asymmetric, and it is this field and the secular variation that Cox believes controls polarity bias. Since the required asymmetry is absent from the differential equations of the dynamo, it must reside in the boundary conditions at the CMB. An example of the type of required asymmetry in boundary conditions is any odd-zonal harmonic. A physical example would be zonal convection through the entire core produced by a difference in the temperature of the lower mantle in the northern and southern hemispheres. Cox further speculated that if lateral variations in temperature and density in the lower mantle partially control the pattern of fluid motions in the core, changes in reversal frequency may reflect shifts in the location of individual convective cells in the lower mantle, whereas changes in polarity bias may reflect changes in the entire pattern of mantle convection. The question of the influence of mantle convection on the Earth's magnetic field is discussed further in Section 8.5.

3.2 Field intensity during a polarity transition

A polarity transition takes place so quickly (on a geological timescale) that it is difficult to find rocks that have preserved a complete and accurate record. Good intensity estimates may be obtained for volcanic rocks but suffer from the fact that there is little chronological control. On the other hand, sedimentary rocks give reasonably good chronological control, but sedimentation rates are often too slow to allow detailed resolution of intermediate fields. Moreover, an initially complete and accurate palaeomagnetic record in sediments may be obliterated or altered by chemical diagenesis and bioturbation occurring after initial deposition.

Observational evidence on changes in intensity during a reversal is conflicting. Although there is general agreement that there is a marked reduction in intensity associated with the reversal (perhaps to as little as 10% of the initial field on occasion), there is some disagreement as to whether the onset of changes in field direction coincides with the reduction in intensity. One of the earliest studies was made by Van Zijl *et al.* (1962) on rocks from the Stormberg volcanic sequence in Africa. They found a major reduction in intensity during the reversal, with some indication of an intensity increase immediately before and after the actual reversal. Similar results were obtained by Soviet workers using sedimentary sections on land (see, e.g., Kaporovich *et al.*, 1966).

Several workers have now succeeded in obtaining a record of the geomagnetic field during a polarity transition. It appears that during a reversal the intensity of the field first decreases by a factor of 3 or 4 for several thousand years while maintaining its direction. The magnetic vector then usually executes several swings of about 30°, before moving along an irregular path to the opposite polarity, the intensity still being reduced, rising to its normal value later. It is not certain whether the field is dipolar during a transition. Moreover, there do not seem to be any precursors of a reversal or any indication later that a reversal has occurred. One of the first detailed records of a field reversal is that of Dunn *et al.* (1971), who examined a single, igneous intrusive body (the Tatoosh intrusion in Mount Rainier National Park, Washington) which cooled comparatively slowly from the outside. If such a body cooled slowly enough, a large time span of the ancient geomagnetic field would be recorded continuously as the Curie point isotherm slowly moved further into the intrusion. This would avoid the discontinuities inherent in lava suites and the poor resolution inherent in sedimentary sequences. It is difficult, however, to establish an absolute timescale, since it depends upon calculations of cooling rates based upon parameters which are themselves poorly known. Dunn *et al.* found that the field intensity decreased by a factor of 10 before any change in field direction occurred, and did not return to normal until after the directional change was completed. The directional change was estimated to have taken 1–4 ka, while the intensity change took 10 ka.

Later work on the Tatoosh intrusion by Dodson *et al.* (1978) confirmed the earlier work of Dunn *et al.* The change in declination was gradual at first but the inclination changed rapidly initially. Moreover, the dispersion of directions, as measured by the Fisher precision parameter *k*, increased markedly during the reversal, reaching a maximum at the time of minimum intensity of magnetization. It must not be forgotten that individual samples record a time-averaged field. The uncertainties regarding the time interval during which the magnetization becomes blocked can be quite large, owing to uncertainties in the shape of the intrusive body and of the way in which it cooled. It is interesting that Dodson *et al.* observed no intensity change for the transition recorded by the Laurel Hill intrusion.

Contradictory results on the behaviour of the geomagnetic field during a reversal were obtained by Opdyke *et al.* (1973) from measurements made on a high-deposition-rate deep-sea core from the southern Indian ocean. The core is reversely magnetized to a depth of 460 cm, is normal from 460 to 940 cm, which is followed by a long unbroken stretch of reversed magnetization to a depth of 2250 cm, where a further polarity change occurs. The rest of the core is normally magnetized. The authors identified the normal interval between 460 and 940 cm as the Jaramillo subchron. Its duration was estimated by Opdyke (1969) to be 56 ka, making the rate of sedimentation during the event 8.6 cm/ka in this core. Because of this high rate of deposition, intermediate directions of magnetization were observed within each transition as well as a sharp decrease in intensity.

Figure 3.5 shows that for the lower Jaramillo reversal a pronounced drop in the intensity of magnetization is coincident with the onset of the direction changes. The time taken for both the intensity and direction of the field to reverse is approximately 4.6 ka, on the basis of a sedimentation rate of 8.6 cm/ka. Figure 3.5 also shows that there are three cycles of intensity changes with periods of

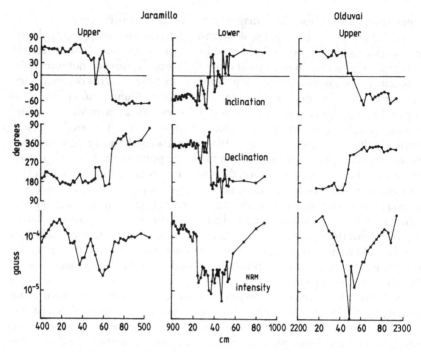

Figure 3.5 Details of the inclination, declination and intensity of magnetically cleaned remanence across the upper and lower Jaramillo and the upper Olduvai polarity transitions. After Opdyke *et al.* (1973).

approximately 1.4 ka during the time of the polarity change. The declination and inclination variations show a similar periodicity. The period of these movements is close to that usually associated with the secular variation of the non-dipole field – supporting the hypothesis that during the transition the dipole field is weak, being of comparable intensity to that of the non-dipole field. The Mono Lake excursion (Lund *et al.*, 1988) also appears to be related to the secular variation (see Section 4.4 for a detailed account of this excursion). Opdyke *et al.* concluded that the decreases in intensity were not caused by variations in the magnetic mineralogy of the sediments but reflected actual decreases in intensity of the Earth's field.

Kawai *et al.* (1976) studied the Brunhes–Matuyama transition, using thin sections of a core from the Melanesia Basin. They found the onset of directional change to be coincident with a sharp drop in intensity. They suggested that this has not been clearly seen before, because with average sedimentation rates the usual length of core samples is too large to reveal fine structure: the core sampled by Opdyke *et al.* (1973) had a relatively high sedimentation rate. Bingham and Evans (1975) also found the duration of the reduction in intensity to be comparable to that for the change in direction in a reversal seen in the Stark formation, Great Slave Lake, Canada, in Precambrian red siltstones (approximate age

1650–1800 Ma). They found, however, some suggestion of very rapid variations in intensity immediately before and after the change in direction. Nevertheless, in the majority of cases studied, the duration of the intensity change appears to be longer than that in direction.

Bol'shakov and Solodovnikov (1980) analysed all world data on intensities measured on continental extrusive and baked rocks (excluding oceanic basalts), ranging in age from the early Triassic to the Brunhes chron. They concluded that the average values of the intensity were the same for both normal and reversed fields from the Pliocene to the late Quaternary, and, moreover, that for the last 250 Ma the Earth's magnetic moment has remained essentially constant. Kristjansson and McDougall (1982) examined the intensity of the geomagnetic field as recorded in late Tertiary lava flows in Iceland, and confirmed that there was no significant difference between normal and reversed flows. Shaw *et al.* (1982) used 199 cores from a sequence of Icelandic lavas to determine 68 intensity values of the magnetic field ranging in age from 2 to 6 Ma (see Figure 3.6). They found large and rapid changes covering about an order of magnitude. The variation does not follow a normal distribution and there is a tendency for positive swings in magnitude to be both larger and sharper than negative swings. The data also suggest that low field values are more likely to occur during reversed periods. The separate averages of the normal and reversed data are not significantly different, but there is some suggestion of asymmetry in the normal and reversed field states in that the mode, median and quartiles of the reversed data are lower than those of the normal data. This question of asymmetry is discussed further in Section 5.8.

The polarity change seen by Bingham and Evans (1975) in the Stark formation, Great Slave Lake, Canada, shows that reversals have been a feature of the geomagnetic field for at least 2000 Ma. The transition lasted about 10^4–10^5 a, which is considerably longer than that for more recent reversals. Ito (1970), in a study of volcanic rocks of Pliocene to Miocene age from south-western Japan, also found that the time of transition was longer (10^4–10^5 a). Some interesting results have been found by Kawai *et al.* (1977) on a core obtained from the Melanesia Basin. They found a number of cases when the intensity of the magnetic field was near zero. The youngest instance was at the Brunhes–Matuyama boundary. Kawai and Nakajima (1975) had earlier estimated that the field was then reduced by two orders of magnitude for about 10 ka. The next instance occurred at the lower Jaramillo boundary, when the field vanished for more than 13 ka. They found yet another case of near-zero field from about 1.05–1.07 Ma ago. In this case the field recovered in the same direction and thus the 'excursion' remained unnoticed. Their oldest example corresponds to what has been called the Réunion event – in this case also the field vanished and recovered in the same direction.

However, palaeointensity measurements are difficult to make – the reliability of any estimate obtained from the analysis of the NRM of rocks depends on both the rock type and the method used. Most palaeointensity results from polarity transitions have been obtained from sedimentary rocks. Each sample carries a magnetization which reflects some time average of the transitional field. The time interval depends on the sedimentation rate if the primary remanence

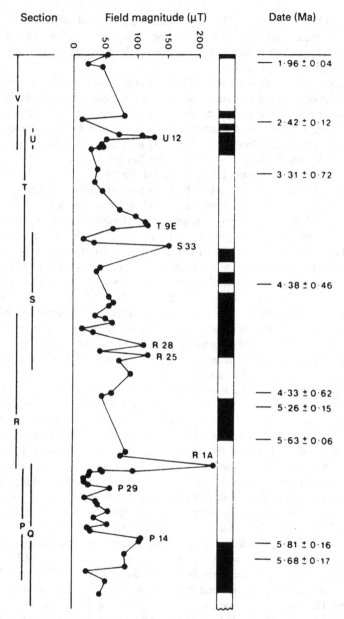

Figure 3.6 Composite diagram of the magnitude of the geomagnetic field with the time axis based on the assumption that on average the lavas are extruded at regular intervals. The time overlaps of the sampling sections are shown together with the individual palaeofield magnitude data and the polarity of all measured samples – dark areas represent normal polarity. After Shaw *et al.* (1982).

is DRM, or on the lock-in time if the magnetization is PDRM. The thinner the sedimentary layer recording a transition, as compared with sample size, the more serious are the averaging effects. The near-zero palaeointensities reported by Kawai *et al.* (1977) were obtained from very narrow transitional zones. Only a few samples were taken from each of these zones, and the weak intensity observed could be the result of the superposition of different time-dependent components with different or even opposite directions.

Herrero-Bervera and Helsley (1983) analysed the palaeomagnetic record in samples collected from the Red Peak Member of the Chugwater Formation of Triassic age in central Wyoming. They found that a slow change in intensity without change in direction begins long before the marked intensity decrease associated with the change in direction. The major decrease in intensity appears to be coincident with the directional change and recovers shortly after the directional change is complete. Thus, although the major portion of the intensity change took the same time as the directional change, the total time involved in the transition may be as much as three times that associated with the period of rapid directional change. Their results indicate that the dipole field 'collapsed' rather than 'flipped over' during the reversal.

In a study of the natural remanent magnetization of cores from Lake Michigan, Dodson *et al.* (1977) found instances of major fluctuations in intensity with little accompanying changes in direction. However, palaeointensities cannot be firmly obtained from sediments, and neither Creer *et al.* (1976) nor Vitorello and Van der Voo (1977), who also worked on Lake Michigan sediments, reported unduly high intensities.

Bogue and Coe (1984) obtained palaeointensity measurements from an R→N transition recorded in lavas from Kauai, Hawaii. The intensity dropped by ~ 75%, while the field remained within 30° of the reversed axial dipole direction. As the reversal neared completion, the field had an intensity of ~ 50% of its final value while still 40° from the final direction. Thus, in this case the relationship of palaeo-intensity to field direction during the early part of the reversal is different from that towards the end. Bogue and Coe interpreted their observations to be the result of an asymmetrical component in a zonal flooding model (see Section 5.4).

A surprising result has been reported by Shaw (1975). During a well-documented transition in a lava flow in western Iceland, he found that the geomagnetic field was large and stable when the VGP was close to the equator (see Figure 3.7). This reinforces an earlier, but far weaker, suggestion of Wilson *et al.* (1972) of rather strong intermediate dipole moments in an otherwise smooth variation. Confirmation of this result by Shaw is stronger, since he used only four lavas, recording a single transition, which is thus much less subject to statistical fluctuations – Wilson *et al.* applied statistical analysis to some 3000 specimens from 1500 lavas (Shaw and Wilson, 1977). The fact that the intermediate palaeofield can remain fixed in one direction for a considerable period of time (43 lavas recorded in western Iceland) and also that it can change in magnitude without changing direction indicates that the Earth's magnetic field may have a third metastable state which occurs much less frequently than the normal and reversed states.

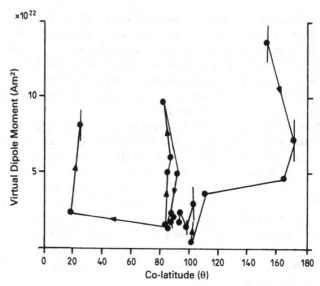

Figure 3.7 Graph of VDM against co-latitude for the R_3 to N_3 transition zone of western Iceland. The error shown is the standard deviation – in most cases it is less than the black circles. Arrows indicate the time progression and show that the large intermediate VDMs grow and decay smoothly. After Shaw (1975).

Shaw (1977) found another case of a strong intermediate palaeofield – in the Lousetown Creek, Nevada, transition. However, Roberts and Shaw (1990) later pointed out that at best the Lousetown Creek formation shows a blurred and partial record of a reversal and that the magnitude of the field was weak during the extrusion of most of the intermediate lava flows. They suggest that it is quite probable that the flows have been severely affected by lightening strikes – also the possibility that the disparate magnetization directions have arisen from CRM cannot be ruled out.

Another case has been found by Barbetti and McElhinney (1976). Archaeomagnetic studies of prehistoric aboriginal fireplaces along the ancient shore of Lake Mungo, a dried-out lake in south-eastern Australia, revealed high field strength during an excursion of the geomagnetic field – an increase by a factor of 5 or 6 over what appears to be the background field strength outside the excursion and an increase by a factor of 3 over the present field strength. It is possible, however, that the excursion seen at Lake Mungo could also be the result of lightning strikes. The Lake Mungo excursion is discussed further in Section 4.3.

Large and rapid changes in both the intensity and direction of the geomagnetic field have been observed during the R→N transition seen in lavas of upper Miocene ages from Steens Mountain, Oregon (Prévot et al., 1985). The directions and intensity approach near-full polarity values within the transition. The Steens Mountain transition is discussed in detail in Section 6.7.

Additional evidence for high transitional intensities comes from geochemical studies. Raisbeck *et al.* (1985) reported a dramatic increase in the concentration of ^{10}Be, a cosmogenic isotope created in the upper atmosphere, across the Matuyama – Brunhes transition as recorded in a deep-sea core. This is to be expected, as the production of ^{10}Be is inversely proportional to the strength of the Earth's magnetic field, so that when less of the atmosphere is shielded from cosmic radiation, as occurs with a drop in the intensity of the Earth's magnetic field, more ^{10}Be is produced. However, Raisbeck *et al.* found that the record of higher ^{10}Be concentration associated with the Matuyama – Brunhes reversal shows a mid-transitional *decrease* in the concentration of ^{10}Be. This they interpret as a return to high magnetic field intensities during the reversal.

The absence of directional changes, if further substantiated, suggests that the fluctuation in intensity is in the dipole field, and that the non-dipole field is relatively weak at that time. Such a suggestion raises two questions – what is the cause of the fluctuations in the dipole field, and why is the non-dipole field so weak? It is interesting to note that in Olson's (1983) model for reversals, high transitional intensities can be observed in certain cases (see Section 5.5).

From an analysis of the variation of the intensity of magnetization with the angular departure from the central axial field, Chauvin *et al.* (1990) suggested that, for reversal angles greater than 30°, the geomagnetic field is purely transitional at the latitude of Polynesia and that the mean intensity of the field remains very weak during the entire transitional period – about one-fifth of its value during stable polarity states. Comparison of data from more than 2000 lava flows from Iceland (Kristjansson and McDougall, 1982; Kristjansson, 1985) shows that the decrease in intensity with increasing angular departure from the axial dipole direction observed in Polynesia is also observed in Iceland with a cut-off level between transitional and stable fields of ~ 30–40° (see Figure 3.8). The intensity of the transitional field also appears to have been higher in Iceland than in Polynesia, suggesting an increase in the average intensity of the intermediate field with latitude. This is opposite to what was suggested earlier by Prévot *et al.* (1985). On the other hand, support comes from data from Kauai (Bogue and Coe, 1984) and Oahu (Coe *et al.*, 1984) which indicate stronger palaeointensities of the transitional field in Hawaii than those observed in Polynesia.

3.3 Field direction during a polarity transition

Another unresolved question is whether the magnetic field vector at a given locality tends to move along the same path during successive polarity transitions. If only the non-dipole field were present over much of a transition, the VGP from many reversals would be expected to be randomly scattered through all longitudes. Some analyses have suggested, however, that there may be preferred meridional bands within which most transition pole paths lie. An account of some of the earlier work is given below. A thorough discussion of this important question is given in Section 6.5. One of the early studies was by Creer and Ispir (1970), who concluded from an analysis of Tertiary polar transitions that during

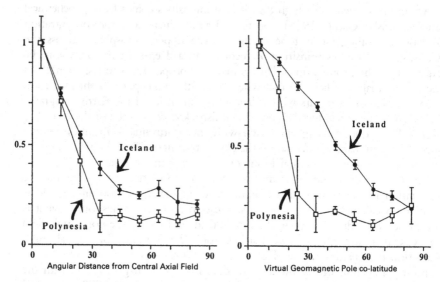

Figure 3.8 Comparison of the variations of the average geometric mean values of the intensity (after 10 mT) from Polynesia and Iceland versus the reversal angle and the VGP latitudes. Data from Iceland after Kristjansson (1985). In order to enable comparison, each data set has been normalized to its respective first value in the 0–10° interval. The arithmetic standard errors have also been normalized. After Chauvin *et al.* (1990).

a transition a significant equatorial dipolar component remains which can reverse independently of the axial dipole and on a shorter timescale. The movement of the dipole from one hemisphere to the other, corresponding to transition field directions recorded at widely separated sites, follows the same path, passing through the Indian Ocean. Petrova and Rassanova (1976) also found preferred locations for pole paths during reversals of different ages, although a later analysis of the transition field by Vadkovskii *et al.* (1980) found no specific trajectories for the movement of the VGP.

Another early study was that of Hammond *et al.* (1979), who analysed the results obtained from two oriented piston cores from the north-west Pacific Ocean (cores K75–01 and K75–02) which recorded the Jaramillo and Olduvai transitions (see Figure 3.9). These cores are approximately 120° in longitude away from the Indian Ocean site of Opdyke *et al.* (1973). The high accumulation rate of core RC14–14 of Opdyke *et al.* allowed them to sample at approximately 200 a intervals. The lower accumulation rate of the cores of Hammond *et al.* meant that the sampling interval during the Jaramillo subchron was about 1300 and 1000 a respectively. The resemblance between the VGP paths for cores K75–01, K75–02, and RC14–14 for corresponding transitions, together with the fact that the cores come from widely separated regions, tentatively supports a transition model in which a dipolar field is predominant during each of the polarity reversals studied. On the other hand, successive VGP paths do not coincide and neither Opdyke *et al.*'s nor Hammond *et al.*'s VGP paths match those determined by Creer and Ispir (1970). Thus, while these transition fields seem to retain predominantly dipolar configurations, the dipole (or, possibly, the non-dipole) field may incorporate large-scale magnetic features that can differ between successive reversals.

Hammond *et al.* also found a decrease in intensity associated with the change in direction which adds support to the non-dipole transition model. The K75–02 intensity values decreased to about 10% of the values before and after the transition, a level consistent with the present non-dipole field intensity throughout most of the Pacific basin region (Doell and Cox, 1972). Moreover, the intensity begins to drop in advance of the directional change and recovers only after the directional change is complete – in agreement with what is usually found, although, as already noted, Opdyke *et al.* (1973) found that, in their deep-sea core, changes in intensity and direction occurred at the same time.

Steinhauser and Vincenz (1973) investigated the longitudinal and latitudinal distribution of palaeopoles during a polarity transition, using data from 23 field reversals ranging in age from Recent to upper Mesozoic. They found two preferential meridional bands of polarity transition centred on planes through 40° E–140° W and 120° E–60° W. Both these bands are separated by two broad regions without transitional poles situated between 6° E–35° W and 85° E–70° W. The 40° E–140° W meridional band of preferential transitions is in agreement

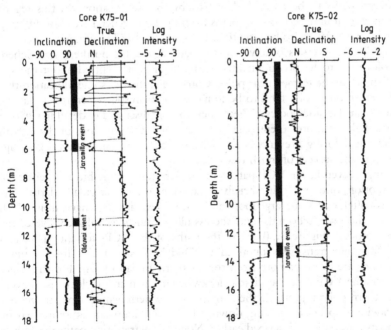

Figure 3.9 Palaeomagnetic data for cores K75–01 and K75–02 from the north-west Pacific Ocean oriented with respect to true north and corrected for corer rotation. Full intervals indicate normal polarity in the core; open intervals indicate reversed polarity. After Hammond *et al.* (1979).

with the results of Creer (1972), who, using 12 polarity changes, determined an average path centred on the plane through 60° E–120° W. It may be significant that the Central Pacific lies within the transition zone between 145° E and 174° W, i.e. within the band between 6° E and 35° W. The secular variation over the Central Pacific has been very small for at least the last million years (Doell and Cox, 1971; McWilliams *et al.*, 1982), suggesting that the non-dipole component of the main field has been very small in this region. Kristjansson and McDougall (1982), in an analysis of the geomagnetic field in the late Tertiary recorded in lava flows from Iceland, found that there was no strong preference for the transitional poles to be found in any one particular longitude interval. The most noticeable tendency was for poles to be found in the two regions which are ± 90° in longitude away from Iceland.

Dagley and Lawley (1974) examined more extensive worldwide data and also found no evidence for a common pole path or even a preferred sector of the globe. This is in agreement with the analysis of Hillhouse *et al.* (1972), who, however, did find a tendency for successive transition paths from the same geographical region to coincide. It must be stressed, however, that the data, in general, consist of a collection of poles in the two hemispheres with but little knowledge of the 'path' between them. Dodson *et al.* (1978), in an examination of all available paths from North America, found that no longitude was strongly preferred, although, as Hoffman (1977) had noted, there appears to be a tendency for several paths to be controlled in part by the site location. In this regard Dodson *et al.* found two reversal paths from the same general area but differing in age by about 10 Ma to be essentially the same. In 1977 Hoffman pointed out that for R→N reversals observed at mid-latitudes in the northern hemisphere the poles tend to lie predominantly in the hemisphere centred about the site meridian. These he termed near poles. Conversely, there is some indication that N→R transitional poles tend to be found in the hemisphere centred about the antimeridian (far-sided poles). These questions are discussed in detail in Chapter 6 in the light of more recent data. The geometry and structure of the Earth's magnetic field during a reversal is one of the most controversial topics in geomagnetism and the subject of much research.

During a reversal, the VGP path is not traversed smoothly but has periods of rapid motion, often returning briefly to positions near its original polarity. The path also often traces out large loops and at times remains stationary, from which position, the pole 'rebounds' and successfully or unsuccessfully completes the reversal process. Figure 3.10 shows the paths of the VGPs for the upper and lower Jaramillo transitions obtained by Opdyke *et al.* (1973) from a high-deposition-rate deep-sea core from the southern Indian Ocean. For the lower transition the VGP describes first a clockwise and then an anticlockwise loop in the southern polar regions. It then enters the northern hemisphere, tracing out a clockwise loop before settling down. These three loops are accompanied by the intensity variations described earlier. None of the transition paths of Opdyke *et al.* pass through the Indian Ocean, as proposed by Creer and Ispir (1970) – the three reversed paths actually occur in different quadrants. Thus, the results of Opdyke *et al.* do not support the model of a toppling dipole: rather they indicate that the intensity of the dipole field drops rapidly to a low but non-zero value, allowing the non-dipole field to predominate. This is reflected in the large looping excursions from the rotation axis seen in the VGP paths (Figure 3.10). Since both clockwise and anticlockwise rotations of the VGP path are seen, it would appear that both westward and eastward drifts of the non-dipole field occurred (Skiles, 1970). Both eastward and westward drifts of the non-dipole field have also been inferred from the sense of VGP loops associated with other reversals, e.g. the upper Miocene Steens Mountain polarity change (see Section 6.7). Detailed studies of VGP paths during a reversal indicate that such looping is quite common. Gurarii (1981) has studied the Jaramillo event in Western Turkmenia. He found that the VGP followed a complex path with a large number of loops. However, the pole positions are confined to a comparatively small region of the Earth's surface – passing through the South Atlantic, the eastern part of South America and the North Atlantic, and across North America.

Clement *et al.* (1982) studied the Brunhes–Matuyama transition in three deep-sea sediment cores from the Pacific, two from mid-northern latitudes less than 200 km apart and one from the equatorial Pacific for comparison with nearby cores studied by Freed (1977). Estimates of the duration of the transition ranged from 49 to 85 ka. It is interesting to note that these estimates are similar to that of 46 ka given by Opdyke *et al.* (1973) for the lower Jaramillo in spite

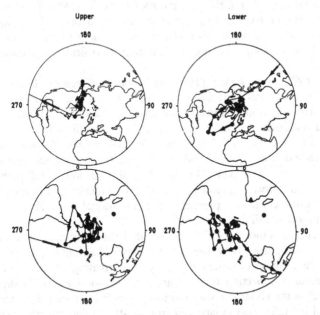

Figure 3.10 Positions of the VGP for upper and lower Jaramillo polarity transitions. VGPs are calculated from remanent directions by suitable adjustment of declinations. Core location shown by circled cross. Position of VGP for intermediate direction from Jaramillo Creek, New Mexico, dated at 0.86 Ma (Doell and Dalrymple, 1966) shown by solid triangle. After Opdyke *et al.* (1973).

of there being almost an order of magnitude difference in sedimentation rates. All three transition VGP paths of Clement *et al.* are characterized by a portion in which the VGP moves from high southerly latitudes to high northerly latitudes along a longitudinally confined path, lying no more than 90° in longitude away from the site, i.e. the paths tend to be near-sided (Hoffman, 1977). However, each path contains one or more loops either preceding or following the longitudinal path. Clement *et al.* point out that there are some indications in the data that the detailed record during the transition may not be simply related to the geomagnetic field. The VGP paths for the two nearby mid-northern latitude cores do not exactly coincide, as might be expected considering their geographic proximity. Again the VGP path for their equatorial core is not coincident with those obtained by Freed from nearby cores. Clement *et al.* suggest that the magnetic record may be distorted by sedimentological factors such as the effects of burrowing organisms and small hiatuses.

Liddicoat (1981, 1982) has examined the Gauss–Matuyama transition in a core from Searles Valley, California. The transition lasted approximately 2 ka. About 2 ka earlier, a short reversal also occurred that lasted about 2 ka. During it and the main reversal, the relative field intensity decreased by at least 70%. The VGP path for the main reversal is confined to a meridional band 20° wide in the Atlantic Ocean and closely matches the VGP path for the Brunhes–Matuyama transition recorded at Lake Tecopa, California (Hillhouse and Cox, 1976)*. On the other hand, the pole path for the Gauss–Matuyama transition recorded at sites in Russia (Burakov *et al.* 1976), about 180° in longitude from Searles Valley, is nearly antipodal at low latitudes to the Searles Valley pole path. This is strong evidence that the transition field was predominantly non-dipolar.

Laj *et al.* (1991) and Clement (1991) have examined records of reversals over the last 10 Ma and found that the poles tend to follow one of two paths – one through North and South America and the other about 180° away through eastern Asia and Australia (see Figure 3.11). There is still much controversy as to whether the transitional field during a reversal remains dipolar. There are cases, as already discussed, which indicate that the transitional field was strongly non-dipolar. On the other hand, the clustering of transitional VGP paths across North and South America found by Laj *et al.* is evidence that in some cases a significant dipolar content is present in the transitional field. This question is discussed in more detail in Section 6.5.

Steinhauser and Vincenz (1973) investigated the latitudinal distribution of transitional palaeopoles. They found a U-shaped distribution, there being a decrease in the number of observed poles with decreasing latitude. They interpreted this result as reflecting an acceleration in the movement of the dipole axis as it approaches the equator – they estimated that it is moving 3.4 times faster in equatorial latitudes than in latitudes around 40°. In contrast, the record of a

---

* Valet *et al.* (1988) resampled the Lake Tecopa section, using improved instrumentation and techniques. Their VGP path is confined within a narrow band of longitude which is significantly different (by 120°) from the path of Hillhouse and Cox.

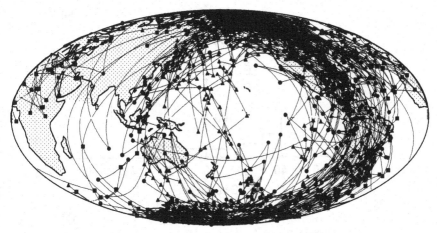

Figure 3.11 VGP trajectories for the Blake event (115–120 ka BP), the Upper Olduvai reversal (~ 1.8 Ma) and two reversals at 6.5 and 11 Ma. After Laj *et al.* (1991).

non-dipole field would give a random latitudinal distribution of poles. They further estimated that the dipole moment is reduced by about one order of magnitude for only about 12% of the transition time, while for two-thirds of the time its magnitude is comparatively high – with field intensities considerably greater than the intensity of the non-dipole field. Kristjansson and McDougall (1982) have also studied the latitude distribution of VGPs, using as data late Tertiary lava flows from Iceland. Their results (Figure 3.12) also illustrate that there is no significant difference between normal and reversed geomagnetic poles. They found that the latitude range where the VGP was most commonly found was around 70° N or S, which is unexpectedly low when compared with the present-day geomagnetic field. The chance that the VGP will be found within 10° of the geographic pole was only 11%. The VGP spent on average 50% of its time below latitude 64.5°, 9% below 35° and about 2% below 10°.

Clement and Kent (1984) obtained palaeomagnetic records of the Matuyama–Brunhes transition from seven low-sedimentation-rate, deep-sea cores from the Pacific Ocean. The cores were taken from near the 180° meridian and spanned latitudes from 45.3° N to 33.4° S. They found that the duration of the transition is dependent on the site latitude, with durations at mid-latitudes being longer by a factor of more than 2 than at equatorial latitudes.

3.4 Changes in the mean frequency of reversals

Figure 3.13 (after Merrill and McFadden, 1990) shows the average frequency of reversals over the past 165 Ma. An earlier estimate by Cox (1975) appeared to indicate that the mean frequency of reversals showed no statistically significant changes from about 75 to 45 Ma and then increased (rather abruptly) by a

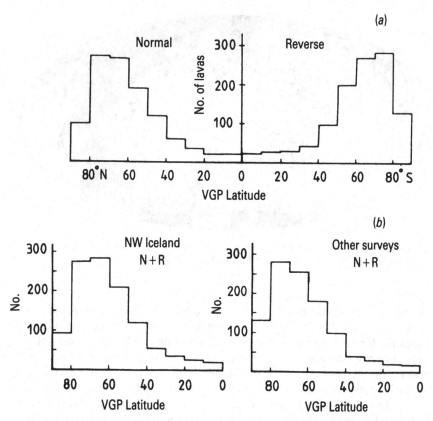

Figure 3.12 (a) Histogram of the distribution of VGP positions in latitude. Normal and reversed flows from north-west Iceland plotted separately. (b) Normal and reversed flows combined. After Kristjannson and McDougall (1982).

factor of more than 2, remaining approximately constant until the present. This interpretation implied that the geomagnetic dynamo was stationary for the 30 Ma prior to this 45 Ma 'discontinuity' and to have been stationary since then. Many papers were written on the supposition that 45 Ma marked a boundary between two intervals during which the statistical properties of the dynamo were distinctly different. This idea has finally been shown to be false. McFadden and Merrill (1984), using a method of analysis developed by McFadden (1984b), found an approximately linear trend on the mean rate of reversals increasing from zero at about 86 Ma ago (just after the end of the long Cretaceous normal superchron) to over 4 per Ma in recent times. They also showed that the pattern of reversals shows an approximately linear decline in reversal frequency from 165 to 119 Ma, when the process of reversals ceased and the field remained in the same polarity state for ∼ 30 Ma (the Cretaceous normal superchron). From about 86 Ma onward, field reversals again occurred at a gradually increasing rate.

Visual inspection of Figure 3.13 suggests that the peaking of the reversal rate near both ends of the record will produce a strong artificial periodicity of ~ 140 Ma, while the absence of reversals near the middle of the record will create a strong first harmonic of ~ 70 Ma. Superimposed on this trend, there seems to be possible oscillations with a period of ~ 30 Ma. Early studies of periodicities in the reversal record are confounded by the fact that different data, different record lengths and different methods of time series analysis have been used.

Lowrie (1982) was the first to question the 'discontinuity' at 45 Ma in the frequency of reversals which was based on data presented by Cox (1975). He estimated the number of reversals per Ma in the Cenozoic and late Cretaceous averaged over intervals of 2, 5 and 10 Ma duration based on the timescale of Lowrie and Alvarez (1981). The 2 Ma averages show large, apparently irregular, fluctuations in reversal frequency; averaging over 5 Ma and especially over 10 Ma intervals indicates that these fluctuations are superposed on an almost linear trend (see Figure 3.14). This suggests that the average reversal frequency of the geomagnetic field has been steadily increasing since the late Cretaceous. Lowrie and Kent (1983) repeated the calculations, using different polarity timescales and an 8 Ma sliding window, and arrived at similar conclusions.

Mazaud *et al.* (1983) analysed the structure of the reversal frequency curve for the last 83 Ma, using the timescale of Lowrie and Alvarez (1981) and that of LaBrecque *et al.* (1977). The frequency of reversals was studied, using a moving rectangular window, with window widths ranging from 2 to 10 Ma. The reversal frequency curves were then analysed in terms of a monotonically increasing component and an oscillating component. The monotonic component was modelled by a least-squares fit to a Lorentzian function. To analyse the oscillating part, they subtracted the corresponding Lorentzian from the reversal

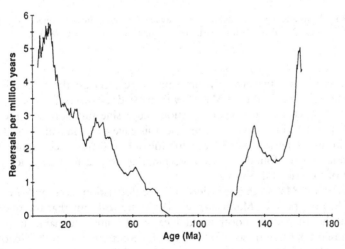

Figure 3.13 Estimated mean reversal rate from the present back to 165 Ma. After Merrill and McFadden (1990).

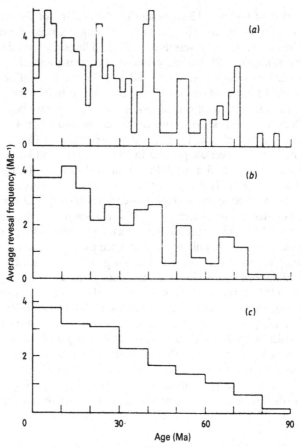

Figure 3.14 The frequency of geomagnetic polarity reversals since the late Cretaceous, averaged over intervals of (a) 2, (b) 5 and (c) 10 Ma, respectively. After Lowrie (1982).

frequency curve and performed a time autocorrelation on the remainder and found a periodicity of about 15 Ma. They then fitted the reversal frequency curve simultaneously with a Lorentzian function and a sine function. For the 4 Ma window and the LaBrecque *et al.* timescale, this gave a 31 Ma half-width for the Lorentzian and a period of 15.1 Ma for the sine function (see Figure 3.15). Similar results were found for windows ranging from 2 to 8 Ma and for the timescale of Lowrie and Alvarez.

McFadden (1984a) does not believe that the fluctuations seen by Lowrie and Kent (1983) or those of Mazaud *et al.* (1983) are real, but that the frequency of reversals has increased smoothly and linearly with time. He maintains that the superimposed fluctuations seen by Lowrie and Kent are due to their choice of smoothing filter (8 Ma) and that those of Mazaud *et al.* are an artefact of their analysis, the filter generating the fluctuations. It is just by chance that the periods

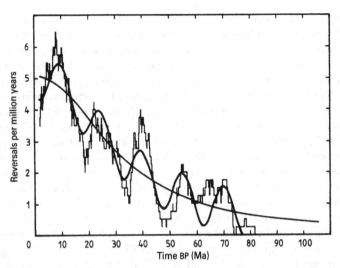

Figure 3.15 Direct least-squares fit of the frequency of reversals (in the LaBrecque *et al.*, 1977, timescale), using a Lorentzian function. After Mazaud *et al.* (1983).

of the fluctuations seen by these two sets of authors are approximately the same.

Negi and Tirwari (1983) had argued earlier for a 32 Ma stationary periodicity in the magnetic field and Raup (1985) later for a 30 Ma signal in phase with, and possibly a harmonic of, the 15 Ma signal claimed by Mazaud *et al.* (1983). Raup used the geomagnetic timescale of Harland *et al.* (1982). Lutz (1985) showed that the 30 Ma signal found by Raup is sensitive to the length of the time series. When the record is truncated by progressively eliminating the most recent events, the spectrum changes, showing that the 30 Ma peak is an accident of the record length. Lutz confirmed the earlier work of McFadden (1984b) and McFadden and Merrill (1984) showing that there is no evidence for periodicity in the reversal record. In a later paper, Lutz and Watson (1988) showed that long-term variations alone can account for most of the spectral evidence that is taken to support a 30 Ma periodicity in the geomagnetic reversal record – a trend in a series analysis can seriously affect spectral estimates. They further pointed out that the reversal record would contain only 5.5 cycles of a 30 Ma periodicity. Moreover, a quiet superchron (such as the Cretaceous normal superchron), roughly one cycle in length and unexplained by the period model, separates the data into two parts, the longest of which would contain fewer than three cycles.

Stothers (1986), however, still believes that there is a statistically significant period of ~ 30 Ma in the reversal record. He believes that the 15 Ma period found by Mazaud *et al.* (1983) in the record for the last 83 Ma is probably a harmonic rather than the basic period (30 Ma). He also claims that the 30 Ma period he finds in his separate analysis for the two intervals 0–83 Ma and 118–165 Ma is not being produced by the 35 Ma quiet interval (the Cretaceous normal superchron). Stother's method of analysis has been questioned by Stigler

(1987) and by McFadden (1987). McFadden points out that the fundamental question has not been addressed, viz. 'What is the source of the structure of the magnetic reversal record – has it been externally imposed upon the reversal sequence or could it have arisen merely by random processes?' To test this is not simple and as yet no properly conditioned test has been performed. McFadden concludes that the question of an externally imposed structure remains open.

Mazaud and Laj (1991) later analysed three different polarity timescales for the last 83 Ma – those of LaBrecque, Kent and Cande (LKC) (1977), Lowrie and Alvarez (LA) (1981) and Berggren, Kent, Flynn and van Couvering (BKFC) (1985). In each case they found large fluctuations in the reversal frequency superimposed on a linear trend. The results were obtained with a 4 Ma rectangular window and the analysis was repeated with different window widths varying from 0.1 Ma to 10.5 Ma. In all cases the window was shifted in 0.1 Ma steps. They also determined autocorrelation and maximum entropy power spectra of the fluctuating part of the reversal record. Their results are shown in Figure 3.16.

Mazaud and Laj found that for the LA and BKFC timescales the frequency peaked between 12.5 and 14 Ma, and at 16.5 Ma for the LKC timescale. The small differences result from small discrepancies in the age calibration of the polarity timescale and from the inclusion of very short events present in the LKC timescale only. They therefore believe that the periodicity observed in the different timescales is real and not an artifact introduced by the use of sliding windows. They then ask the question raised by McFadden (1987) – is the periodicity inherent in the mechanism responsible for triggering reversals or does it arise from fluctuations of a random process? Mazaud and Laj attempt to answer this question by following up a suggestion of McFadden which consists in comparing the geomagnetic record with a large number (200 000) of synthetic sequences produced by a random process, characterized by a linear time variation of its mean activity. The synthetic sequences were analysed, using sliding windows in the same way as for the geomagnetic record. The Fourier spectra of the frequency signals were then compared. Mazaud and Laj concluded that the periodicity detected in the magnetic record is not a simple statistical fluctuation of an aperiodic generator and, hence, a 13–16 Ma periodicity in the geodynamo must be seriously considered.

Marzocchi and Mulargia (1990) analysed 11 different reversal timescales for the last 83 Ma and found that the rate of reversals has a uniform exponential trend with no significant periodicities superimposed. In a later paper (1992), using a periodogram analysis for the Harland *et al.* (1990) timescale, they came to a different conclusion. They found as before an exponential trend in the rate of occurrence of reversals, but with a 15 Ma periodicity superimposed – in agreement with the work of Mazaud and Laj (1991). Marzocchi and Mulargia point out that periodicity studies based on intervals between reversals and those in the rate of occurrence are equivalent only asymptotically for infinitely long series, a fact which is often overlooked. The former verifies the dependence of the length of an interval on previous ones, while the latter permits a test for the existence of periodic time variations in the rate of occurrence. Marzocchi and Mulargia stress that the independence between the length of the intervals is the

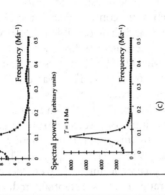

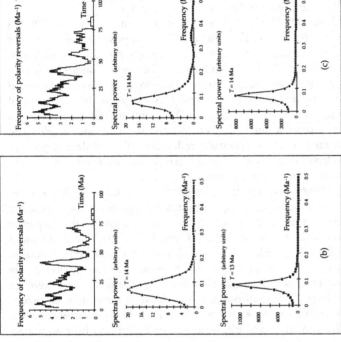

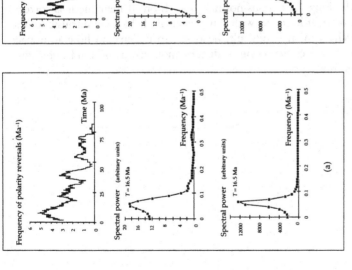

Figure 3.16 Analysis of the LKC polarity timescale (a), of the LA polarity timescale (b) and of the BKFC polarity timescale (c). In all cases the upper graph represents the frequency of the geomagnetic reversals obtained using a 4 Ma sampling window, the middle shows the Fourier transform (autocorrelation method) of the fluctuating part of the above frequency signal and the lower represents maximum entropy analysis of these fluctuations. After Mazaud and Laj (1991).

central feature of a process without memory, while periodicities in the rate of occurrence together with a trend are the two basic features of the time-dependent mean in a generalized random process. They found no periodicity in the intervals between reversals, although there is a significant periodicity of 15 Ma in the rate of reversal occurrence. The question of periodicities in the reversal record is still controversial – even more controversial is the question of possible periodicities in major extinctions (and their possible relation with like periodicities in reversals). These further questions are discussed in Section 8.6.

McFadden and Merrill (1984) suggested that there may be a 150–200 Ma periodicity in reversal frequency. They based this on the post-Jurassic pattern of reversals and the long normal Cretaceous superchron and the long reversed Kiaman superchron. Gallet *et al.* (1992) investigated the magnetostratigraphy of a late Triassic pelagic limestone section from south-western Turkey and found rough agreement with this suggestion. There also appears to be a number of short-term fluctuations in reversal frequency. The significance of these is hard to judge, since many magnetic intervals were defined by only one sample. The reversal frequency range is also substantially reduced if only well-defined polarity intervals are considered. Another major problem is the uncertainty in the absolute chronology since the Kiaman superchron.

Pal and Creer (1986) presume that the geomagnetic field should be most susceptible to extraterrestrial catastrophic events during its relatively unstable periods of frequent reversals. They claim to have found 'spurts' in the frequency of geomagnetic reversals at ~ 30 Ma intervals, which they attribute to enhanced core turbulence during episodes of bombardment by extraterrestrial bodies. Spurts in the frequency of geomagnetic reversals occurred approximately every 30 Ma since 165 Ma during mixed-polarity superchrons. There were no spurts during the long Cretaceous normal superchron. If their thesis is correct, reversal spurts have little to do with any stochastic behaviour intrinsic to the geodynamo mechanism. It is difficult, however, to understand how an impact on the Earth's surface could seriously disturb motions in the fluid outer core (see also Section 8.4).

4
Geomagnetic excursions

4.1 Introduction

In addition to polarity changes, the Earth's magnetic field has often departed for brief periods from its usual near-axial configuration, without establishing, and perhaps not even instantaneously approaching, a reversed direction. This type of behaviour has been called a geomagnetic excursion. Geomagnetic excursions have been reported in lava flows of various ages in different parts of the world and from some deep-sea and lake sediments. Excursions are generally observed to commence with a sudden and often fairly smooth movement of the virtual geomagnetic pole (VGP) towards equatorial latitudes. The VGP may then return almost immediately, or it may cross the equator and move through latitudes in the opposite hemisphere before swinging back again to resume a near-axial position. Barbetti and McElhinny (1976) define the term 'excursion' to describe a VGP movement of more than 40° from the geographic pole (following the suggestion of Wilson *et al.* (1972) for intermediate pole positions) which terminates with a return of the Earth's field to its pre-existing polarity, without the dynamo being observed to establish itself in the opposite polarity. Defined in this way, excursions are differentiated from the secular variation (when the VGP co-latitude $\theta < 40°$) and from short polarity events – a term Barbetti and McElhinny apply only when the opposite polarity ($\theta < 40°$ or $\theta > 140°$) persists long enough for at least one oscillation in the strength of the main dipole field (about 10^4 a). It is possible that excursions represent abortive reversals.

It is not always easy to distinguish between an excursion and a short polarity event. Indeed, as we shall see later, it is a moot point whether there are any real differences in the physical processes that initiate reversals. In this chapter the term 'excursion' will be used in those cases which have been generally accepted as such (e.g. the Laschamp excursion and the Mono Lake excursion). In those cases which are believed to be genuine reversals of short duration, the term 'subchron' will be used (e.g. the Jaramillo subchron). In those cases where there is still some doubt, the term 'event' will be retained on the understanding that it may be an excursion or a subchron (e.g. the Blake event).

Figure 4.1 shows the declination and inclination recorded in the Jiuzhoutai, China, loess section. Whether the fluctuations in the magnetic field seen in the

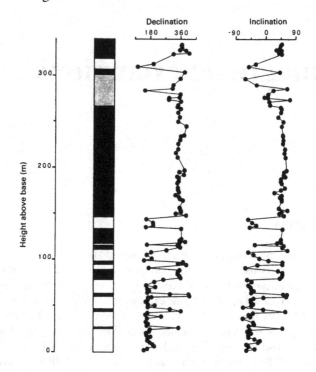

Figure 4.1 Declination and inclination results from the Jiuzhoutai loess section and the inferred magnetic polarity stratigraphy. After Rolph *et al.* (1989).

first 100 m above the base of the section can be classified as excursions is not clear. Another example of short-period events (probably excursions) is shown in the inclination record of a hydraulic piston core from the Gulf of California (Figure 4.2).

The youngest excursion proposed is the Starno event (Noel and Tarling, 1975), which they inferred from a 30° shallowing in the inclination of demagnetized post-glacial sediments from Blekinge in southern Sweden. The low inclinations were seen in two samples from one core and in one sample from a second core at another site. Age correlation between the two sites is poor: about 1.88 ka for the first site (Starno) and about 3.7 ka for the second (Stilleryd). The mean of these two ages, 2.8 ka, corresponds to that of a possible excursion suggested by Ransom (1973) on the basis of archaeomagnetic data from Italy and Greece. Although the archaeomagnetic data imply an inversion in the sign of the inclination, the lake sediment data of Noel and Tarling do not show this. Verosub and Banerjee (1977) have expressed grave doubts as to the reality of this excursion. It is extremely difficult to establish the reality of an excursion in every case.

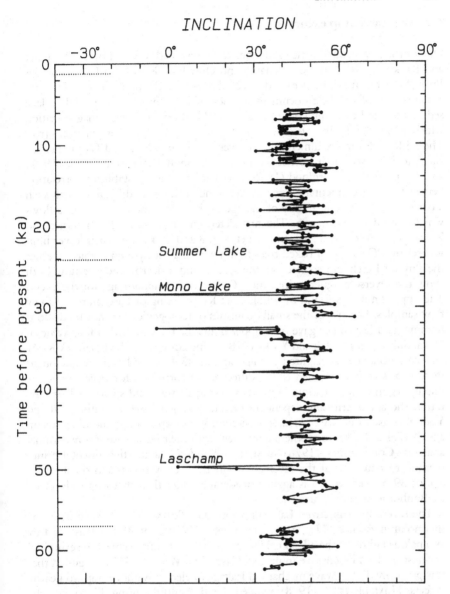

Figure 4.2 Inclination versus age profile for a DSDP hydraulic piston core from the Gulf of California, from 5–62 ka. The horizontal dotted segments at the left age ordinate indicate the calibration points for the age estimates. After Levi and Karlin (1989).

4.2 The Laschamp excursion

One of the first geomagnetic excursions to be reported was that by Bonhommet and Babkine (1967) in the Laschamp and Olby lava flows from La Chaîne des Puys (Massif Central, France) and since called the Laschamp excursion. This still remains one of the best-documented cases of a recent excursion. Unlike lake sediments, the Laschamp volcanics have a stable thermoremanent magnetization which is not susceptible to structural degradation or detrital inclination error. The evidence for it in sediments is, however, conflicting. Noel and Tarling (1975) claim to have seen it in recent sediments, although Denham and Cox (1971) and Thompson and Berglund (1976) have been unable to establish a correlation between it and sedimentary core samples. One of the main difficulties with such a correlation is the uncertainty of the age of the Laschamp and Olby lava flows which record the excursion. The first determination was by Bonhommet and Zahringer (1969), who gave an age between 8 and 20 ka ago. Their lower limit was set by ^{14}C dating of baked trees found within a trachyte unit which overlies the Puy de Laschamp scoriae. As the scoriae and a nearby andesite flow both show the reverse magnetization, the ^{14}C date is a minimum age for the event. The upper limit was found by whole-rock K–Ar dating of Laschamp and Olby flow samples. Because of the small quantities of radiogenic ^{40}Ar measured, Bonhommet and Zahringer gave only upper limits for the age, with a final value of 20 ka. Subsequently Hall and York (1978), using conventional whole-rock K–Ar analyses, obtained a weighted average age of 45.4 ± 2.5 ka. They attributed these consistently older ages to the difference in correction for atmospheric (or initial) argon contamination. With such young samples and such low ^{40}Ar/^{36}Ar ratios, the age is critically dependent on the assumed ^{40}Ar/^{36}Ar ratio. Hall and York also used the ^{40}Ar/^{39}Ar dating method with step heating and obtained an age of 47.4 ± 1.9 ka, in excellent agreement with their results using conventional analyses. Condomines (1978) used the ^{230}Th/^{238}U radioactive disequilibrium method to date one of the reversely magnetized Olby flows, and obtained an age of 39 ± 6 ka, which is again considerably older than that originally given by Bonhommet and Zahringer.

The reversely magnetized Laschamp and Olby flows have also been dated by thermoluminescence (TL). An earlier attempt (Wintle 1973) was frustrated by anomalous fading of the TL signal. Such anomalous fading seems to be exhibited by most of the TL-sensitive minerals in lavas, and Wintle (1977) suggested that it is probably due to wave mechanical leakage of electrons out of traps to nearby defects. Huxtable *et al.* (1978) avoided this difficulty by using TL to date the clay and sediment baked by the lava flows. They obtained an age of 25.8 ± 1.7 ka in fair agreement with a TL age of 33 ± 4 ka obtained by Valladas *et al.* (1977) based on quartz extracted from a granitic inclusion. Guérin and Valladas (1980) have also obtained ages for the Laschamp and Olby lava flows, using TL techniques on plagioclase feldspars. They obtained ages of 32.5 ± 3.1 ka, for Laschamp and 37.3 ± 3.5 ka for Olby. More recently, they have dated these lava flows, using high-temperature (800–1000 K) TL properties of plagioclase and K–Ar techniques. Their TL ages for the Laschamp and Olby events are 32.5 ±

3.1 ka and 37.3 ± 3.5 ka, respectively, and their K–Ar ages 38 ± 8 and 42 ± 9 ka.

There have been still more attempts to determine the age of the Laschamp and Olby reversely magnetized lava flows. Using whole-rock samples, Gillot *et al.* (1979) obtained a K–Ar age of 43 ±5 ka for the Laschamp flow and 50 ± 7.5 ka for the Olby flow, in excellent agreement with the ages obtained by Hall and York (1978). They also estimated the ages by TL. For the Laschamp flow they used quartz from the granitic inclusion and obtained an age of 35 ± 3 ka; the age of the Olby flow was measured from five quartz pebbles from the baked palaeosoil and gave an age of 38 ± 6 ka. They also obtained TL ages of plagioclases obtained from groundrock samples, and found an age of 33.5 ± 5 ka for the Laschamp flow and 44.1 ± 6.5 ka, for the Olby flow. Both these ages are greater than the other TL ages obtained by Huxtable *et al.* (1978). Gillot *et al.* (1979) also obtained ^{14}C ages for these flows – the most probable age of the Olby organic layer was found to be greater than 36 ka.

Barbetti and Flude (1979) have made measurements of the geomagnetic field strength of sediments baked by the lava flows from La Chaîne des Puys. For Royat the dipole moment was only 30% of its present value. The TL age for the Royat flow obtained by Huxtable *et al.* (1978) was 25.8 ± 1.7 ka. However, Hall *et al.* (1979) obtained a ^{40}Ar/^{39}Ar age of about 40 ka, which is not significantly different from that of the Laschamp and Olby flows.

Roperch *et al.* (1988) measured the intensity of the Earth's magnetic field during the reversed directions of the Laschamp and Olby flows, and obtained a value of 7.7 μT (i.e. less than one-sixth of the present field). They suggested that this low value is more characteristic of transitional behaviour and, hence, that the directions of the Laschamp and Olby flows were not acquired during a stable reversed polarity interval – a more likely explanation is that the Laschamp excursion represents an unsuccessful or aborted reversal. Chauvin *et al.* (1989) measured the intensity of the Earth's magnetic field in the Louchadière lava flows. K–Ar age determinations confirm that these flows, which are in the same volcanic province as the Laschamp and Olby flows, are contemporaneous with them. They obtained a value of 12.9 μT (± 3.3) which is one-third to one-quarter of the present-day field. The VDM associated with this flow is 2.2×10^{22} Am^2, which is considerably less than the average VDM for the last 10 ka, and even for the last 5 Ma, which is about 8.7×10^{22} Am^2 (McFadden and McElhinny, 1982). Kristjansson and Gudmundsson (1980) found evidence for a geomagnetic reversal in lavas from three different hills in the active volcanic zone of the Reykjanes peninsula, south-west Iceland (called by them the Skalamaelifell excursion). There is no firm age for the excursion. The mean magnetic field direction is similar to that occurring in the transitional lava outcrop near Maelifell reported by Peirce and Clark (1978). It is also consistent with the shallow geomagnetic inclinations found in Danish marine clays between 23 and 40 ka by Abrahamsen and Knudson (1979).

More extensive fieldwork by Levi *et al.* (1990) has identified the same excursional palaeomagnetic directions at four additional outcrops in the Reykjanes peninsula. K–Ar dating of the excursion gives a mean age for 19 determinations of 42.9 ± 7.8 ka. Levi *et al.* obtained a new estimate for the Laschamp excursion of 46.6 ± 2.4 ka, using 30 K–Ar ages from three laboratories, and suggested that the Icelandic lavas recorded the same excursion as the Laschamp and Olby flows in France. They also obtained a value for the intensity of the palaeomagnetic field of 4.2 ± 0.2 μT, more than an order of magnitude weaker than the present magnetic field in Iceland and even less than that found by Roperch *et al.* (1988) for the Laschamp and Olby flows in France. Marshall *et al.* (1988) obtained an almost identical value of 4.3 μT. A well-defined intermediate direction has been observed at 63 sites in eight distinct lava flows of the Albuquerque volcanoes, New Mexico (Geissman *et al.*, 1990). Preliminary K–Ar studies indicate a late Pleistocene age (20–100 ka) as the time of the extrusion, but the dating is too imprecise to correlate the event with other excursions.

This intense amount of work on the dating of these flows underlines one of the main difficulties in correlating reported excursions of the Earth's magnetic field – namely that of determining reliable ages. In any theory of magnetic excursions it is essential to know whether they are worldwide events or are observed over only a small part of the globe. When the Laschamp excursion was originally dated as lying between 8 and 20 ka (Bonhommet and Zahringer, 1969), many attempts were made to correlate it with other excursions observed in other parts of the world, such as the Gothenburg (Mörner and Lanser, 1974), Gulf of Mexico (Clark and Kennett, 1973) and Lake Biwa (Nakajima *et al.*, 1973). Denham and Cox showed in 1971 that if the Olby–Laschamp excursion had affected the entire Earth's magnetic field, it could not have happened between 13.3 and 30.4 ka ago or lasted less than 1.7 ka. Now that it is known that the event is much older, correlations have been sought with older reported excursions. Other magnetic field disturbances observed in the period 30–50 ka are the Lake Mungo excursion (Barbetti and McElhinny, 1972, 1976), which lasted from 28 ka to at least 31 ka (see Section 4.3); the Lake Biwa excursion (Yaskawa *et al.*, 1973; Yaskawa, 1974), dated by extrapolation at 49 ka; and an excursion, the age of which is estimated at 40 ka, in sediments from the Indian Ocean (Opdyke *et al.*, 1974).

Evidence for the Laschamp excursion has also been found in tuffaceous sediments from Japan (Hirooka *et al.*, 1977) and from alluvial deposits of the river Obj in the USSR (Kulikova and Pospelova, 1976). Løvlie *et al.* (1986) found two zones with shallow to steep negative palaeomagnetic directions in three high-latitude cores from the Arctic ocean. Amino acid ratios and oxygen isotope values suggest that the sediments in which the reversed polarity zones were observed were deposited within oxygen isotope stages 2–3, i.e. less than 60 ka ago. Løvlie *et al.* suggest that the two reversed polarity zones may reflect the Lake Mungo and Laschamp excursions.

Levi and Karlin (1989) obtained an inclination record for the past 60 ka in sediments in the Gulf of California. They identified the Laschamp excursion between ~ 52.3 and 48.8 ka, somewhat older than the age reported in the Massif Central range in France and of the Skalamaelifell excursion in Iceland. However, the very low palaeointensities of these flows (Roperch *et al.*, 1988; Levi *et al.*, 1990) suggest that the Laschamp excursion might represent an aborted geomagnetic reversal during which the field at the Earth's surface was not dipolar. Hence, the age and morphology of the excursion might vary for widely separated areas. Moreover, in Iceland and France the excursion was observed in extrusive lavas, representing geologically an instant snapshot of the field. In contrast, sediments record a more continuous time-averaged signal, whose amplitudes represent a lower limit of the actual geomagnetic fluctuations. Thus, it would not be surprising to observe different geomagnetic expressions of the Laschamp excursions at different geographic locations and recorded by different remanent acquisiton processes.

Creer *et al.* (1990) examined the palaeomagnetic secular variation record in sediments from Lac du Bouchet – a maar lake about 100 km from the Laschamp–Olby sites. They found no evidence of any strongly anomalous directions. They concluded that the only way in which the Laschamp–Olby anomalous directions could have been caused by geomagnetic field variations would be if they were 'spot' records of very strong inclination/intensity minimum. If this were the case, the 'excursions' could not have lasted more than a few hundred years.

A similar conclusion was reached by Tric *et al.* (1992) in their study of the palaeointensity of the Earth's magnetic field during the last 80 ka, using records from five marine cores. They found that the lowest dipole moment occurred 39 ka ago, about the time when weak intensities and excursional directions were seen in the Laschamp and Olby flows in France and the Skalamaelifell excursion in Iceland. However, Tric *et al.* found no departures of the geomagnetic vector from the normal direction during this period in the samples from their marine cores. Although discontinuities and/or breaks in the sedimentation rate could be responsible, it is difficult to imagine that these phenomena occurred synchronously in different oceans, as well as in the lacustrine sediments that Creer *et al.* (1990) studied from Lac du Bouchet. Tric *et al.* conclude with Creer *et al.* that the duration of the Laschamp event could not have exceeded a few hundred years.

In a later paper Thouveny and Creer (1992) discuss in more detail possible reasons for the discrepancy between the Laschamp–Olby volcanic and the Lac du Bouchet sedimentary records. They attribute it to the different mechanisms by which the rocks became magnetized. The lavas instantaneously freeze the ambient magnetic field (TRM), whereas in sediments the porosity allows magnetic particles to realign progressively (PDRM) until the locking-in depth is reached. The time for this depth to be reached is controlled by the sedimentation rates. The effects of this PDRM filter would attenuate the amplitude of high-frequency secular variations, introduce a phase lag and decrease the intensity of the magnetization. They showed that wavelengths of 0.5 and 1 m (typical of the Lac du Bouchet record) would have locking-in depths of 5 and 10 cm, respectively, and suffer attenuations of ~ 20%. With an average sedimentation rate of 0.27 mm/a, such depths correspond to delays of 200 and 400 a. A full reversal could thus appear as a secular variation signal if the duration of the reversal was less than ~ 200 a.

Heller (1980) and Heller and Petersen (1982) have carried out thermal laboratory experiments and observed complete or partial self-reversal of the NRM in many of the Olby samples and to a lesser extent in some of the Laschamp flows. After magnetic and optical examination of the samples, they suggested that the self-reversal mechanism is caused by a negative magnetostatic coupling between titanomagnetite phases of widely varying degrees of oxidation. Heller and Petersen also studied the magnetostratigraphy of a contemporaneous loess section at Steinheim in southern Germany and found no reversed polarities. They did observe some abrupt changes in declination of very short duration (less than 2 ka), which they suggested are probably caused by mechanical disturbances (such as solifluction) rather than representing geomagnetic field variations. No positive evidence for the Laschamp event was detected in other loess profiles from Czechoslovakia and northern China.

This work indicates the need for thorough rock magnetic investigations before ascribing every excursion to a reversal of the geomagnetic field.

4.3 The Lake Mungo excursion

Barbetti and McElhinny (1972, 1976) carried out archaeomagnetic studies of prehistoric aboriginal fireplaces along the ancient shore of Lake Mungo, a dried-out lake in south-eastern Australia. Directions of magnetization preserved in oven stones and baked hearths showed that wide departures of up to 120° from the direction of the axial dipole field occurred about 30 ka ago. The geomagnetic excursion recorded between at least 30.78 ± 0.52 and 28.14 ± 0.37 ka is associated with very high field strengths between 1 and 2×10^{-4} T (see Section 3.2). The field strength subsequently decreased to between 0.2 and 0.3×10^{-4} T after the excursion. This main excursion is referred to as the Lake Mungo excursion. There is some evidence that a second excursion, associated with low field strengths of $0.1-0.2 \times 10^{-4}$ T, occurred around 26 ka ago. These dates were obtained from ^{14}C measurements in charcoal collected from the fireplaces and are consistent with the stratigraphic evidence. Huxtable and Aitken (1977), using quartz grains extracted from the baked clay–sand matrix, obtained TL ages for the first excursion of 35 ± 4.3 ka. This age is not significantly different from the average of the ^{14}C dates.

Because most of the excursions proposed so far have been found in sediments and have not been convincingly documented over a sizable portion of the globe, several workers (Verosub, 1975; Verosub and Banerjee, 1977; Thompson and Berglund, 1976) have suggested that some or most of them may reflect sedimentological rather than geomagnetic phenomena. The record of the Lake Mungo excursion, however, is contained in sedimentary material that was baked in prehistoric aboriginal fireplaces. Thus, sedimentological phenomena cannot be invoked to explain the anomalous directions found in this case, because the NMR is not detrital or diagenetic in origin but rather is thermoremanent.

Løvlie and Sandnes (1987) recorded two excursions in mid-Weichselian cave sediments from Skjonghelleren, Valderøj, in West Norway. They identified the younger excursion (30 ka) with the Lake Mungo excursion. Remarkable agreement between synchronous VGP positions from the two almost antipodal localities (Skjonghelleren cave and Lake Mungo) suggest that the Lake Mungo event is global and does not represent a local or regional feature. Løvlie *et al.* (1986) had earlier tentatively correlated a reversal polarity zone seen in three Arctic ocean cores with the Lake Mungo excursion.

Mörner (1986), using data from multiple cores from lakes in France (particularly the Grande Pile in the southern Vosges), found that the VGP paths from France and Lake Mungo were almost identical, again suggesting that the Lake Mungo event is global. It is interesting, however, that Oberg and Evans (1977) found no evidence of the Lake Mungo excursion in a 7 m sedimentary sequence in southern British Columbia. It should be noted that it has been suggested that the anomalous field seen at Lake Mungo is the result of lightning strikes.

It is possible to simulate the record of a reversal in a number of ways, e.g. the dipole field terms can be reduced to zero and then allowed to grow with opposite polarity, while some attempt is made to simulate the behaviour of the non-dipole field. Larson *et al.* (1971) carried out such a simulation, using a number of subsidiary dipoles placed close to the CMB. These were allowed to drift past the site to simulate the westward drift of the secular variation. In this way they were able to generate a sequence of field directions at the site which were similar to those observed in the reverse-to-normal transition recorded in a sequence of late Miocene volcanics dated at about 15 Ma in the Santa Rosa Range, north-central Nevada. In a similar model of the Tatoosh reversal, Dodson *et al.* (1977) found that to achieve a reasonable simulation it was necessary to maintain a fixed horizontal dipole term, which did not reverse at the same time as the axial dipole. Also, to obtain a path which is confined in longitude it was necessary to maintain some part of the non-dipole field with constant polarity throughout the reversal of the main dipole.

Harrison and Ramirez (1975) have modelled pseudo-reversals (i.e. ones not involving the main dipole field) by assuming that they are caused by a dipole at the CMB, whose magnetic field is opposite to that of the main dipole field. They showed that the areal coverage of disturbed magnetic field around such a pseudo-reversal can be quite small, so that observations made only a few thousand kilometres away would show no anomalous directions. If this model is correct, they further showed that pseudo-reversals should be much less common at low latitudes than at high latitudes. Barbetti and McElhinny (1976) used such a model to explain the Lake Mungo excursion. A radial dipole (offset from the centre of the Earth by 0.5 Earth radii) would need a moment of only one-eighth that of the main geocentric dipole to produce at the point on the surface nearest to it a field greater than or equal to that of the main dipole, whereas on the far side of the Earth it would produce a field only 1/27 as large.

Coe (1977) considered various simple sources, all in the Earth's outer core, that could account for the Lake Mungo excursion – an eccentric radial dipole or current loop, an eccentric horizontal dipole and a pair of eccentric radial current loops, of opposite sign. A horizontal eccentric source or current loop is the most efficient in terms of causing the least disturbance of the field elsewhere, but there is little basis for expecting one to exist, since the modern non-dipole field is dominated by a few large features that are most simply modelled by radial sources.

Coe showed, however, that single radial sources would produce extremely high fields over very large areas and significantly high anomalous fields over the entire globe. Thus, they are improbable unless simultaneous worldwide effects are discovered. The best model seems to be a combination of two radial eccentric sources of opposite sign. In this case horizontal components add and the area dominated by their combined fields is much less than that from a single source. Coe showed that in their most efficient configuration such a pair would cause a maximum field of about 2.3×10^{-4} T while producing the required non-dipole field at Lake Mungo. However, significant effects would occur as far away as south-eastern Asia and India, southern Africa, and much of Antarctica and the south Pacific Basin.

4.4 The Mono Lake excursion

Denham and Cox (1971) carried out palaeomagnetic measurements on late Pleistocene sediments in Mono Lake, California, and found no evidence for a reversal between 30.4 and 13.3 ka with ages controlled by ^{14}C dating. If the Laschamp excursion occurred during this time, its duration can have been no longer than 1700 a (the largest sampling gap in their data). However, the sediments did record a well-defined excursion in the direction of the field about 24 ka ago with a peak-to-peak amplitude of 25° and a period of about 600 a. They attributed the anticlockwise looping motion of the magnetic field vector to a local eastward drift of the non-dipole field. Denham later (1974) investigated the records in more detail and showed that the major features of the loop could be explained by an inwardly directed radial dipole located at 0.5 Earth radii drifting eastwards along a path 15° south of Mono Lake. Its drift velocity was 0.10–0.19°/a as compared with the present westward drift of about 0.2°/a and its strength was 0.12–0.21 relative to the main dipole moment. He estimated the lifetime of the disturbance to be of the order of 850 a.

Verosub (1977a) has pointed out that the presence of anomalous directions in only the inclination or declination but not in both is more likely to be caused by distortion in the palaeomagnetic recording process or by errors in the sampling procedure than by fluctuations in the geomagnetic field. A major problem is the difficulty in finding consistent records of geomagnetic excursions. In some cases a proposed magnetic excursion can be found at the same stratigraphic horizon but with greatly varied magnetic signature (Freed and Healy, 1974): in other cases the magnetic feature may be entirely absent from nearby, contemporaneous sediments (Creer *et al.*, 1976a,b).

The most convincing evidence for the existence of the Mono Lake excursion is that it has been found at two sites 17 km apart within the same sedimentary environment (Denham, 1974). On the other hand, sediments from Clear Lake, spanning the time interval 21–29 ka, do not record any anomalous magnetic features (Verosub, 1977a). Since Clear Lake is only 320 km from Mono Lake and since each sample from Clear Lake represents 26 years of sedimentation, the magnetic signature of the Mono Lake excursion should be recorded in detail in the Clear Lake samples. The absence of the Mono Lake feature from the palaeomagnetic record of Clear Lake makes it difficult to accept the Mono Lake record as a real geomagnetic excursion. If the reality of the Mono Lake geomagnetic excursion is not confirmed, it indicates that even the existence of magnetically consistent records from two widely separated sites within the same sedimentary environment is not sufficient to establish the existence of a geomagnetic excursion. On the other hand, if the reality of the Mono Lake excursion is confirmed, its absence from the palaeomagnetic record of Clear Lake indicates that unfortunately geomagnetic excursions can have only limited reliability as magnetostratigraphic horizons for correlation between different sedimentary environments.

Liddicoat and Coe (1979) later re-examined the Mono Lake excursion in more detail – extensive sampling revealed a new feature of the excursion. They found the previously known eastward swing in declination and steepening of inclination to be preceded by an even greater swing to westerly declination and shallow inclination. The duration of the entire excursion is about 1000 a. Excellent agreement of palaeomagnetic directions between four sites shows that the excursion is a real expression of changes of the geomagnetic field. They modelled the source by a radial eccentric dipole at high northern latitudes pointing outward during the first part of the excursion, and near the equator and pointing inward during the latter part. The eccentric dipole was offset from the centre of the Earth by a distance of 0.28 Earth radii (approximately 1784 km). Movement of the source appears to be localized, displaying a complex pattern of eastward, westward and even northward drift. The average moment of the hypothetical eccentric dipole during the excursion is comparable with the largest calculated for the 1945 field, and the maximum moment is almost twice as great. The ages of all the samples range from 30.7 to 23.35 ka, on the basis of ^{14}C dates and the assumption that the sedimentation rate was uniform. Their data are in general agreement with Denham and Cox's (1971) result of relatively quiet behaviour of field direction from about 30.7 to 25.0 ka followed by an excursion between about 25 and 24 ka. The answer to the Mono Lake – Clear Lake problem may be that part of the record is missing. Uniform sedimentation rates were assumed but some of the deposition may have been destroyed by, e.g., bottom water currents. More probably, however, the answer lies in errors in ^{14}C dating (see, e.g., Barton and Polach, 1971, and Section 4.9).

Lajoie and Liddicoat (1980) later found large swings in declination (up to 60°) and inclination (up to 50°) in widely spaced samples from the post-excursion portion of the stratigraphic section record, confirming the broad swings reported previously. Five and a half clockwise loops span the set of smoothed VGPs in the interval 29–13 ka. The loops have an average duration of 2·9 ka and their movement is comparable with the westward drift of the non-dipole field.

Verosub et al. (1980) have examined a sequence of deep lake clays exposed on the shores of Pyramid Lake, Nevada, 230 km from Mono Lake, covering the time interval 25–36 ka. These dates were obtained from ^{14}C measurements on two wood fragments and from one sample of disseminated organic material. The spread in ages is consistent with the assumption of a uniform and continuous

sedimentation rate. The measured ranges of inclination and declination are 40°
and 75°, respectively. Thus, the palaeomagnetic record from Pyramid Lake, like
that from Clear Lake, contains no evidence for a geomagnetic excursion. Taken
together, it would seem that Northern California and Western Nevada were not
affected by an excursion during the time period 21–36 ka. This is in direct
conflict with the data from Mono Lake. Verosub *et al.* (1980) have discussed
this problem in detail. If the absence of the excursion at both Clear Lake and
Pyramid Lake is attributed to non-deposition or erosion, such processes would
have to have occurred simultaneously at each lake and simultaneously with the
geomagnetic excursion. Such a coincidence is highly unlikely. The simplest explanation is that the original data of the Mono Lake excursion are in error. It should
also be noted that studies of contemporaneous lake sediments by Oberg and
Evans (1977) in southern British Columbia and by Doh and Steele (1981) in
Fargher Lake, south-west Washington, failed to confirm the existence of an
excursion. In a later paper, Doh and Steele (1983) confirmed that for the most
part the sediments of Fargher Lake have a strong, stable remanent magnetization
showing no evidence of reversals. They concluded that none occurred during
deposition except perhaps during two short gaps in their records totalling about
1700 a for which they were unable to recover any sediments. If the Heussers'
(1980) timescale is correct, no reversals occurred in south-western Washington
during the time interval 14.1–30.3 ka unless it occurred during the intervals
15.8–16.8 ka or 21.1–21.8 ka.

Palmer *et al.* (1979) have carried out palaeomagnetic and sedimentological
studies at Lake Tahoe, California–Nevada, and found fair agreement between
the Mono Lake and Lake Tahoe records in both declination and inclination (see
Figure 4.3). The correlation is based on the similarity of longer wavelength
variations. On the basis of this correlation, the one short-wavelength high-
amplitude feature on the Mono Lake record is aligned with a similar short-
wavelength feature common to all three cores from Lake Tahoe. There is,
however, no definite age determination on the cores from Lake Tahoe, and the
Mono Lake excursion was fixed at 24 ka.

Liddicoat has looked at the Pyramid Lake records and suggested that perhaps
there is a hint of an excursion near the middle of the record where it should be
on the basis of the Mono Lake records. Later, Liddicoat *et al.* (1982) identified
the Mono Lake excursion in the Lake Lahontan Sehoo Formation in Carson
Sink, Nevada, 200 km north-east of Mono Basin, where it was originally documented in the Wilson Creek beds near the north-west shore of Mono Lake. The
age of the Mono Lake excursion is placed at 28–26 ka, on the basis of interpolation from a linear regression of 16 ^{14}C dates on nodular and platy algal tufa
in the Wilson Creek beds. This age is consistent with a 24.48 ± 0.43 ka ^{14}C date
on wood associated with the Wono tephra bed in Lake Lahontan lacustrine
deposits near Pyramid Lake.

Turner *et al.* (1982) sampled an 18 m sedimentary sequence at Bessette Creek,
British Columbia (about 12° north of Mono Lake). Radiocarbon ages suggested
that the sequence spans the interval 31.2–19.5 ka. No evidence for any large
geomagnetic excursions was found, although a distinctive pattern of 'normal'

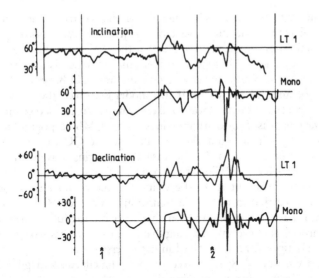

Figure 4.3 Proposed correlations of NRM inclination and declination data from Lake Tahoe and Mono Lake. Correlation is based upon the similarity of longer wavelength variations between points 1 (13.3 ± 0.5 ka) and 2 (23.3 ± 0.3 ka). After Palmer *et al.* (1979).

secular variation was observed with declination and inclination swings of 45° and 25° peak-to-peak amplitude, respectively. For the most part the secular variation consisted of low-amplitude oscillations about the field vector of a geomagnetic axial dipole expected at the site latitude, but three relatively large perturbations occurred at approximately 4 ka intervals.

Mörner (1986) also found no record of the Mono Lake excursion in cores from lacustrine deposits from Grande Pile in the southern Vosges, although many other excursions were identified during the last 140 ka covered by the cores. Wang Jingtai *et al.* (1986) found an inclination anomaly in sediments from Dabusan Lake, Qaidam Basin, Central Asia, dated about 20–30 ka but only in two partially reversed samples – they attached little importance to it.

Lund *et al.* (1988) have obtained a new palaeomagnetic secular variation (PSV) record from the late Quaternary Wilson Creek beds of Mono Lake. They found a distinctive periodic vector waveform that follows, and is almost certainly related to, the Mono Lake excursion. A comparison of all published PSV results from the Wilson Creek beds suggests that the magnetic field at Mono Lake went through an interval (36–28 ka) of very-low-amplitude PSV followed by the Mono Lake excursion (28–27 ka) and four subsequent recurrences (27–12.5 ka) of the excursion waveform with relatively diminished amplitudes (lower than the excursion amplitude, but higher than PSV amplitudes at Mono Lake since 12.5 ka). This suggests that the core dynamo process responsible for PSV is capable of near impulse onset ($\leq 10^3$ a), a very slow dissipation rate ($\sim 10^4$ a) and quasi-periodic behaviour that is non-wavelength-dispersive in time.

Lund *et al.* suggest two possible models for the recurring waveform of the PSV. In the first the generalized waveform is just the spatial non-dipole field. Zonal drift of this non-dipole field would cause the spatial waveform to repeatedly drift past Mono Lake. This type of waveform recurrence is also seen in the younger Lake St. Croix palaeomagnetic record (Lund and Banerjee, 1985), but is not well documented elsewhere. If this model is correct, the same variation should be seen at any site at Mono Lake's latitude. In the second model the recurring waveform is due to irregularities at the CMB (topographic bumps, temperature, . . .) which could alter the flow in the OC to produce several 'standing' non-dipole sources. In this case the recurring waveform would be related to the long-term stability of the pattern of fluid flow past Mono Lake, limiting the waveform pattern to the vicinity of the Mono Lake region. It is interesting to note that in a detailed transition record from Crete Valet *et al.* (1986) found that the secular variation characterizing the full polarity intervals continues through the reversal, implying that the process responsible for generating secular variation was not involved in the reversal.

Hanna and Verosub (1989) reviewed 16 lake sediment palaeomagnetic records from western North America spanning the last 40 ka. Five studies contain evidence for the Mono Lake excursion, but six studies do not show any record of it. The Mono Lake excursion has been confirmed by three studies at Mono Lake, California, at Summer Lake, Nevada, and, tentatively, at Lake Tahoe, California–Nevada. On the other hand, as already discussed, there are no indications of any geomagnetic excursions at Clear Lake, California; Pyramid Lake, Nevada; Devil's Park and Triangle Park, Colorado; Fargher Lake, Washington; and Bessette Creek, British Columbia. Clear Lake is only 320 km from Mono Lake and, if the above observations are true, would place serious constraints on models of geomagnetic field behaviour. (If excursions are a result of fluctuations in the core, then the expressions in the surface field should be of the order of 3200 km in areal extent, i.e. the Mono Lake excursion should have been recorded at least at Clear Lake and Pyramid Lake.) Verosub *et al.* have suggested (1980) that the inconsistencies are a result of radiocarbon dating errors in the Mono Lake record. Later Hanna and Verosub (1989) reviewed all the ^{14}C dating and concluded that any geomagnetic excursions in western North America would be constrained to have occurred between 13.3 and 14.1 ka, between 15.8 and 16.8 ka or after 31.2 ka. On the other hand, the discovery of the Mono Lake excursion between 24.9 and 28.9 ka at Summer Lake substantiates the evidence that the Mono Lake excursion occurred during the time intervals covered by the Clear Lake and Pyramid Lake records. Hanna and Verosub concluded that the matter is not completely settled and that the Mono Lake excursion raises questions about the accuracy of ^{14}C dating and also about the palaeomagnetic recording process in sediments.

Levi and Karlin (1989) analysed a 60 ka palaeomagnetic record from sediments in the Gulf of California. They believe that the Mono Lake excursion was recorded between ~ 29 and 26 ka. This time span is about 3 ka, as compared with an estimated 1 ka at Mono Lake. They also observed a narrow zone near 23 ka with a very similar palaeomagnetic signature to that of the excursion seen at Summer Lake. They suggest that the Summer Lake excursion is distinct from and younger than the Mono Lake excursion by some 3–5 ka and is of considerably shorter duration, lasting no more than a few hundred years.

The most recent study of the Mono Lake excursion is that of Liddicoat (1992), who sampled two new localities, Mill Creek (MC) and Warm Springs (WS), 20 km apart in the Mono Basin, and also in exposed sediments from Pleistocene Lake Lahonton in north-western Nevada, 250 km to the north, where the sampling was done at Pyramid Lake and in the Carson Sink Bed (CSB). The most interesting result of this study is that the Mono Lake excursion can be seen at Pyramid Lake. Liddicoat and Coe (1979) had shown earlier that the excursion consists of two parts separated by a volcanic ash bed. At Pyramid Lake and with CSB, the older half of the excursion is not recorded, but the younger half can be recognized. Figure 4.4 shows the magnetic record at three sites in the Mono Basin – the first from Wilson Creek (WC), which is site A of Liddicoat and Coe's earlier work (1979), and the other two at MC and WS.

The older part of the excursion (the portion below the Wilson Creek ash bed 15) is a rapid swing in declination to $\sim 300°$ that is immediately followed by an 80° change in inclination (from about 50° to $-30°$). There is a fivefold decrease in the relative intensity and the VGPs describe a large clockwise loop (Liddicoat and Coe, 1979). The younger half (the portion above the Wilson Creek ash bed 15) has an inclination that steepens to 80° as the declination moves quickly to the east. The relative intensity recovers to the pre-excursion level and the VGPs make a small counterclockwise loop. On the basis of ^{14}C dating and tephrachronology and the assumption that the sedimentation rate was constant at about 25 cm/ka, the middle of the excursion is estimated to be ~ 28 ka, in agreement with the age given by Levi and Karlin (1989) from their study of sediments in the Gulf of California. The full excursion lasted less than 2 ka.

Figure 4.5 (Liddicoat, 1992) shows the record from three sites in the Carson sink. Below the Carson sink ash bed there is easterly declination, inclination $\sim 30°–40°$ and low relative intensity, whereas for the Mono Lake excursion below the Wilson Creek ash bed 15 a westerly declination is followed by negative inclinations and reduced relative intensity. However, above the Carson sink ash bed the Mono Lake excursion can be recognized – easterly declination, nearly vertical positive inclinations and high relative intensity. There is also similar looping of the VGPs.

The record from Pyramid Lake (Figure 4.6) does not appear to show the Mono Lake excursion when specimens are AF demagnetized at 15 mT or 20 mT. However, when AF magnetization is increased to 60 mT, the younger half of the excursion can be seen – as in the Carson Sink, the older half of the excursion is absent. Liddicoat (1992) suggested two possibilities for only partial recording of the Mono Lake excursion at Pyramid Lake and Carson Sink – the low relative strength of the palaeomagnetic field during the older half of the excursion and the slower sedimentation rate in Lake Lahontan, which has been estimated to be about one-half of the rate of deposition of the lacustrine sediments in the Mono Basin.

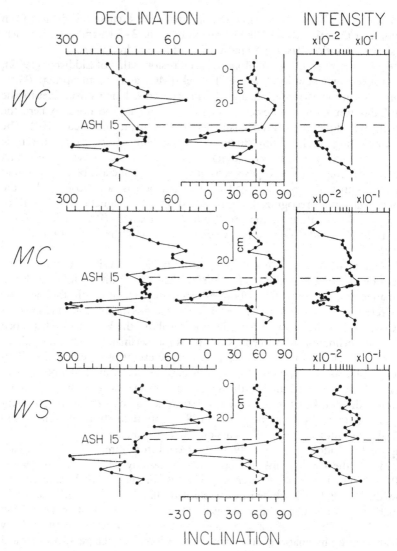

Figure 4.4 Palaeomagnetic curves for the Mono Lake Excursion from approximately 30 cm above to 25 cm below the Wilson Creek Ash Bed 15 (horizontal dashed line) at Wilson Creek (WC), Mill Creek (MC) and Warm Springs (WS) in the Mono Basin, California, for AF demagnetization at 20 mT. The data for Wilson Creek are from Liddicoat and Coe (1979); the data for Mill Creek and Warm Springs are from Liddicoat (1992). Note that westerly declination and negative inclination occur during low relative intensity below the Wilson Creek Ash Bed 15, and that above the ash bed easterly declination and steep positive inclination occur during high relative intensity. The low and high relative intensity have about the same duration if the sedimentation rate was constant within this interval. Vertical line in the plot of inclination is the inclination (57.4°) of an axial dipole field. Intensity is in A/m. After Liddicoat (1992).

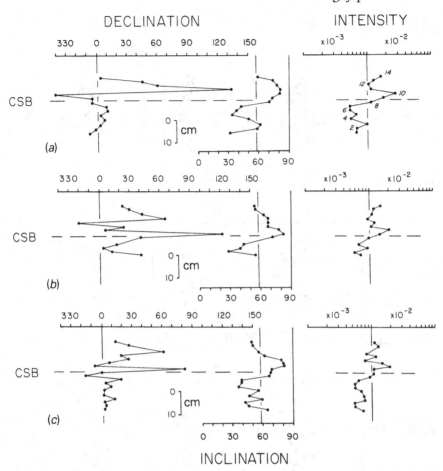

Figure 4.5 Palaeomagnetic curves for Carson Sink Sites A, B and C. The horizontal dashed line is the position of the Carson Sink Bed (CSB). The data are for 20 mT AF demagnetization. Note that the younger half of the Mono Lake Excursion (the portion above the CSB) is recorded in part at each site but that the older half (the portion below the CSB) is absent. Vertical line in the plot of the inclination is the inclination (58.9°) of an axial dipole field. Intensity is in A/m. After Liddicoat (1992).

4.5 The Gothenburg 'flip'

Mörner *et al.* (1971) and Mörner and Lanser (1974) reported a very rapid 'excursion' in the magnetic field which they subsequently named the Gothenburg 'flip'. They first recognized it in a core in the Botanical Gardens, Gothenburg, Sweden. The upper boundary was fixed very precisely at the boundary between the Fjaras Stadial and the Bolling Interstadial, dated at 12.35 ka. The change

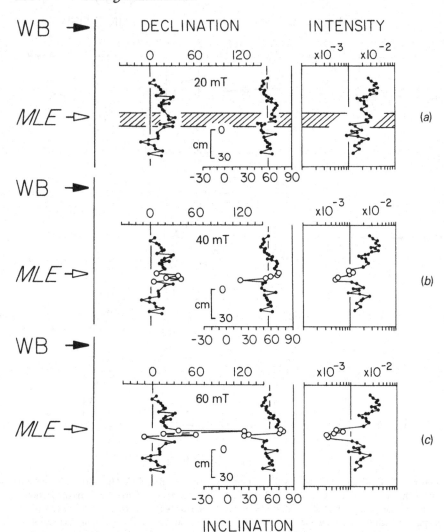

Figure 4.6 Palaeomagnetic curves for the Sehoo Formation at Site A at Pyramid Lake, Nevada. Plot (a) is for 20 mT AF demagnetization. Plots (b) and (c) are for the same curves except in the zone 97 – 105 cm below the Wono Bed (WB, black arrow), where the data (open circles) are for single specimens spaced back-to-back in one hand sample; that sample is from the shaded part of plot (a). The sample contains the younger half of the Mono Lake Excursion (MLE, white arrow), which appeared when AF demagnetization was increased to 40 mT (plot b) and 60 mT (plot c). Note in plot (c) that declination is about 120° when the inclination approaches 80°. Vertical line in the plot of the inclination is the inclination (59.2°) of an axial dipole field. Intensity is in A/m. After Liddicoat (1992).

from reversed to normal polarity is very rapid, occurring within a few years! Furthermore, there is no major intensity change related to the polarity change. They speculated that there may have been a long period (1–2 ka) of unstable magnetism, with several flips between normal and reversed polarity, or a short period (about 100 a) of reversed polarity at around the Fjaras Stadial (12.4–12.35 ka) which followed a millennium of irregular (but not fully reversed) magnetism. Mörner (1977) later claimed that the excursion is also present at the same stratigraphic level in four other Swedish cores.

Thompson and Berglund (1976) believe that the intermediate palaeomagnetic directions observed by Mörner and Lanser are due to slumping rather than a real change in the magnetic field, which they maintain was of normal polarity from 13–11 ka in Sweden. They base their view on an analysis of 408 subsamples taken from two cores in southern Sweden. Their data suggest that previously reported reversed palaeomagnetic directions are not reliable indicators of the ancient geomagnetic field, but have been distorted by mechanical sedimentation processes, slumping or weathering. They further suggested that the proliferation of unusual palaeomagnetic directions in Scandinavia around 12 ka is a reflection of changing climatic conditions. In many localities fluctuations of climate produced sediments of very variable mechanical properties, particularly at times of peri-glacial activity, which were poor recorders of the direction of the ambient mag-netic field. Banerjee *et al.* (1979) confirm this conclusion. An earlier report that the Gothenburg 'flip' had been observed in New Zealand has also since been refuted (Sukroo *et al.*, 1978).

Barbetti *et al.* (1980) have successfully measured the ancient magnetic direction and field strength in baked stones from late Pleistocene hearths at Étiolles and Marsangy, France. The dates for two hearths at Étiolles have been estimated as 12 ± 0.22 ka, obtained by ^{14}C dating of a mammoth scapula excavated from a lower level in another part of the site. The only means of dating the stones at Marsangy is by archaeological seriation based mainly on the stone industry. Barbetti *et al.* assumed an age of about 12 ± 0.5 ka. They found the VGP to be very close to, and not significantly different from, the Earth's geographic pole, and the corresponding VDM to be only slightly less than the present-day dipole moment. These results strengthen the conclusions of Thompson and Berglund (1976), casting doubt on the reality of the Gothenburg excursion.

Mörner and Lanser (1975) later analysed core A179–15 – a high-deposition-rate core from the southern North Atlantic – and found a sudden change in the declination at 12.35 ka which coincides exactly with the end of their Gothenburg flip. Opdyke (1976) does not accept their findings, believing that the signal seen in declination only is noise arising from the sampling procedures used in this old core. However, Mörner (1976) does not accept Opdyke's criticism and is still convinced of the reality of the Gothenburg flip. Abrahamsen (1982) exam-ined three sedimentary cores from Solberga, Brastad and Moltemyr, north of Gothenburg, Sweden. In the Solberga core magnetic directions between depths 17 and 12 m showed great scatter, indicating a possible excursion of the magnetic field. However, no well-documented excursions of the geomagnetic field in Holo-cene time are yet known, and Abrahamsen suggested that the directional scatter could most simply be explained as due to some kind of post-depositional disturbance such as sliding, slumping, bioturbation or compaction.

There are several signs in the Solberga core around 18–19 m depth of a change in the environment from saline to more brackish water conditions. This may be related to the climatic amelioration of the Pleistocene–Holocene transition, with an increase in the meltwater flux and with the drainage of the Baltic Ice Lake. A change in the wet density is seen at 18.95 m and also a jump in the NRM intensity at that depth and again at 19.35 m. A change in declination is observed at 17.95 and 18.50 m but not in the cleaned records. The inclination shows a smooth variation below 17 m with a gradual increase between 20 and 17 m, indicating that, at least up to this level, a post-depositional disturbance of the sediment is unlikely to have occurred.

The 'Gothenburg excursion' of the geomagnetic field, which Mörner *et al.* (1971) postulated to end at 12.35 ± 0.05 ka is too early to appear in the Solberga core with an age at the bottom of 11.2 ± 0.4 ka. The Brastad core, however, is likely to reach further back in time. Below 13.5 m the declinations and inclinations are very scattered, and at the very bottom of the core, four specimens with low inclinations and ordinary declinations were found. During magnetic cleaning the scatter is not significantly altered in the bottom metre, indicating that viscous components do not influence the scattered directions. The directional scatter must therefore be due to orientational scatter in the sediment rather than a differential response to viscous overprints in the ambient field. Abrahamsen concluded that a geomagnetic excursion is probably seen at the bottom of the Brastad core.

During progressive AF demagnetization, an interval of the Holocene part of both the Solberga and Moltemyr cores shows decreasing inclinations probably due to viscous magnetic overprinting, which may indicate either unusual magneto-sedimentological properties related to an increase in deposition rate, or a hitherto unrecognized post-glacial geomagnetic low-inclination excursion, occurring shortly after 10 ka. The Gothenburg excursion, dated to end around 12.35 ka, is not seen in these records, although the Brastad and Moltemyr cores do probably reach further back.

Sandgren (1986) could find no evidence for the Gothenburg excursion in sediments from the Torreberga Basin, south Sweden, and concluded that Mörner's results are a reflection of unsuitable palaeomagnetic conditions. Wang Jingtai *et al.* (1986) found an inclination departure in sediments from Dabusan Lake, Qaidam Basin, central Asia, dated about 9–15 ka, but in only one sample – they attached little importance to it.

Mörner (1986), however, continues to insist that the Gothenburg flip is real. He claims that some records show a reversed inclination (with the VGP in the central equatorial Pacific), while others show a reversed declination (with the VGP in the eastern equatorial Pacific). In a French set of cores a rapid inclination flip is immediately followed by a 180° declination switch, implying a rapid switch between two opposite equatorial VGP positions. Mörner believes that the length of the flip is very short – only a few years and certainly less than 50 a. It is difficult to accept the validity of these statements. He does receive some support, however, from Petrova and Pospelova (1990), who claim that deviations of geomagnetic directions in the interval 13–11 ka have been found in many regions of the Soviet Union, both in marine and lacustrine bottom sediments and in continental sediments.

4.6 The Blake event

A transition in the magnetic field which may be an excursion or a true reversal is the Blake event, which was first discovered in seven cores from the Caribbean and Indian Ocean by Smith and Foster (1969), who dated it at between 105 and 114 ka and estimated its duration as 5–7 ka from their most reliable sedimentation rates, although the sedimentation rate in one core indicates a duration of 12 ka. The nature of the transition is primarily a change in sign in the inclination, and it is not clear that the excursion involved a full 180° rotation of the geomagnetic axis. Denham (1976) and Denham *et al.* (1977) observed the Blake event in two large-diameter cores from the Greater Antilles outer ridge and reported that average sedimentation rates gave durations of 16.5 and 30 ka, while another method led to an estimated duration of 50 ka. Denham did not consider such long durations to be realistic, since the Blake event has been observed so rarely. He suggested that the VGP shift is consistent with a regional excursion caused by fluctuating non-dipole field intensities. Because of its apparent long duration, the Blake event may be distinct from other excursions. Sasajima *et al.* (1980) found an excursion in pyroclastic flows at Keno and Kogashira, Japan, and also in a rhyolitic tuff from Sigulagava, North Sumatra, which they tentatively correlated with the Blake event. If this is correct, it is the first time that the Blake event has been identified in igneous terrestrial rocks and would suggest that the Blake event is a global reversal and not a local event. The Blake event has also been reported from lacustrine sediments in Lake Biwa, Japan (Yaskawa *et al.*, 1973; Yaskawa, 1974), in offshore clays from Italy (Creer *et al.*, 1980), and in loess deposits from Czechoslovakia (Kukla and Koci, 1972) and Poland (Tucholka, 1977). Aksu (1983) recorded an excursion in sediment cores from Baffin Bay which he correlated with the Blake event. He estimated the age to be between 129 and 161 ka on the basis of sedimentation rate and correlation with oxygen isotope curves. Eardley *et al.* (1973) found several metres of reversed polarity near the top of a long drill core from Lake Bonneville, Utah. They identified it with the Blake event and estimated its age to be 125 ka, assuming a constant sedimentation rate.

Tucholka *et al.* (1987) found evidence of the Blake event in five cores in the Mediterranean. They estimated the age to be ~ 117 ka and the duration ~ 6ka. Wang Jingtai *et al.* (1986) believe that they have found evidence for the Blake event in sediments from mainland China, and Herrero-Bervera *et al.* (1989) in lacustrine sediments from Pringle Falls, Oregon, again strengthening the arguments for its global existence. However, Mankinen *et al.* (1986) found no evidence for the Blake event in a study of volcanic rocks from Long Valley caldera in California with K–Ar ages ranging from 95 to 115 ka. They suggested that perhaps the Blake event occurred prior to this time interval or that the event is so short that it was not recorded by episodic volcanism. On the other hand, Champion *et al.* (1988) believe that the Blake event may be older than previously thought. Data they collected from the Laguna basalt flow in the Bondera Field of west central New Mexico indicate a K–Ar age of 128 ± 33 ka. This is the first evidence from volcanic rocks for the age of the Blake event. Rais *et al.* (1991) have since carried out a palaeomagnetic study of 72 lava flows from the Piton des Neiges in Réunion Island with K–Ar ages ranging from 88 to 129 ka encompassing the Blake event. They found strong deviations from normal polarity directions, including a complete reversal, the field displaying character- istics of the Blake event observed elsewhere. They also made palaeointensity measurements which showed large fluctuations – 20–80 μT, the average being 47 μT, which is higher than the present-day value of 35 μT.

Creer *et al.* (1980) carried out palaeomagnetic and palaeontological studies on a long core from Gioia Tauro, Italy, and claimed to have seen the Blake event. As recorded by the inclination, the Blake event is split into two parts of estimated duration 16 and 23 ka, separated by a short sequence of normal inclination of estimated duration 10 ka. Manabe (1977) claimed to have seen the Blake event in a marine terrace in Japan, the dating of which is based primarily on palaeoclim- atological evidence. Like the Italian record and two of the western North Atlantic records, the Blake event in the Japanese record appears as two reversed intervals separated by a short normal interval. Herrero-Bervera *et al.* (1989) also found that their record from Pringle Falls Oregon is split into two parts separated by a short normal interval.

Tric *et al.* (1991) obtained two high-resolution records of the Blake event from marine cores in the Mediterranean. An upper limit for the duration of the event based on oxygen isotope stratigraphy and tephro-chronology is ~ 4 ka. The directional changes do not occur at constant speed, but rather in a stop-and- go manner, as in detailed records of full reversals (see, e.g., the discussion of the Steens Mountain reversal, Section 6.7).

During the first phase of the reversal the VGPs move rapidly from the North Pole to intermediate southern latitudes and then back to the North Pole. In both cases the path is confined to a longitudinal band over the Americas. During the second phase the VGP undergoes a similar, but slower, motion over the Americas, followed by a period of standstill at high southern latitudes. In the final phase of the transition back to normal polarity the VGP passes first over South America up to mid-southern latitudes and then switches suddenly (by ~ 180° in longitude) to a path over Australia and south-eastern Asia. The two bands of longitude coincide with those of many recent reversals, e.g. in the Pringle Falls record of Herrero-Bervera *et al.* (1989). In the first phase of the transition both VGP paths are confined to the same longitudinal band over the Americas, and the first loop from northern polarity to southern latitudes and back, followed by a second north to south pole shift, is also confined to the same band of longitudes. The main feature of the last phase of the reversal (the jump of the VGP path to longitudes almost antipodal to the Americas) is also seen in both records (see Figure 4.7). This suggests that a large dipolar component may

(a)

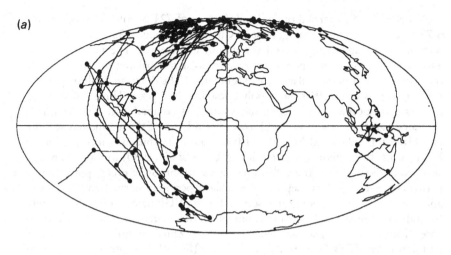

(b)

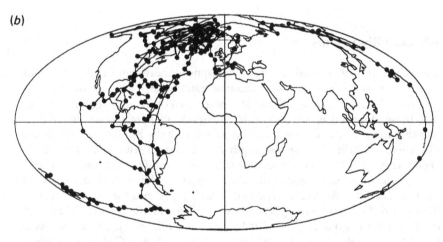

Figure 4.7 Plots of the VGP path during (a) the Blake event recorded in a core from the Mediterranean (Tric *et al.*, 1991) and (b) the geomagnetic event from Pringle Falls obtained by Herrero-Bervera *et al.* (1989). After Tric *et al.* (1991).

have been present during these transitional fields. This question is discussed further in Section 6.5. The coincidence of the two bands of longitude seen also in many other reversals suggests that the physical processes leading to reversals or events are basically the same and that the same mechanism is statistically observed over long periods of time. In conclusion, it seems that the Blake event is a worldwide feature of the geomagnetic field, but whether it should be considered as a true reversal or an excursion is not clear.

Zhu *et al.* (1993) carried out a detailed study of 721 oriented samples from a 5.7 m loess section near Xining, western China. They found a 55.2 cm anomalous section in an intervening loess layer within a palaeosol. The palaeomagnetic anomaly consists of *three* periods of reverse polarity separated by two relatively short periods of normal polarity (see Figure 4.8). The duration of all six polarity transitions was extremely short – only a few hundred years. Three different age models corresponding to different modes of loess accumulation gave essentially the same values for the upper and lower boundaries of the anomalous section, viz. 116.50 ± 1.2 ka and 111.40 ± 1.0 ka. Zhu *et al.* thus identified it with the Blake event. It is interesting to recall that Creer *et al.* (1980) in their study of a core from Gioia Tauro, Italy, claim to have seen the Blake event split into two periods of reverse polarity separated by a short period of normal polarity. There appears to be preferred longitude sectors for the VGP paths which are different from those observed in lacustrine sediments in the western USA and in deep-sea cores from the Mediterranean (see Section 6.5). Zhu *et al.* find that in northern latitudes most VGPs lie between 120° W and 180° W. In southern latitudes the VGPs are relatively clustered with a longitudinal distribution centred around 100° E.

4.7 Lake Biwa events

Yaskawa *et al.* (1973) carried out palaeomagnetic measurements in a 197.2 m core from Lake Biwa, Japan. Sedimentation rates in large lakes are usually high – in Lake Biwa as much as 500 m/Ma – nearly two orders of magnitude greater than those of most deep-sea sediments. The results of a preliminary survey indicate that there have been five short reversals during the Brunhes normal chron. The curve of inclination against depth correlates well with that obtained by Wollin *et al.* (1971) in a deep-sea core from the North Pacific. From this correlation Yaskawa *et al.* suggested that the youngest of the three clear reversals is the same as the Blake 'event'. The two other clear reversals in inclination have been called the Biwa I event, between about 176 and 186 ka, and the Biwa II event, between about 292 and 298 ka (Kawai *et al.*, 1972). These ages were based on extrapolation of a depth–time plot derived from ^{14}C ages in the upper part of the core. Fission track ages on zircons from ash layers within the core have given more reliable ages. Kawai (1984) suggested mean ages of the reversal events which are of 6–13 ka in duration to be 100, 160, 310 and a fourth interval of negative inclination below Biwa II at about 380 ka – this he called the Biwa III event.

Hayashida (1980) has confirmed the existence of the Biwa I excursion which he found in the Takashina Formation of the Kobiwako Group. The Kobiwako Group is considered to consist of the same sediments as those of the ancient Lake Biwa. Two volcanic ash layers in the Takashina Formation showed reversed polarity in an otherwise normal polarity section and were correlated in age with the Biwa I excursion seen in the lake core. Hayashida (1984) later identified the

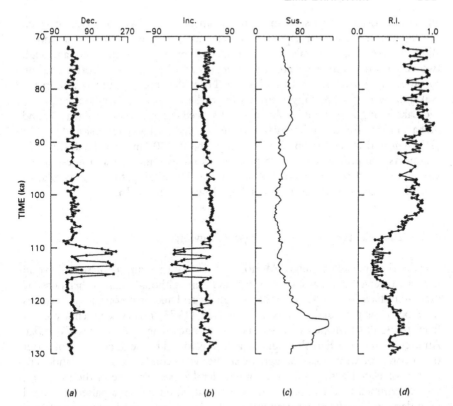

Figure 4.8 Palaeomagnetic data from a loess section near Xining, western China: (a) declination, (b) inclination, (c) susceptibility, (d) relative intensity (normalized by ARM at 300°C). After Zhu *et al.* (1993).

Biwa I excursion in terrace-forming deposits in the north-eastern area of Ashanti city, about 150 km west of Lake Biwa. A short reversal about the time of the Biwa II excursion has been observed by Harrison (1974) in deep-sea cores in the north-western Pacific. Creer *et al.* (1980) suggested that the alpha excursion of Gioia Tauro, Italy, correlates with Biwa II and the beta excursion with Biwa III.

Torii *et al.* (1974) found fully reversed directions in the Kasuri ash in Plio-Pleistocene deposits of the Osaka group – apatite in the ash gave fission track ages of 370–380 ka, in agreement with the estimated age of Biwa III. Palaeomagnetic results from glacial loess deposits of East Germany and Western USSR record up to five polarity events within the Brunhes chron (Bucha, 1983) – in particular, Bucha correlated reversed polarity directions from deposits of the late Okam glaciation with Biwa III. Champion *et al.* (1988) suggested that anomalous declinations seen in a core from Summer Lake by Negrini *et al.* (1988) just above an ash deposit dated at 360–370 ka may also correspond to Biwa III. No clear reversals were seen in the upper part of the core, although two very short excursions were seen at depths of 13 m and 26 m corresponding to times of approximately 18 and 49 ka. From the similarity of the pole paths Nakajima *et al.* (1973) suggested that the earliest excursion may be the same as the Lake Mungo excursion. Nakajima *et al.* placed the second excursion at around 38 ka and speculated that it may be the same as the Laschamp excursion. Hirooka (1976) found an excursion near Ina City, about 200 km from Lake Biwa, in sediments composed of layers of pumice, scoria, volcanic ash and fine-grained sediments dated between about 60 and 35 ka. The VGP path is very similar to that of the second excursion seen at Lake Biwa around 49 ka.

4.8 Other excursions and short-period events

The Laschamp, Lake Mungo, Mono Lake and Gothenburg excursions have in the past received most attention in the literature, although many other possible excursions have been reported. Palaeomagnetic and micropalaeontological studies have been carried out by Clark and Kennett (1973) on 28 sedimentary cores from the Gulf of Mexico with sedimentation rates ranging from 9 to 20cm/ka. An excursion of the Earth's magnetic field was found in the upper parts of 8 out of 15 cores for which palaeomagnetic studies were conducted and was independently correlated with planktonic foraminiferal zones. The age of the excursion was determined by extrapolation of sedimentation rates from a palaeontological boundary and occurred between 12·5 and 17 ka. Clarke and Kennett attributed the excursion in several cores from the western Gulf of Mexico to the dominance of calcium carbonate. Almost all the cores from the eastern Gulf of Mexico that were used for palaeomagnetic measurements show the excursion. These cores were, except for the uppermost few centimetres, almost devoid of calcium carbonate. Freed and Healy (1974) have also examined deep-sea sediment cores from the Gulf of Mexico. They found two excursions, one at 17 ± 1.5 ka and the other at 32 ± 1.5 ka. The excursions were dated by microfaunal analysis.

A regional excursion (the Erieau excursion) has been proposed by Creer *et al.* (1976b) on the basis of their study of two cores, 19 km apart, of late Wisconsinan sediments in Lake Erie. This excursion was dated indirectly as starting at about 14 ka ago and ending at a horizon whose age is somewhere between 10.5 and 7.6 ka. The excursion manifests itself as a change in inclination from about −90° to +90°. They suggested that the event could be the same as that reported by Yaskawa *et al.* (1973) in the core from Lake Biwa (estimated age approximately 18 ka) and seen by Clark and Kennett (1973) in cores from the Gulf of Mexico, dated between 12.5 and 17 ka.

Banerjee *et al.* (1979) obtained high-resolution data of variations in the inclination recorded in the sediments of two post-glacial lakes in Minnesota. The first, from Lake St Croix, covered the last 9.6 ka and the second, from Kylen Lake, from about 9 to 16 ka. They found no evidence for any abnormal behaviour or excursions of the geomagnetic field over the past 16 ka in Minnesota, although several authors (e.g. Creer *et al.*, 1976b) have claimed that there have been sharp excursions in field direction during this time interval. Verosub and Banerjee (1977) have suggested that only a few of these proposed excursions represent real geomagnetic fluctuations, the rest, especially those recorded in sediments, probably being caused by disturbance of the sediment during its recovery or by imperfect recording of the geomagnetic field during deposition.

Abrahamsen and Knudsen (1979) carried out palaeomagnetic measurements on a single piston core in marine clay at Rubjerg, Denmark. They found an apparent excursion of the geomagnetic field in the undisturbed older marine deposits (the older Yolida Clay). They obtained 17 specimens which exhibited the excursion in a section at least 1.2 m thick. The age is not well determined – foraminifera fauna and stratigraphic position indicate an age between 23 and 40 ka. They speculated that it may be correlated with the excursions reported at Laschamp, Mono Lake and Lake Mungo, and in the Gulf of Mexico. Further work in Denmark has been carried out by Abrahamsen and Readman (1980) in clays at Nørre Lyngly. In the older Yolida Clay they confirmed the existence of the excursion seen in Rubjerg. In addition, they found an excursion in declination in the Younger Yolida Clay, aged about 14 ka. This age was obtained from foraminifera fauna and by ^{14}C dating of *in situ* shells. They speculated that this excursion may be the same as that observed in the Gulf of Mexico (Clark and Kennett, 1973) and in Lake Michigan (Dodson *et al.*, 1977).

Shibuya *et al.* (1992) recorded anomalous geomagnetic directions in alkali basalt lavas from the Auckland volcanic field, New Zealand. This is the first reported evidence of excursions seen in igneous rocks from the southern hemisphere. The authors place the ages between approximately 25 and 50 ka, and tentatively correlate them with the Lake Mungo, Mono Lake and Laschamp excursions. If substantiated by better age control, these results would be important in establishing that these excursions were worldwide phenomena.

Champion *et al.* (1988) have reviewed all the data documenting short reversal records from both volcanic and sedimentary rocks, and found evidence for eight changes in polarity in the Brunhes and two in the late Matuyama, besides the Jaramillo. These include a new reversal from lava flows in the Snake River area of Idaho which they have called the Big Lost reversed polarity event, dated at 565 ± 14 ka. Petrova and Pospelova (1990) have claimed that there were 12 excursions in the Brunhes chron, and no fewer than seven within the 580 ka interval after the Matuyama–Brunhes transition.

Cox (1968) proposed that the frequency distribution of polarity chrons and subchrons was continuous and that there were numerous undiscovered 'events' with durations shorter than 50 ka. The ages of the principal polarity chron boundaries for the last 5 Ma are now fairly well determined, as are the times and durations of the major subchrons. Discovery and confirmation of very short events, however, has been slow, because they are of such short duration that evidence for their existence is not widely observed.

Figure 4.9 shows the position of the ten reversed polarity events documented by Champion *et al.* (1988), who believe that they are true subchrons and not excursions. They may or may not be associated with low palaeointensities, although low field strengths might explain why the reversal process aborts. They will be discussed in more detail later in this section. Champion *et al.* (1988) point that the ten reversed polarities of Figure 4.9 occur at regular intervals. The regularity is not strictly periodic in character, although it could represent a periodic function masked by another stochastic process. A simple mean of the polarity intervals is 90 ka. This number is very close to the 100 ka orbital eccentricity period of the Earth and raises the question of possible connections between geomagnetic variations and climate (see Section 8.1).

Ryan (1972) identified four reversed events in the Brunhes in cores from the Caribbean and Mediterranean. He correlated the uppermost reversal with the Blake event and named the others the Jamaica, Levantine and Emperor events. He estimated the ages for the Jamaica and Levantine to be ~ 200 and 300 ka. Other short reversals or excursions in the Brunhes which may be the same as those found by Ryan have been given different names (see Figure 4.9).

A short-duration full reversal has been found at several sites in loess and silt sections in Alaska and the Yukon (Westgate *et al.*, 1985). The reversals occur about 3 m below the Old Crow tephra, which was originally dated at ~ 86 ka and were originally correlated with the Blake event. More recent estimates give the age of the Old Crow tephra as 149 ± 13 ka (Westgate, 1988). Champion *et al.* (1988) therefore suggest that the reversed strata may well be the Jamaica/ Biwa I events rather than the Blake. Hayashida (1984) had previously correlated the Jamaica and Biwa I events.

Evidence for the Levantine reversal of Ryan (1972) dated at 230 ka is the Biwa II event at ~ 310 ka (Kawai, 1984). Short polarity intervals dated at ~ 300 ka have also been found at Tulelake, northern California (Rieck *et al.*, 1992) and at Summer Lake, Oregon (Negrini *et al.*, 1988). Kochegura and Zubakov (1978) reported a number of anomalous or reversed directions in strata from the Pleistocene deposits of the Ponto-Caspian region of eastern Europe dated at ~ 270–280 ka – they called it the Chegan event. Bucha (1983) quoted polarity events for deposits of the Saale–Dneiper glaciation dated at 275 ka. Champion *et al.* (1988) recalculated the age of the α event of Creer *et al.* (1980) to be 260 ka. Rampino (1981) suggested that the double Blake events of Creer *et al.* (1980) may be the Blake and Jamaica events juxtaposed by irregular sedimentation rates.

The oldest of the four reversal events found by Ryan (1972) in cores from the Caribbean and Mediterranean was called by them the Emperor and dated between 400 and 500 ka – probably ~ 460 ka. Wilson and Hey (1981) collected a series of marine magnetic profiles across the Galapagos spreading centre and found consistent short-wavelength anomalies which they interpreted as the Emperor event at 490 ± 50 ka, with a duration of ~ 10 ka. Kochegura and Zubakov (1978) in their work on the Pleistocene deposits of the Ponto-Caspian region reported a reversed polarity event beginning at ~ 550 ka and lasting about 100 ka – they called it Ureki. Bucha (1983) quoted reversed polarity sediments associated with Elster II and late Okan (Dainar) glacial deposits, with an estimated age somewhere in the mid-400 ka time range. Champion *et al.* (1981)

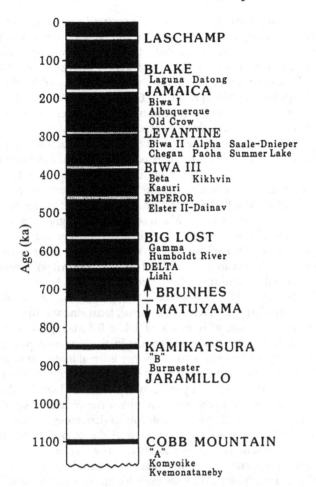

Figure 4.9 Positions of well-identified reversed polarity subchrons in the Brunhes and late Matuyama polarity chrons. Events that have been identified and age-dated from volcanic rocks shown as solid bars, those from sedimentary rocks as stippled bars. Typeface and type size of upper case preferred subchron names represent the degree of confidence in the exact age or existence of a given magnetic subchron. Corroborative reversal records are listed in capital and lower case letters beneath the subchron that they support. After Champion *et al.* (1988).

found evidence for a brief (5–10 ka) reversal of the field in cores through a sequence of basalt flows in the eastern Snake River Plain, Idaho. They could not find material in the reversed flows suitable for dating, and assigned an age of 465 ± 50 ka, using K–Ar ages of bracketing normally magnetized flows. They correlated it with the Emperor event. Later Champion *et al.* (1988) analysed the data from a new core in the Snake River plain in which they were able to date the reversed lavas. They obtained an age of 565 ± 14 ka, which is considerably older than the Emperor event, and named the reversed event Big Lost. The age of the γ event of Creer *et al.* (1980) agrees with that of Big Lost. A further correlative of the Big Lost event may have been seen by Negrini *et al.* (1987), who found a short reversed interval in Pleistocene sediments in the Humboldt river canyon. They estimated an age of less than 495 ka obtained for the Dibeku-lewa ash. This age has since been revised to 610 ka Sarna-Wojcicki *et al.* (1988).

Mankinen *et al.* (1978) found a short normal polarity event in the Matuyama reversed chron recorded in a rhyolite unit exposed on Cobb Mountain in northern California. K–Ar dating gave an age of 1.12 ± 0.02 Ma, which is definitely older than the Jaramillo normal subchron. Mankinen and Grommé (1982) later found two basalt flows in the Coso Range, California, which erupted ~ 1.08 Ma ago to have normal polarity – a nearby rhyolite dome (dated at 1.04 Ma) showed reversed polarity. Rea and Blakely (1975) had earlier found evidence of a small positive anomaly dated at about 1.1 Ma in some of the magnetic profiles across the East Pacific Rise. Watkins (1968) had suggested even earlier the possibility of a short event at about 1.07 Ma from a study of some marine sedimentary cores from the Southern ocean. Another brief period of normal polarity preceding the Jaramillo normal subchron has been detected in cores from the Melanesia Basin by Kawai *et al.* (1977). Maenaka (1983) and Maenaka *et al.* (1977) reported a normal polarity ash layer in the fluvial, lacustrine and marine deposits of the Osaka Group, Japan, with an age of 1.1 ± 0.1 Ma, again predating the Jaramillo subchron. Marine and loess deposits in the Ponto-Caspian region of the USSR also contained a brief normal polarity interval immediately preceding the Jaramillo (Kochegura and Zubakov, 1978). The data seem to indicate that the Cobb Mountain normal polarity event was global: the high VGP positions also suggest that it was not an excursion and that a complete reversal of polarity occurred. The duration of the event is difficult to determine – it probably was less than 10 ka. It has been seen in only a few of the deep-sea cores that have been studied: it is difficult to identify polarity events with durations shorter than 20 ka if sedimentation rates are slower than 1 cm/ka. However, Clement and Kent (1986/7) in a study of the records from North Atlantic sediments confirm that the Cobb Mountain is a short (~25 ka) full normal polarity interval, and not an excursion.

Hsu *et al.* (1990) also obtained records of the Cobb Mountain event from two sites in the Celebes and Sulu Seas. Although the two sites were very close together, the records of the transitional fields were found to be quite different – and also very different from the record obtained by Clement and Kent (1986/7) from the North Atlantic. The authors concluded that the transition field for the Cobb Mountain event was non-axisymmetric. Clement and Martinson (1992) have also compared the records from two deep-sea cores – this time from two sites in the North Atlantic, but some 1300 km apart. Although there is little similarity between the two intensity records, the directional records display a number of very similar features – especially in the inclination. There is, however, a lack of fluctuations in the declination records associated with large inclination rebounds. (It is interesting in this connection that Hsu *et al.*, 1990, in their investigation of the two records from the Celebes and Sulu Seas found a rebound at one of their sites, but not at the other.) The Cobb Mountain event is discussed in more detail in Section 6.4.

Other short reversals have been observed in the Matuyama chron besides the Cobb Mountain event already discussed. Maenaka (1983) found an interval of normal polarity younger than the Jaramillo subchron in sediments of the Osaka Group centred around the Kamikatsura ash. He suggested an age of 850 ± 30 ka based on that obtained from a basalt from another area of south-west Japan. Mankinen *et al.* (1981) gave an age of 840 ka (± 30) for a short normal polarity interval on sanidine from a dacite from the Clear Lake area of California. Eardly *et al.* (1973) had earlier found a short normal interval (~ 850 ka) in their drill core in deposits of Lake Bonniville, Utah. Numerous sections of loess in China with an age of ~ 850 ka have been found to contain short normal polarity intervals by Wang Jingtai *et al.* (1986). A later more detailed geomagnetic record of Chinese loess has been given by Rolph *et al.* (1989), who found seven normal polarity intervals in the Matuyama chron (see Figure 4.10). They attempt to correlate these events with the polarity timescale of Mankinen and Dalrymple (1979) adapted by the addition of the Cobb Mountain and Gilsa events.

Hyodo *et al.* (1992) obtained records of a geomagnetic excursion in sediments from Sangiran, central Java, and Moyokerto, east Java. At both sites, which are 200 km apart, the declinations have a maximum westerly swing of more than 70° from the axial dipole field direction, while the inclinations remain low. The VGP path for the Sangiran record is confined to a band of longitude 20° E ± 10° which is almost antipodal to the band of longitude (150° W ± 20°) to which the VGP path for a short sequence of the upper Olduvai transition field observed at Mojokerto is confined. The duration of the excursion is estimated to be ~ 130 ka (1.52–1.65 Ma), assuming a constant sedimentation rate – this is quite long for the majority of excursions. No other excursion corresponding to that seen in Java has been observed, although many short events have been reported between the Jaramillo and Olduvai subchrons. Hyodo *et al.* model the excursion with a fixed central axial dipole and an eccentric radial dipole at the CMB (35° S, 135° E) which is stationary for a long time. The ratio between the moment of the eccentric radial dipole and the central axial dipole varies up to 0.25.

Rieck *et al.* (1992) carried out magnetic and tephra studies on five long drill cores (to a depth of 331 m) in lacustrine sediments at Tulelake, northern California, spanning the past 3 Ma. The Brunhes, Matuyama and Gauss chrons were recognized as well as the Jaramillo, Olduvai, Réunion (?) and Kaena (?) subchrons. In addition, six short intervals showed anomalous inclinations – five of which they tentatively correlated with other excursions or short subchrons: Lake Biwa (~ 18 ka), Mono Lake (~ 27 ka), Blake event (~ 114 ka), Kamikatsura (~ 850 ka), Cobb Mountain (1.10 Ma). The ages were based on identified magnetostratigraphic horizons and tephra chronologic horizons dated independently outside the Tulelake basin. Major fluctuations in the sedimentation rate occurred – periods of slow or sporadic accumulation occurred between about 620 and 200 ka and between about 2.5 and 2.1 Ma. Rapid deposition occurred between about 170 and 125 ka.

Twenty-five successive reversals have been observed by workers from the

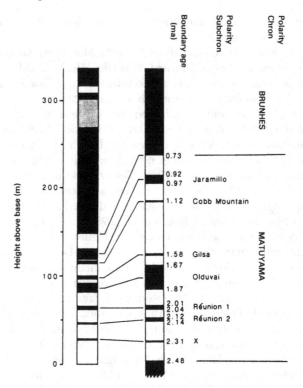

Figure 4.10 A comparison between the magnetic polarity stratigraphy derived by Rolph *et al.* (1989) and the magnetic polarity timescale of Mankinen and Dalrymple (1979). After Rolph *et al.* (1989).

Palaeomagnetic Laboratory, Ultrecht, from 7.5 to 1.5 Ma in the Upper Miocene of Crete and in the Pliocene of Sicily and Calabria (for references, see van Hoof and Langereis, 1992). Not all records are suitable for detailed investigation and many may be short polarity events rather than excursions. Moreover, if they are of short duration, the precision in dating is not sufficient to decide whether events observed in different locations are contemporaneous.

MacDougal and Chamalaun (1966) reported the possible occurrence of a short event around 2 Ma ago based upon palaeomagnetic and K–Ar age investigations of igneous rocks from Réunion Island. They also claimed that a very short reversed episode occurred at a late stage of the Olduvai subchron, which would thus be split into two normal events. Other short polarity episodes have been found in deep-sea cores but most of them have not been considered seriously, because of a lack of consistency between independent data.

Sueishi *et al.* (1979) found two short normal polarity events in the Matuyama chron, in two deep-sea cores taken in the western equatorial Pacific. One is dated at about 1.07 Ma and the other at about 1.94 Ma. Both were accompanied by a pronounced drop in field intensity. The age of the later one is in agreement with that of a short reversal in declination seen by Ninkovich *et al.* (1966) in a core from the North Pacific and by Watkins (1968) in four cores from the Southern Ocean. Sueishi *et al.* believe that the observed 'reversal' in palaeomagnetic direction is not due to a polarity event but is more likely to be related to an excursion of the pole, because the observed declination fluctuates very rapidly within this core. They suggested that the most plausible explanation is that the moment of the Earth's dipole field decreased so much during this episode that palaeomagnetic directions, including repeated reversals, observed at various sites are mainly determined by the non-dipole field. The age of the earlier episode is in rough agreement with that of the Réunion event. The length of the transition is about 5 ka, which would seem to classify it as an excursion. Heller and Liu (1982) examined a long core from a borehole near Lochuan, Shaanxi Province, China, which penetrated a 136-m-thick loess deposit. After thermal demagnetization at 35 °C, several clearly defined polarity zones were observed which they correlated with the Brunhes–Matuyama boundary and the Jaramillo and Olduvai subchrons. No definite evidence of shorter-term geomagnetic excursions was found (three samples out of a total of 231 do not fit the polarity sequence – two of them would fit the Réunion events). Intensity variations were especially pronounced in the upper part of the core (the last 1.2 Ma) and are interpreted to record rather drastic climatic changes coinciding with major episodes of loess deposition. Heller and Liu believe that the rock magnetic properties are largely controlled by chemical alteration during rock formation reflecting climatic fluctuations during late Pliocene to Pleistocene times.

Bingham and Stone (1976) examined 36 late Tertiary lava flows (3–4 Ma) from three separate sequences from the Wrangell Mountains of south-central Alaska. Twenty-seven out of the 36 flows gave mean directions more than 35° from an axial dipole or present field direction, suggesting that they were extruded during an excursion of the geomagnetic field. The data indicate that there were two different excursions around 3.4 Ma (since this age is near the Gilber–Gauss boundary, the field behaviour may have occurred during polarity transitions). Since the probability of sampling three separate sections and finding that two of them contain different field excursions is rather low, Bingham and Stone suggested that such excursions may have occurred more frequently at the time of the extrusion of these lavas than appears to be the case in Quaternary times. This conclusion, however, rests on the very small number of sites compared. Later Hamilton and Evans (1983) examined 50 flows during the last 6.5 Ma at Level Mountain, a composite volcano in northern British Columbia. Four of the flows yielded a tight group of divergent directions which they interpreted as being erupted during a geomagnetic excursion or polarity transition near the base of the Gauss chron. This lends some support to the suggestion of Bingham and Stone that the geomagnetic field was abnormally disturbed about 3.5 Ma ago – at least locally in north-western North America.

Steiner (1983) examined two deep-sea cores from the western Pacific containing the reversed polarity interval bounding the younger end of the Cretaceous superchron (the long interval of normal polarity). Near the younger end of the reversal interval an excursion of the inclination was found in both cores. No change in declination accompanied the change in inclination. Because of the stability of the samples to both AF and thermal demagnetization and the fact that the change in sign of the inclination occurred at the same depth in sediments of the same age within two cores 500 m apart, Steiner believes that the cores contain a true record of the behaviour of the geomagnetic field. Approximate sedimentation rates indicate that the excursion had a duration of between 46 and 54 ka. Because of this rather long duration and its proximity (about 236–303 ka) to a field reversal, Steiner tentatively suggests that the excursion may have been an aborted attempt of the field to reverse.

Turner and Vaughan (1977) examined a core drilled off the north-east coast of England covering a complete sequence of the Marl Slate, and obtained evidence of rapid changes in the Permian magnetic field during the Zechstein marine transgression. During most of the Permian the magnetic field was almost always reversed, with very few cases of normal polarity. In one section of the core Turner and Vaughan found a transition from normal polarity to reversed polarity – the transition from positive to negative inclination was extremely short, although there was no sharp transition in declination. Both inclination and declination showed much less variation in the upper, negative, part of the core. The authors believe that the magnetization of the Marl Slate was acquired during deposition or early diagenesis, and that the lower, positive, zone is not a simple normal polarity zone but rather evidence of rapid field changes similar to an excursion.

Bingham and Evans (1975) found evidence for a reversal and an earlier apparent excursion recorded in red siltstones of Precambrian age (approximately 1650–1800 Ma) in the Stark formation, Great Slave Lake, Canada. The excursion is seen in a 5 m section approximately 150 m below the reversal, the palaeopole moving through about 80° from the stable direction above and below it. It is interesting that the path taken by the VGP during the excursion is close to that described by the later reversal, although there is a significant time difference between the two occurrences. This lends some support to the hypothesis that excursions represent aborted reversals.

4.9 Further comments

Although records have been obtained from a wide variety of environments and places, in no case has it yet been shown that a proposed excursion has been recorded with the same palaeomagnetic signature in two nearby but distinct environments. Again in no case has it been shown that two independently dated proposed excursions at widely separated sites are of precisely the same age. Demonstration of such a temporal consistency is critical in establishing a global as opposed to a regional nature of an excursion. Most geomagnetic excursions have been recorded in lake or deep-sea sediments and it is extremely difficult to assess the reliability of ages assigned to them. The horizons are usually dated by assuming uniform sedimentation rates between or beyond ^{14}C dated horizons. Varve counting techniques as used in dating the excursions recorded in Sweden (Noel and Tarling, 1975) are generally regarded as being more reliable than ^{14}C dating. The problem of interpreting the data depends largely upon the extent to which different ages may be taken as being significantly different from one another.

The basic assumption of the radiocarbon dating method is that the atmospheric ^{14}C/^{12}C ratio (Δ^{14}C) has remained constant through time. However, there is increasing evidence that changes in Δ^{14}C have occurred in the past. Significant differences have been found between ^{14}C dates and dating of tree rings and lake sediments dated by varves. A comparison of U–Th and ^{14}C dating of coral reefs from Barbados (Bard *et al.*, 1990a) has now shown that there have been large Δ^{14}C changes over the past 20 ka. A number of possible causes have been suggested which could produce changes in atmospheric Δ^{14}C, such as changes in oceanic ventilation, associated with climatic changes or changes in the rate of production in the upper atmosphere, arising from changes in the flux of cosmic rays, changes in solar activity, or variations in shielding by the Earth's magnetosphere. Mazaud *et al.* (1991) have used the most recent estimate of the geomagnetic dipole field strength over the last 80 ka (Tric *et al.*, 1992) to examine the extent to which changes in the intensity of the geomagnetic field have affected the radiocarbon timescale. They found that the geomagnetic field has been the main factor governing the production of cosmogenic ^{14}C and that Δ^{14}C changes due to other causes have been of lesser importance. Low values of the geomagnetic dipole moment which was ~ 50% of its present-day value in the interval 18–40 ka gives rise to a positive Δ^{14}C of 350%. In terms of ^{14}C ages, the low values of the VADM in this time interval introduces a shift of the ^{14}C age of about 2–3 ka towards younger ages, in good agreement with recent U–Th and ^{14}C age comparisons (see Figure 4.11). A neglible difference between true and ^{14}C ages is predicted for the interval 45–50 ka.

True reversals caused by a reversal of the dipole field should be capable of being seen worldwide, provided that measurements in the correct time interval can be made. Excursions of the field could be caused by the main dipole field tilting at a large angle to the rotation axis and then returning to the same orientation as before. Such excursions should also be capable of being seen worldwide and the VGP paths from different locations should agree. The presence of the non-dipole field could allow excursions of the field to be seen at one location which are not observed at other locations far away, provided that the sources of the non-dipole field are sufficiently large and provided that they are not located close to the centre of the Earth.

Marino and Ellwood (1978) have provided additional evidence that some apparent geomagnetic field excursions recorded in sediments may not represent short-term changes in the Earth's magnetic field. An excursion had been reported from Imuruk Lake, Alaska (Noltimier and Colinvaux, 1976) dated at some 18 ka (^{14}C dates span 10–25 ka). Marino and Ellwood (1978) measured the magnetic fabric of both the excursion and normal sediments in the core, using the magnetic susceptibility anisotropy method. They found that normal sediments exhibited a characteristic primary magnetic fabric, whereas excursion sediments exhibited an anomalous, distorted fabric. They suggested several causes for the distorted fabric, such as slumping, deposition by turbidity currents, secondary mineralization and

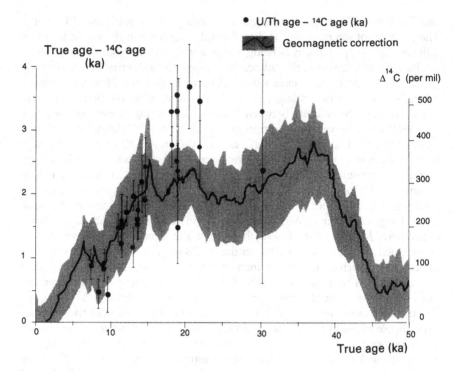

Figure 4.11 Calibration of the radiocarbon timescale, together with spot calibrations of ^{14}C ages obtained from U–Th age determinations. [14]C ages determined by β-counting ^{14}C disintegrations (Bard *et al.*, 1990a) and by accelerator mass spectrometry (Bard *et al.*, 1990b). Shaded area corresponds to the uncertainties of the magnetic intensities given by Tric *et al.* (1992). Vertical axis on right gives atmospheric Δ [14]C. After Mazaud *et al.* (1991).

increase in biogenic activity. Since the magnetic fabric of the excursion is anomalous, the magnetic directions recorded may be suspect, and they suggested that without independent corroborative evidence such excursions should be regarded with suspicion.

Løvlie and Holtedahl (1980) have given a further example of an 'apparent' excursion – an almost exact 180° change in declination being observed in a piston core at a depth of 90 cm from the continental margin off the east coast of Norway. This inversion in declination is associated with changes in the inclination from a tightly grouped low dipping distribution above this depth to a more scattered distribution, broadly coinciding with the expected inclination of the present geomagnetic field, below 90 cm. The overconsolidated top section, defined by anomalously low inclination directions, shows a magnetic fabric typical of slurries or deposits affected by water currents, as opposed to the fabric in the lower section associated with magnetic directions coinciding with the present geomagnetic field. The sampling area shows seismic features indicative of slumping, erosion and redeposition, which are also reflected by micropalaeontological evidence. They concluded that the pre-consolidated top section has been transported by slumping, the anomalously low dipping directions resulting from processes acting during the initial consolidation. The effect of post-depositional compaction on the acquisition of a remanent magnetization in fine-grained sediments has been experimentally shown to induce a reduction of the inclination (Blow and Hamilton, 1978), which is comparable in magnitude to the decrease observed in the pre-consolidated section, assuming a Quaternary age for the latter. Løvlie and Holtedahl tentatively concluded that the anomalous direction of magnetization reflects processes acting during consolidation. As compactive DRM is likely to play an important role in sediments overlain by ice during glacial periods (Blow and Hamilton, 1975), the reported indication of a low inclination excursion at Rubjerg (Abrahamsen and Knudsen, 1979) may possibly also be related to this effect.

Laboratory experiments have shown that misalignments affecting inclination and declination can be caused by the deposition process itself, by water currents and by the slope of the bedding plane. These errors may be present to varying degrees in different sedimentary environments (for a review, see Verosub, 1977b). However, the maximum deviation which can arise is generally less than 30°, and cannot account for the large-scale anomalous directions that are sometimes observed.

As discussed in Section 2.2, studies of deep-sea sediments have shown that, from a palaeomagnetic point of view, there are 'stable' and 'unstable' cores. In an unstable deep-sea core, as much as 30–70% of the magnetic directions are anomalous, so that the core may really consist of thin layers of stable and unstable materials. The boundaries between these layers may not show up as distinct lithological or chemical changes; rather, they may represent changes in mineralogy in a self-reversal region or a viscous magnetization region. If a core contains only one layer or a few unstable layers, the anomalous directions in these layers will have the characteristics of palaeomagnetic excursions. No palaeomagnetic excursion can be based on the evidence from a single core or site. At the very least there must be 'internal consistency' whereby the proposed palaeomagnetic excursion is found repeatedly in sediments of the same age in a given sedimentary environment. Bioturbation of the sea-floor can also degrade and even erase portions of the palaeomagnetic record. Thus, a regional geomagnetic fluctuation may not be recorded with the same signature at every site and may even be erased at some. Failure to detect a palaeomagnetic excursion in an adjacent lake or ocean basin is in itself an important result.

Finally, proposed palaeomagnetic excursions must show 'temporal consistency'. An event which only lasted a few hundred years cannot be recorded in two well-dated cores at horizons which are several thousand years apart. In this regard, one must beware of the reinforcement syndrome (Watkins, 1972): 'The initial report of a palaeomagnetic excursion will encourage other workers to re-examine previously unexplained or disregarded "curious" results and to reinterpret sedimentation rates so that the anomalous behaviour seen by them is contemporaneous with the palaeomagnetic excursions. Subsequent work will also focus on sediments of the same age. Reported excursions will then tend to cluster around a single date, whereas negative results showing no anomalous behaviour will tend to remain unpublished because they are not interesting.'

5
Models for reversals

5.1 Introduction

Theories on the cause of reversals of the Earth's magnetic field fall into one of two general classes. First there are those which maintain that reversals result from magnetic hydrodynamic instabilities, triggered by finite-amplitude perturbations of an otherwise stable dynamo. The perturbations are usually assumed to have their origin within the core, such as instabilities in the pattern of convective flow (Cox, 1968; Parker, 1969; Levy, 1972a–c; Olson, 1983). An external origin has been suggested (e.g. Muller and Morris, 1986), although, as discussed later, this seems unlikely. This class of dynamos is considered in Section 5.2. The second class of theories maintains that reversals are the result of irregular oscillations of a non-linear a.c. dynamo (e.g. homopolar disc dynamos – Rikitake, 1958; Robbins, 1977). Such dynamos are considered in Section 5.3. They are intrinsically oscillatory and, unlike models in the first category, perturbations are not required to initiate a reversal. Most linear a.c. kinematic dynamos have sinusoidal behaviour in time, although the Earth's field does not. The palaeomagnetic field has long periods of one polarity, in which the field intensity fluctuates about an average value, separated by relatively brief transitions. The Earth's field reverses its polarity at irregular intervals, the average time between reversals at present being a few hundred thousand years. This is much shorter than the timescales of mantle convection and plate tectonic processes ($\sim 10^8$ a), but much longer than the timescale for the free decay of electrical currents in the OC ($\sim 1.5 \times 10^4$ a). There are no known dynamical processes in the Earth's deep interior which operate on the reversal timescale.

There are essentially two classes of models that have been proposed to account for reversals. In the first model (the 'standing field' model of Hillhouse and Cox, 1976), the main dipole field together with part of the non-dipole field decay to zero and regenerate with the opposite polarity. The transition field at any site is thus the sum of a steady non-dipole component and an axial dipole component that changes only in magnitude and sign. This is consistent with the longitudinal confinement observed for many VGP paths (see Section 6.5), provided that the non-dipole part of the field that persists during the transition does not drift on a large scale. In the second model (the 'flooding' model of Hoffman, 1977,

1979), the reversal process begins in localized zones. These zones of reversed flux grow and eventually flood through the entire core. Hoffman's model is discussed in Section 5.4. The structure and geometry of the transition field are discussed in Chapter 6.

In some models the non-dipole field remains a small part of the total field but the dipole is considered to have two or three components, the strength, polarity and possible orientation of which change during the transition. Bochev (1969) suggested such a model with three dipoles – for epoch 1960, he found that two dipoles are approximately parallel to the Earth's axis of rotation, while the third is aligned obliquely to the axis. Creer and Ispir (1970) suggested that these three dipoles correspond to different dynamo processes and that each dynamo might drift or undergo periodic changes in position – each might also fluctuate in strength and even reverse its polarity independently of the others.

Verosub (1975a) made a similar suggestion to account for reversals. In his model the field arises from two separate sources, the observed dipole component of the Earth's field being the vector sum of the dipole components of each source, i.e.

$$M_0 = M_1 + M_2 \tag{5.1}$$

If the dipole components of each source are oppositely directed, one pointing essentially towards geographical north and the other south, then

$$M_0 \simeq M_1 - M_2 \tag{5.2}$$

The prevailing magnetic polarity reflects the polarity of the dominant source component: a magnetic reversal, which corresponds to a change in the sign of M_0, represents a shift in the relative sizes of M_1 and M_2. If both M_1 and M_2 are very large compared with M_0, then only a small fluctuation in M_1, M_2 or both will result in a reversal. If the Earth's field does arise from two sources, then differences could be expected between normal and reversed polarity states, indicating that a different source is dominant in each state. Some authors claim to have seen such effects. Wilson (1972) found significant differences in the time-averaged mean pole positions of normal and reversed populations, while Dagley and Lawley (1974) found a clear predominance of reversed over normal transitions and concluded that the reversed state is less stable than the normal one. However, as mentioned in Section 3.1, Kristjansson and McDougal (1982), as a result of more detailed studies, concluded that normal and reversely magnetized states of the geomagnetic field were equally probable during the last 14 Ma. They found complete equivalence between normal and reversely magnetized lavas (in Iceland) in all statistical properties of their intensity distribution. The question of differences between normal and reversed states is discussed further in Section 5.9.

Verosub (1975b) developed a new method of modelling the Earth's field by fitting a set of spherical harmonic coefficients with a geocentric axial dipole and an eccentric (but not necessarily axial) dipole. The free parameters are the relative strengths of the two dipoles and the co-ordinates of the eccentric dipole. When applied to the first eight coefficients of the International Geomagnetic Reference

Field for 1965, the best fit consists, not of two more or less parallel components whose sum is the present field, but rather of two large, more or less antiparallel components whose difference is the Earth's field. This is precisely the configuration described by Equation (5.2). Verosub suggested that the two sources are located in the inner and outer cores (IC, OC). The source field in the liquid OC could be generated by MHD action, but it is much more difficult to postulate a source mechanism in the solid IC.

There have been a number of attempts in the past to obtain a numerical solution of the dynamo problem (e.g. Stevenson and Wolfson, 1966; Kropachev, 1971a, b). All these models include rotation and are driven by a radial body force. None of them have been successful, because of the severe restriction on the size of the allowable time step. Gubbins (1975) avoided this difficulty by neglecting rotation and assuming the fluid to have an appreciable viscosity. He found that if the dynamo was driven well above some critical limit, the magnetic field underwent large oscillations which might ultimately lead to a reversal. However, the time required to complete a reversal was very long ($\sim 3 \times 10^4$ a).

Watanabe (1981) investigated the non-steady state of a simplified $\alpha\omega$ dynamo allowing for Coriolis, Lorenz and viscous forces. He found that the dipole field undergoes relatively gradual variations with occasional reversals. The reversal of the dipole field is caused essentially by the disruption of the non-uniform rotation, the time required for a complete decay being about 3000 a. The quadrupole field also changes sign and it appears that the toroidal field, from which poloidal fields (such as g_1^0, g_2^0, g_3^0) are produced by the effect, also reverses.

5.2 The models of Cox and Parker

Cox (1968) developed a probabilistic model in which he assumed that polarity changes occur as a result of an interaction between steady oscillations and random processes. The steady oscillator is the dipole component of the field and the random variations are the components of the non-dipole field. The random variations serve as a triggering mechanism that produces a reversal whenever the ratio between the non-dipole and the dipole fields exceeds a critical value (see Figure 5.1). Parker (1969) developed an alternative model for reversals. He showed that a fluctuation in the distribution of the cyclonic convective cells* in the core can produce an abrupt reversal of the geomagnetic field.

The simplest fluctuation leading to a reversal is a general absence of cyclones below about latitude 25° for a time comparable to the lifetime (≈ 1000 a) of an individual cell. Parker's model has been further developed by Levy (1972a–c), Cox (1975a) and Hoffman (1977, 1979). Parker's model is not, like Rikitake's, two coupled disc dynamos (see Section 5.3) which are dependent on an exchange of energy between magnetic and mechanical forms – in fact Parker's model, has no explicit inertial component. It depends instead on the interaction of two

---

* Because the fluid in convective cells undergoes a twisting motion as it rises and falls due to Coriolis forces, the individual cells have been compared to cyclonic weather patterns.

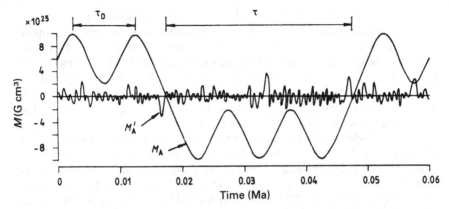

Figure 5.1 Probabilistic model for reversals. τ_D is the period of the dipole field and τ the length of a polarity interval. A polarity change occurs whenever the quantity ($M_A + M'_A$) changes sign, where M_A is the axial moment of the dipole field and M'_A is a measure of the non-dipole field. After Cox (1968).

almost independent self-sustaining dynamos magnetically coupled. They are assumed to vary individually in regenerating strength, either statistically or with secular change of physical parameters. One is idealized in the calculations as a pair of rings of upward streams, or cyclones, at high latitude near the pole in each hemisphere. The other is such a pair of rings at low latitude near the equator. Normally, when they are both operating, or when the distribution of upwellings is random, they are parts of the same global dynamo. However, if one of them ceases to function, the direction of the magnetic field at its position is determined by the toroidal field generated by the other. This can locally have reversed direction and, when the dormant ring starts generating again, it will be excited in the reverse direction and may take over if the other ring weakens.

In both Cox's and Parker's models, cyclonic convection cells in the core produce reversals by a two-step mechanism. At any instant they are randomly distributed throughout the core; reversals occur when, through random processes, they arrive at certain configurations. However, there is a fundamental difference between the two models. In that of Parker the occurrence of a reversal depends only on the spatial distribution of the cyclones, and not on the intensity of the dipole field. In the model of Cox the occurrence of reversals depends upon both the distribution of the cyclones and the field strength of the cyclone disturbances (i.e. the non-dipole field) relative to the dipole field.

Laj *et al.* (1979) pointed out that there is a mechanical difficulty in Cox's model arising from the different time constants of the random fluctuations of the non-dipole field and the steady oscillations of the dipole field. Oscillations of the dipole field typically have a time constant of the order of 2×10^4 a. Characteristic times for the non-dipole field range up to about 10^3a; however, most of the power is concentrated in the range 2–4×10^2 a. It is quite difficult to couple two systems with very different time constants. Laj *et al.* maintain that, if Cox's model is correct, the field would reverse but would maintain the opposite polarity for a very short time – of the order of the time constant of the non-dipole field. This problem of the differences between the time constants of the dipole and non-dipole fields also occurs in other models. In Parker's (1969) model the quantitative nature of the cyclones is not sufficiently well known to enable the calculation of the minimum amount of time during which they have to be absent from low latitudes in order to induce a reversal. Parker believes that reversals occurring as often as 10^5 a are not likely.

In Cox's model there is a finite probability that the sum of the axial component of the non-dipole (ND) contributions will overcome the dipole (D) field causing a reversal. Hence, the probability of reversal should be high if the ratio ND/D is large, and vice versa. If the dipole field is much larger than the sum of the axial components of the ND field, then the field is stable. As the ND fields increase, their axial components may overwhelm the dipole field, and the geomagnetic dynamo then amplifies the field in the opposite sense. Thus, the magnitude of the palaeosecular variation should be related to the frequency with which reversals occur: when reversals are frequent, the palaeosecular variation should be large, and vice versa. Brock (1971) showed that such a relationship is generally borne out by palaeomagnetic data. He also showed that the magnitude of the palaeosecular variation in pre-Cenozoic times was on the average about 15% less than during the Cenozoic and suggested that this difference is reflected in the reversal rate in the past. Irving and Pullaiah (1976) carried out a study of the palaeosecular variation in the later Phanerozoic (0–350 Ma), using a database about twice as large as that available to Brock, and confirmed his conclusions.

A mechanism for reversals based on cyclonic convection in the core as proposed by Parker (1969) has been further developed by Levy (1972a–c). A steady dipole field is maintained by a two-stage process. Non-uniform rotation of the liquid core first draws the poloidal field into a toroidal magnetic field (see Figure 5.2). The toroidal field is then twisted into meridional loops of poloidal field by cyclonic convective motions in the core. If a meridional loop has the same sense as the large-scale dipole, then the dipole field is regenerated. If a loop has the opposite sense, the dipole field is degraded. Levy showed that there are generally two types of toroidal zones in the core: regions of normal toroidal flux (where convection reinforces the main dipole) and regions of reverse toroidal flux (where convection degrades the poloidal field). Levy suggested two possible reversal schemes. In the first the dipole field is maintained by cyclonic convection concentrated at low latitudes. A fluctuation resulting in a large burst of cyclones in high latitudes suffuses the low-latitude region with reverse toroidal flux. Normal regenerative low-latitude cyclones then create a dipole field with sign opposite to the original field (see Figure 5.3). In the second scheme the field is maintained normally by cyclones concentrated at high latitudes. In this case, the low-latitude region of the core normally contains reverse toroidal flux. A fluctuation in which a large burst of cyclones appears at low latitudes directly produces poloidal field with sense opposite to that of the large-scale dipole field, thereby causing the dipole field to change sign (see Figure 5.4).

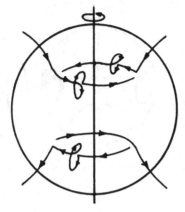

Figure 5.2 In the normal regeneration of the geomagnetic field, large scale velocity shear stretches the poloidal field into an azimuthal, toroidal field B_ϕ. Rising cyclones twist B_ϕ into poloidal loops of field. If these loops reinforce the original dipole field, then a regenerative dynamo can exist. After Levy (1972b).

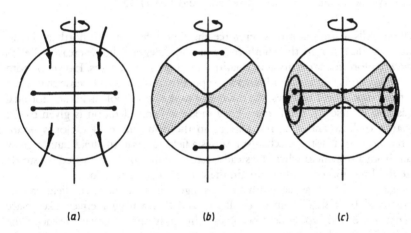

Figure 5.3 (a) A dipole field maintained by cyclones concentrated at low latitudes produces normal toroidal field throughout the core. (b) A fluctuation in the distribution of cyclones, consisting of a large burst of high-latitude cyclones, produces a region of reverse toroidal flux at low latitude (shaded area). (c) Subsequent low-latitude cyclones then generate poloidal field with sense opposite to the original dipole, thus reversing the dipole field. After Levy (1972b).

Although neither of these two schemes may actually occur, Levy maintains that a fluctuation in the distribution of cyclones with latitude can reverse the dipole field. Cox's (1975a) analysis of global palaeomagnetic data supports Levy's model in the sense that large non-dipole anomalies generally tend to maintain the same sign along a given circle of latitude. This observation is consistent with the existence of Levy's zones of reversed and normal toroidal flux. The symmetry of Levy's and Cox's models is different in the sense that in Cox's model the

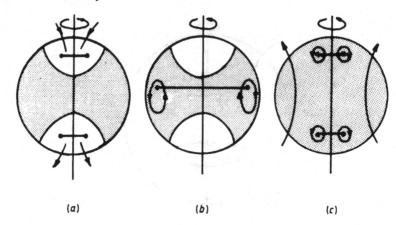

(a)   (b)   (c)

Figure 5.4 (a) A dipole maintained by cyclones concentrated at high latitudes produces both normal and reverse toroidal flux in the core. Shaded region represents reverse toroidal flux. (b) A burst of low-latitude cyclones produces poloidal field with sense opposite to the original dipole. (c) It also floods the entire core with reverse toroidal flux, so that subsequent high-latitude cyclones then maintain the reversed field. After Levy (1972b).

cyclonic cells at mid-latitudes reinforce the dipole field in the northern hemisphere, while those in the southern hemisphere degrade it, whereas in Levy's model reinforcement occurs at mid-latitudes in both hemispheres. However, both models are antisymmetric with respect to polarity reversal in the sense that when the main dipole field reverses, the directions of the non-dipole fields and toroidal fields also reverse. In the Parker–Levy model no consideration is given to the question of how and why fluctuations in the distribution of cyclones occur. Moreover, even if some mechanism causes a field reversal, the field cannot grow again from a stray field when the steady state fluid motion is recovered – a steady state fluid motion can only maintain the existing magnetic field.

Nagata (1969) disagrees with Cox's suggestion that the main field can be represented by a steady dipole oscillator, and that a trigger effect takes place whenever the non-dipole field becomes sufficiently large relative to that of the dipole. He offered an alternative interpretation of reversals, based on the hypothesis that the main dipole field is steadily maintained only so long as the convection pattern in the core is asymmetric, as is the case in the Bullard–Gellman–Lilley dynamo (Lilley 1970a), but collapses when the convection pattern becomes symmetric, as in the original Bullard–Gellman model (1954). Lilley (1970b) suggested that the deviation of the geomagnetic axial dipole from true north is an expression of asymmetric motion in the Earth's core. A wandering of the magnetic poles to coincide with the geographic poles may thus be an indication of the flow in the core becoming symmetric. If this should happen, dynamo action would be lost and the dipole field would decay, perhaps to grow again in the opposite direction (i.e. reverse) when the flow pattern once more becomes asymmetric. As a corollary, strong dipole fields may accompany a large deviation of magnetic north from true north. In this regard it is interesting to recall that cases have been reported when the geomagnetic field was large and stable when the VGP was close to the equator (see Section 3.2).

Kono (1972) and McFadden and McElhinny (1982) have shown that the Cox (1968) model is incompatible with the palaeomagnetic data. In addition, his model implicitly attributes separate sources to the dipole and non-dipole fields, whereas real current sources will contribute to both.

Mazaud and Laj (1989) developed a model of reversals based on the dynamo models of Parker (1969) and Levy (1972a,b,e). The field source consists of N interacting point dipoles with finite lifetimes. At regular intervals a dipole is suppressed and a new one immediately generated. Changes in the total moment of the system may occur because the polarity of a new dipole is statistically determined by its coupling with the others. The evolution of the system depends only on N, the number of dipoles, and c, the strength of the coupling. For $N = 16$ (typical of the number of convective cells in Parker's and Levy's models), and for various values of the parameter c, the morphology and temporal distributions of reversals and excursions can be reproduced, including cases of varying rate of field change. Larger values of N generate more continuous reversals, while smaller values of N generate rapid, impulsive changes. The modelled VGP paths are confined to narrow bands of longitude, but are not systematically close to the site or its antipode.

5.3 The disc dynamo

Because of the complexity of the non-linear equations that describe the geodynamo, various models have been proposed. One of the simplest analogies is the homopolar disc dynamo (see Figure 5.5) first suggested and studied by Bullard (1955). It consists of an electrically conducting disc which can be made to rotate on an axle by an applied couple. If it rotates in an axial magnetic field, a radial EMF will be produced between the axle and the edge of the disc. If this were all, the EMF would be balanced by an electric charge on the edge of the disc and no current would flow. If one end of a stationary coil coaxial with the disc and axle is joined to the edge of the disc by a sliding contact (a brush) and the other end is joined to another sliding contact on the axle, a current will flow through the coil and an axial magnetic field will be produced. No external source of field or current is required and no part of the machine is ferromagnetic. This system becomes a dynamo when this induced field becomes equal to the field required to produce it. It must not be forgotten, however, that the homopolar dynamo is not simply connected. It requires complicated conduction paths and slipping brushes in order to operate; and it is not at all obvious that a homogenous conducting fluid can be put into motion in such a way as to initiate dynamo action.

It is easier to formulate the equations for the system if the current in the disc is axially symmetric; this can be achieved by substituting a highly conducting ring for the brush. The current in the disc can then be represented by a single variable and is not a function of azimuth.

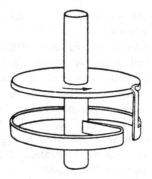

Figure 5.5 Disc dynamo. After Bullard (1955).

If the disc is driven by a constant couple G, the equation of motion is

$$C\dot{\omega} = G - MI^2 \tag{5.3}$$

where C is the moment of inertia of the disc, ω its angular velocity, I the current and $2\pi M$ the mutual inductance of the coil and disc. The equation governing the current is

$$L\dot{I} + RI = M\omega I \tag{5.4}$$

where L and R are the inductance and resistance of the coil. The solution of these equations has been discussed by Bullard (1955). There are two non-linear terms, I^2 in Equation (5.3) and ωI in Equation (5.4). Since Equation (5.3) is quadratic in I and Equation (5.4) is linear and homogeneous in I, the dynamo can produce a current (and therefore a magnetic field) in either direction. It cannot, however, switch from one direction to the other, since, if I is zero, so is $\dot{I}$. In order to show reversals, another term must be introduced into Equation (5.4) which is not proportional to either I or $\dot{I}$. This can be achieved by using two coupled dynamos (Rikitake, 1958), or by adding an impedance between the brush and the coil and a shunt connected across the coil of a single dynamo (Malkus, 1972). This latter arrangement is shown in Figure 5.6. If R_s and L_s and R_b and L_b are the resistance and inductance of the shunt and series impedance, respectively, and I_s the current carried by the shunt, the equations of the system are

$$C\dot{\omega} = G - MI(I + I_s) \tag{5.5}$$

$$L_s\dot{I}_s + R_sI_s = M\omega I \tag{5.6}$$

$$(L + L_b)\dot{I} + (M + R_b)I + L_b\dot{I}_s + R_bI_s = M\omega I \tag{5.7}$$

The properties of these equations have been investigated by Robbins (1977), who showed that there are four possible regimes, depending on the values of the different parameters. If $L/R > L_s/R_s$ and $L_b/R_b > L_s/R_s$, irregular reversing solutions occur. Robbins has studied this regime in some detail for the simplest

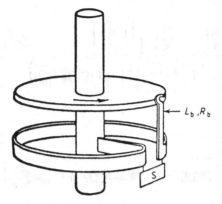

Figure 5.6 Disc dynamo with an impedance between the brush and the coil and a shunt connected across the coil. After Bullard, (1978).

case, in which $L_s = 0$. She showed that there exist solutions whose path in phase space stays in the neighbourhood of one equilibrium point and then switches to the neighbourhood of the other without ever being captured by either. The currents in the coil and shunt show growing oscillations around one equilibrium state, culminating in a reversal of current and a transition to oscillations about the other. An example of oscillations produced in this way is shown in Figure 5.7. The number of oscillations between reversals is very variable – in the example shown it varied from 1 to 66. Other solutions exist where the erratic behaviour does not persist and where, after 18 reversals, the solution is captured by one of the equilibrium points.

If we suppose that the fluid motion in the Earth's core consists of a number of eddies and that each eddy motion is represented by a disc, then we can imagine a series of disc dynamos each coupling with the next stage. The field produced by the last stage may be fed to the first stage. Rikitake (1958) discussed the case of two identical coupled-disc dynamos in which the current from each disc energizes the coil of the other (see Figure 5.8). He found by numerical integration that it was possible for the currents to reverse sign. Later Allan (1958, 1962) showed that the reversals had an apparently random distribution in time. Further numerical and analytical investigations have been carried out by Mathews and Gardner (1963), Sommerville (1967) and Cook and Roberts (1970).

If the two dynamos are similar and the couples applied to them are equal, the equations are

$$L\dot{I}_1 + RI_1 = M\omega_1 I_2 \tag{5.8}$$

$$L\dot{I}_2 + RI_2 = M\omega_2 I_1 \tag{5.9}$$

$$C\dot{\omega}_1 = G - MI_1 I_2 \tag{5.10}$$

$$C\dot{\omega}_2 = G - MI_1 I_2 \tag{5.11}$$

where the suffixes 1 and 2 refer to the first and second dynamos. The equations can be put into non-dimensional form by measuring time t in units of

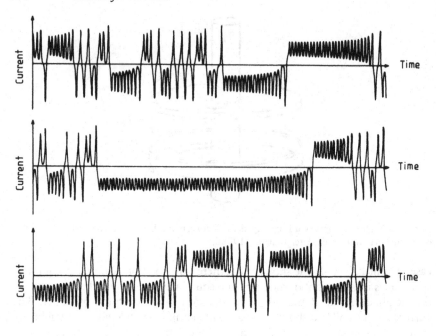

Figure 5.7 Oscillations of the current in the coil of the disc dynamo with shunt and series impedance. After Robbins (1977).

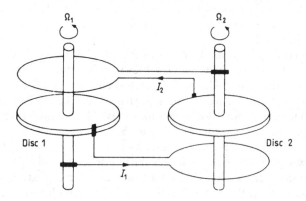

Figure 5.8 Coupled system of two disc dynamos (after Rikitake, 1958).

$(CL/GM)^{1/2}$, currents in units of $(G/M)^{1/2}$ and angular velocities in units of $(GL/CM)^{1/2}$. Writing

$$I_i = (G/M)^{1/2}X_i \qquad \omega_i = (GL/CM)^{1/2}Y_i \qquad i=1, 2 \tag{5.12}$$

the governing equations (5.8)–(5.11) become

$$\dot{X}_1 + \mu X_1 = Y_1 X_2 \tag{5.13}$$

$$\dot{X}_2 + \mu X_2 = Y_2 X_1 \tag{5.14}$$

$$\dot{Y}_1 = \dot{Y}_2 = 1 - X_1 X_2 \tag{5.15}$$

where

$$\mu = (CR^2/GLM)^{\frac{1}{2}} \tag{5.16}$$

is the ratio between the mechanical acceleration time (in the absence of magnetic field) and the electromagnetic decay time (in the absence of motion). Equations (5.13)–(5.15) have, in general, no known analytic solutions.

It follows from Equation (5.15) that $Y_1 - Y_2 = A$ (a constant), i.e. the difference in angular velocities is constant. Figure 5.9 illustrates the behaviour of the system for $\mu = 1$, $A = 3.75$ and clearly shows reversals taking place at time $t = 11, 21, 44, 51, 53, 71, 85$, etc.

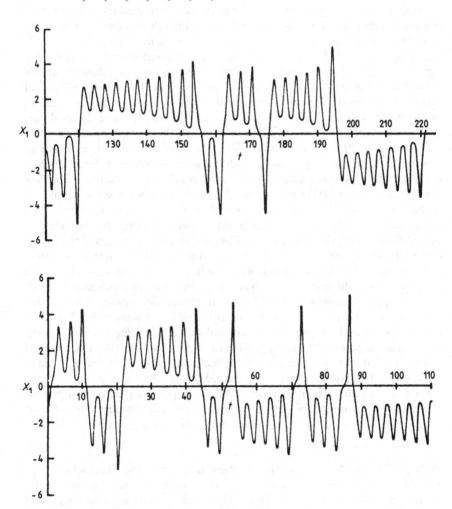

Figure 5.9 A typical evolution of the current X_1, as a function of time (in seconds). After Cook and Roberts (1970).

Although it is clear that such two-disc dynamos are unstable and oscillate and reverse, it is very doubtful whether any quantitative results have been obtained which are applicable to the Earth. The behaviour of such models depends very much on the value of μ. Both Allan (1962) and Lowes (unpublished) have estimated μ as lying between about 10^{-2} and 10^{-3} for the core of the Earth. All earlier published results were for values of μ of about unity. Because of computational difficulties no solutions have been obtained for values less than 10^{-2}.

Rikitake's treatment of two coupled-disc dynamos has been generalized by Nozières (1978). He showed that the non-linear dynamical equations involve two very different timescales – a fast MHD scale and a slow inductive scale. For a certain range of the parameters involved the magnetic field suddenly reverses, the intervals between reversals being the slow timescale and the duration of the reversal the fast timescale. Such a relaxation mechanism explains quantitatively the comparative rapidity of an actual reversal compared with the time interval between reversals. In his treatment Nozières allowed only one mechanical variable – the magnitude of the velocity field at any time, i.e. he assumed a constant geometry for the convection pattern. In Bullard's homopolar dynamo there is, in addition to one mechanical variable (the angular velocity), one electrical variable (the current). Nozières broadened the scale of the problem by allowing two electromagnetic degrees of freedom. Using the ratio of the two timescales as an expansion parameter, he obtained an analytical solution of the non-linear differential equations which describe his model, and investigated the stability of the slow motion against fast MHD oscillations. Among the various possible regimes he found a 'ratchet-like' relaxation behaviour which may account for the very sudden reversals of the field. An overall cyclic slow motion is driven by the inductive forces. Periodically, the MHD fast oscillations become unstable, and the system jumps to a new equilibrium configuration, in which *J* and *B* are roughly reversed.

There exists a wide class of systems of ordinary differential equations which represent forced dissipative hydrodynamic systems such as geodynamos, which have non-periodic solutions. Such systems may oscillate randomly between two states of fixed points: i.e. they may have two unstable polarity states. Rikitake's double-disc model belongs to such a class and such a model displays non-periodic reversals in the absence of any triggering mechanism. Lorenz (1963) first pointed out such behaviour in a discussion of the feasibility of very-long-range weather forecasting. The phenomenon (since named 'chaos') has been found in many other fields, e.g. models of population dynamics, physiological control systems and chemical reactions. The Rikitake system has two equilibrium points N and R ($\pm k$, $\pm k^{-1}$, μk) in (X_1, X_2, Y_1) phase space, where k is given by

$$A = \mu(k^2 - k^{-2}) \tag{5.17}$$

Both are unstable foci – around them there is an 'attracting' plane which traps all orbits starting from any point except those on the Y_1 axis. An orbit circles around N or R on this plane, irregularly travelling from an orbit round one point to one around the other (Figure 5.10). This corresponds to a polarity reversal of the magnetic field. Ito (1980) has investigated the statistical properties of this system. Figure 5.11 is a phase diagram in the parameter space (μ, k) showing various regions of periodic regime $(P_1, P_2, \ldots)$ and a chaotic regime. The transition from the periodic to the chaotic regime is characterized by a succession of period-doubling bifurcations (see Figure 5.12). Ito found in the centre of the chaotic regime a parameter region in which reversals seldom occur and the dynamics is less disordered. The Markov entropy of the Lorenz map for the system has a sharp minimum in this parameter region which Ito calls the minimum entropy regime. The smallness and non-uniformity of the frequency of reversals as shown by the palaeomagnetic data suggest that the present geodynamo is in such a state of minimum entropy.

Merrill *et al.* had earlier suggested (1979) that hydromagnetic dynamos might exist in which there are at least two different acceptable velocity fields and in which changes in polarity occur only when there is a change from one velocity field to the other. The existence of two velocity states suggests bifurcating solutions that are frequently encountered in fluid dynamical systems (e.g. Chandrasekhar, 1961; Huppert and Moore, 1976; Roberts, 1978). A characteristic of many finite-amplitude systems is that more than one state is stable or metastable for a given set of externally imposed conditions. It has been speculated

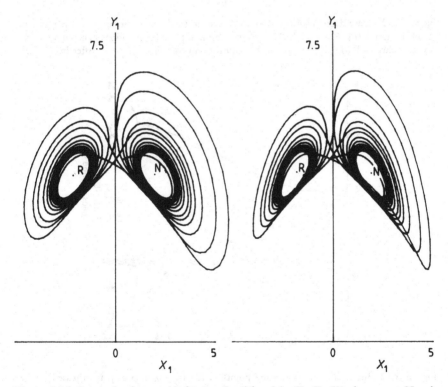

Figure 5.10 Stereo plot for an orbit for $\mu = 1$ and $k = 2$ in (X_1, X_2, Y_1) phase space. N and R are unstable fixed points. The orbit circles round N or R on an attracting plane which looks like two curved discs. After Ito (1980).

(for example, by Roberts, 1978) that reversals could be associated with 'cata-strophic' transitions between primarily geostrophic and primarily magneto-strophic states. If this interpretation is correct, then it is perhaps surprising that the normal and reversed fields are apparently so similar.

Hoshi and Kono (1988) have investigated the Rikitake two-disc dynamo in

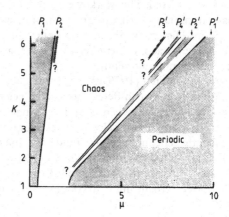

Figure 5.11 Phase diagram of the Rikitake system in (μ, k) space. Solutions are periodic in stippled regions $(P_1, P_2, \ldots P_1', P_2' \ldots)$, but are non-periodic in regions denoted as chaos. The sequence of P_1, P_2, is supposed to be a mirror sequence of P_1', P_2'. . . . After Ito (1980).

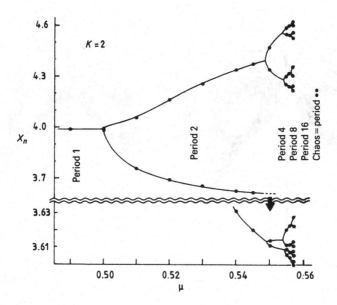

Figure 5.12 Bifurcation of stable fixed points in the Lorenz map as μ is increased in the transitional zone of the periodic regime P_1 at $k = 2$. The ordinate X_n is the value of the nth maximum of $|X_1|$. The figure illustrates a cascade of period doubling bifurcations. After Ito (1980).

more detail for a range of values of the parameters (μ, k). They showed that the current intensity fluctuates round one of two stationary states, the amplitude of the oscillations growing monotonically with time. When the amplitude exceeds some threshold value, the system flips to the opposite polarity and the oscillations resume round the new stationary state. If large amplitudes are attained just before a reversal, the oscillations start with small amplitudes in the reversed state and a very long polarity interval ensues. The system is quite deterministic in that the last amplitude before the reversal controls the oscillations after the reversal. It should be pointed out that Kono (1987) claims that the Rikitake two-disc dynamo is not compatible with known palaeomagnetic data.

The equations describing simple dynamo models are known to exhibit instabilities with respect to starting conditions and integration method, and it has been shown (e.g. Sparrow, 1982) that the existence of strange attractors and chaotic behaviour is inherent in the equations. The precise initial conditions as well as numerical round-off error affect the details of chaotic solutions, and in all physically realistic situations the addition of some noise must intrude to influence the evolution of the system. For fluid motions in the Earth's core, irregularities, in either material properties, physical or chemical processes or fluid dynamical turbulence, will provide an adequate source of noise.

It has already been shown that the Rikitake double-disc dynamo shows all the attributes of chaotic behaviour. The stable points of the Rikitake dynamo are, however, repulsive, i.e. all trajectories which start arbitrarily close to the stable points eventually spiral away, for every value of (k, μ), and the dynamo has no parameter regime with a stable state. Therefore, this model is not a good candidate for additional stochastic excitation. Thus, Crossley *et al.* (1986) considered the Robbins dynamo (Robbins, 1977), which is governed by the equations

$$\dot{\omega} = R - zy - v\omega \tag{5.18}$$

$$\dot{z} = \omega y - z \tag{5.19}$$

$$\dot{y} = \sigma (z - y) \tag{5.20}$$

in which y is the current in the coil (velocity of the fluid), z the current in the disc (horizontal temperature distribution) and ω the angular velocity of the disc (vertical temperature distribution). The non-dimensional parameters are R, the driving torque (mechanical convective force, e.g. temperature gradient), v, the bearing friction (viscous dissipation), and σ (the ratio between disc current decay time and coil current decay time). The variables were interpreted by Robbins according to the thermal convection model of dynamo action and should be reassigned according to either gravitationally driven convection or to other models as appropriate. Equations (5.18)–(5.20), involve three parameters, compared with two in the Rikitake model, and the solutions exhibit different behaviour in different regimes of the parameter R.

To illustrate the effect of round-off error in the solutions, Figure 5.13 shows three plots of $y(t)$ for different integration error tolerances. The parameters chosen were $v = 1$, $\sigma = 5$ and $R = 13$. It can be seen that as the error tolerance improves, the trajectory departs after a time (denoted by the asterisks) from the previous plot and the subsequent evolution after this time is quite different. By induction, one can argue that, regardless of the numerical accuracy of any particular integration scheme or the power of any computer, the eventual behaviour of any one realization (including reversals) of this type of dynamo is governed by numerical inaccuracies. On the other hand, as the initial part of each plot demonstrates, reversals obviously do occur as a result of chaos in the non-linear equations and are caused by more than numerical approximations.

The lack of predictability associated with the sensitivity to initial conditions means that one should not and, indeed, cannot account for every observed fluctuation – individual reversals do not require an 'explanation'. One must consider how long a period might occur without a reversal (such as the Cretaceous normal superchron) as a natural fluctuation rather than as something requiring a special explanation. Again, to what extent do trends in the geomagnetic data (e.g. variations in the average frequency of reversals) reflect major changes in conditions in the Earth's core, or to what extent are they produced by only minor changes to which the statistics are sensitive or, indeed, without any causal change at all (Tritton, 1989)?

Figure 5.13 also illustrates other pertinent behaviour. In particular, the time prior to a reversal is always preceded by a growing instability of the primary oscillation, as observed by Bullard (1978). Second, once decaying oscillations begin they continue until the stable solutions are obtained and there is no subsequent reappearance of growth towards a reversal. At the present time we do not know whether these conditions are met by the actual geodynamo, because of the limited palaeomagnetic record.

Crossley *et al.* (1986) introduced an 'external' stochastic component into the equations of motion to simulate what they assume must be irregularities in

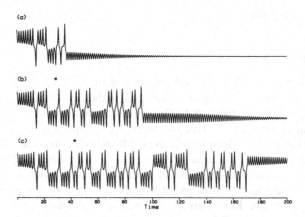

Figure 5.13 Plots of $y(t)$ for Robbins' dynamo with $v = 1$, $\sigma = 5$, $R = 13$ for three values of error tolerance: (a) 10^{-4}, (b) 10^{-6}, (c) 10^{-8}. The asterisks indicate where adjacent plots first differ in time (increasing). The time step for plotting the variables is 0.1 units. After Crossley *et al.* (1988).

the underlying hydrodynamics of the Earth's core. Under some form of core convection, especially that due to gravitational settling (e.g. Loper, 1978; Gubbins and Masters, 1979), one may readily imagine clumps of denser or lighter fluid forming locally and giving rise to small-scale irregularities in a convective velocity field. This component would, through the Lorentz force in the Navier–Stokes equation, generate local irregularities in the magnetic field as well. Alternatively, perhaps the core conductivity, owing to compositional variations, can vary locally, thus affecting the balance of diffusive to generated magnetic flux in the induction equation. Perhaps instead there is a change of conditions at the CMB due to earthquakes or mantle torques. All these effects are lumped into a single driving term which has predetermined statistics and is inserted into the equations of motion.

For the Robbins dynamo, Equation (5.18) is modified by writing

$$R = \bar{R} + r(t) \tag{5.21}$$

where $r(t)$ is some form of noise. The case of a pseudo-random uncorrelated Gaussian process (white noise), which is stationary with zero mean and specified variance σ_1^2, is shown in Figure 5.14 for different combinations of $\bar{R}$ and σ_1.

The integrations are begun at the stable points, where they would remain throughout in the absence of excitation. One can see in Figure 5.14(a) a long sequence of one polarity with varying amounts of stimulation, including growing and decaying fields without reversals, followed by a burst of reversal activity in the middle of the sequence. Figure 5.14(b) demonstrates the effect of decreasing $\bar{R}$ and increasing the excitation, and shows how isolated polarity excursions can occur both with and without long-term changes of polarity. Again, however, the reversals are initiated by growing instability in the primary oscillation.

More extreme examples of excitation are shown in Figure 5.14(c) and 5.14(d), where the value of $\bar{R}$ is further decreased and that of σ_1 increased. One can see that, as $\bar{R}$ is decreased, the oscillations and reversals become more irregular and more frequent and, in particular, the behaviour prior to a reversal loses the characteristic of a gradual build-up of amplitude. At the level of Figure 5.14(d) the random part of R is almost 100% of the mean. It is clear that, with the addition of white noise excitation, the Robbins dynamo now exhibits a wider range of behaviour than a purely chaotic dynamo.

One of the features of white noise excitation is that it is uncorrelated. This can lead to unphysical requirements in the dynamical system, especially when there are very large and rapid excursions in driving torque. An alternative model for the noise inherent in physical systems, which has become fashionable as a result of the work of Mandelbrot (1983), is that of self-similar or 'fractal', 'one-over-f' noise. This stochastic process is called 'flicker noise' because it is analogous to the phenomenon of that name in solid state electronic systems which dominates all other sources of noise at very low frequencies. Flicker noise is correlated over all times with a magnitude that decreases as the correlation interval increases such that the process ideally possesses a $1/f$ power spectrum. Physically, flicker noise allows one to describe a random dynamo which has the longest possible

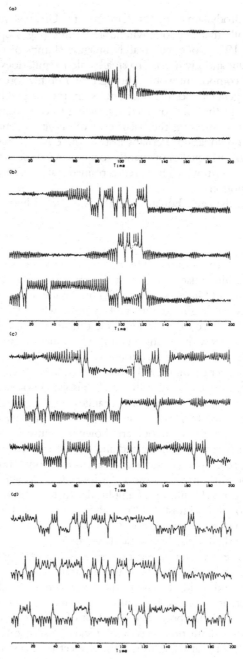

Figure 5.14 Plots of $y(t)$ for Robbins' dynamo (a) $\bar{R} = 14$, $\sigma_1 = 1$; (b) $\bar{R} = 13$, $\sigma_1 = 2$; (c) $\bar{R} = 11$, $\sigma_1 = 4$; (d) $R = 8$, $\sigma_1 = 7$. Error tolerance of 10^{-5}; other parameters as Figure 5.13. The timescale applies to only the first trace of each plot; for subsequent traces add 200 per line. After Crossley *et al.* (1986).

memory and the widest possible spatial correlation, while not evolving through geological time.

To demonstrate the effect of self-correlation in the excitation of a driven dynamo, three examples of Robbins's dynamo forced by flicker noise are shown in Figure 5.15. It can be seen that the first trace in Figure 5.15 yields no reversals, owing to the level of excitation being generally lower than the mean. The second trace shows many reversals from a flicker noise sequence (not shown) that is persistently above the mean, while the third trace is somewhere between the other two in response. If this model were correct, one might expect to find weak evidence of correlation in the reversal record, decreasing with the length of time considered.

Finally, an example of the Robbins dynamo with 'brown noise' excitation is shown in Figure 5.16. This type of process, known also as a random walk, differs from flicker noise in being non-stationary (as well as highly correlated) and provides magnetic field response similar to the flicker noise process. Owing to the very long correlation lengths possible, the excitation could wander quite far from the mean $\bar{R}$ for considerable periods of time, thus emphasizing any tendency for the reversals to appear as clusters in time.

One cannot on the basis of all these examples decide *a priori* whether the palaeomagnetic results favour a chaotic dynamo or a noisy dynamo, or, indeed, some combination of the two. However, like the empirical models of Cox (e.g. 1981), a physically derived stochastic process is invoked to trigger the reversals.

Similar results have been obtained by Honkura and Matsushima (1988), who considered the single-disc dynamo of Bullard (1955) modified by the inclusion of external perturbations. Their models are different from those of Crossley *et al.* (1986) in that the perturbations are considered to be due to magnetic fields exerted externally instead of by the driving torque. They found that excursion-like variations were equally likely as full polarity reversals. Lebovitz (1969) and Ito

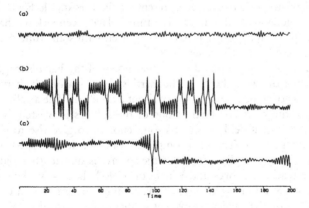

Figure 5.15 Plot of $y(t)$ for Robbins' dynamo with flicker noise: (a) seed = 0.1, (b) seed = 0.2, (c) seed = 0.3 for $\bar{R} = 13$, $\sigma_1 = 3$. The traces may be regarded as part of the same magnetic field sequence at different 'epochs'. After Crossley *et al.* (1986).

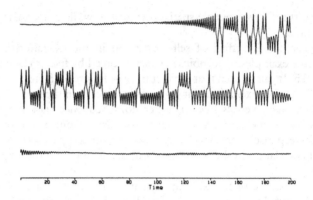

Figure 5.16 Plot of $y(t)$ for Robbins's dynamo with brown noise excitation. Unlike Figure 5.15, the three traces form a continuous sequence. After Crossley *et al.* (1986).

(1988) have considered the case of N interacting disc dynamos. Ito (1988) suggests that reversals can result from large deviations in stochastic systems driven by small, random forces, in systems where 10–20 eddies in the Earth's core evolve stochastically, coupled together through a mean-field-like interaction. He claims that such a model satisfies the observed data better than the view which regards the reversal phenomena as chaos in dynamical disc dynamo models.

Shimizu and Honkura (1985) have also studied the statistical nature of reversals of the Earth's magnetic field with a model of N-disc dynamos. All their models exhibit chaotic behaviour, the statistical nature of reversals being similar to those for the Earth's magnetic field. The best results were obtained for 'inhomogeneous' N-disc systems. Such a system is characterized by a dominant disc which is added to an otherwise homogeneous N-disc system. In the case of the Earth's magnetic field, the dominant disc may be regarded as simulating the dipole field and the other discs as representing the non-dipole fields. A feature of the inhomogeneous N-disc model is a time-delayed feedback mechanism, i.e. the effect of the dominant disc is fed back to itself with some time delay arising from interactions with the smaller discs.

In 1958 Herzenberg proposed a dynamo model of the Earth's magnetic field which consisted of two spheres in the Earth's core each of which rotates as a rigid body at a constant angular velocity about a fixed axis. The axially symmetric component of the magnetic field of one of the spheres is twisted by rotation, resulting in a toroidal field which is strong enough to give rise to a magnetic field in the other sphere. The axial component of this field is twisted as well and fed to the first sphere. If the rotation of the spheres is sufficiently rapid, a steady state may be reached. Lowes and Wilkinson (1963) built a working model of what is effectively a homogeneous self-maintaining dynamo based on Herzenberg's theory. For mechanical convenience they used, instead of spheres, two cylinders placed side by side with their axes at right angles to one another so that the induced field of each is directed along the axis of the other. If the directions of rotation are chosen correctly, any applied field along one axis will lead, after two stages of induction, to a parallel induced field. If the velocities are large enough, the induced field will be larger than the applied field which is no longer needed, i.e. the system would be self-sustaining. Wilkinson has since carried out many more experiments using as many as four cylinders and has been able to produce reversing magnetic fields, the reversal characteristics depending on the experimental configuration used. In most cases the reversal waveform is periodic, unlike the reversal record of the Earth's magnetic field, and yields a simple spectrum when analysed into its Fourier components. In some arrangements a more complex reversal pattern develops, and analyses of these records reveal complicated spectra (Kerridge and Wilkinson, 1983).

5.4 The flooding model of Hoffman

The flooding model of Hoffman (1977, 1979) is based on the idea that the reversal process floods through the fluid core from a localized zone or point of initiation. The entire dipole source remains active, but normal and reversed flux is produced simultaneously in different regions. The VGP moves along the great circle defined by the geographical poles and the observer's site. Depending on the sense of the reversal, where in the core source region the reversal process begins and the hemisphere of the observer, the VGP path for the reversal will be 'near-sided' (i.e. the same longitude as the site) or 'far-sided' (180° away). It is not known whether the non-dipole field is predominantly quadrupolar or octupolar. Figure 5.17 shows schematically a sequence of five field configurations for an R→N reversal. Figure 5.17(a) is for a pure octupolar transition field geometry and 5.17(b) for a pure quadrupolar field. In the upper section of each figure the lines of force of the poloidal fields are shown – the shaded area represents the region of new polarity and illustrates its growth throughout the reversal. In the lower part of each diagram schematic dipoles are used to illustrate the sources of the observed fields. The shading in the lower section of the diagrams indicates the region of the core in which the reversal was initiated.

It is evident from Figure 5.17(b) that the sequence of inclination values for the quadrupolar geometry, whether observed at northern or southern hemisphere sites, includes a vertically downward inclination, implying that the VGP is immediately beneath the site. Hence, each of these records would be near-sided. In contrast, the sequences of inclination values for the octupolar geometry do not both include a +90° inclination. The northern hemisphere site does, and its path will be near-sided; however, the southern hemisphere site has a −90° inclination and would therefore be far-sided, with a VGP antipodal to the site. Fuller *et al.* (1979) considered other models in which the reversal is initiated in different areas of the core and which may not depend upon the sense of the reversal. Figure 5.18 lists various possibilities. The examples illustrated in Figures 5.17(a) and 5.17(b) are given in the top left-hand quadrants of Figures 5.18(a) and 5.18(b), respectively. Both northern and southern hemisphere records of a single reversal are needed to distinguish between quadrupolar and octupolar geometries. Which of the two models for reversals (the flooding or the standing field model)

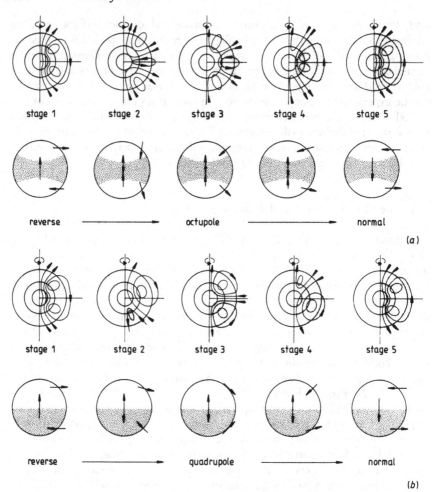

stage 1   stage 2   stage 3   stage 4   stage 5

reverse   ⟶   octupole   ⟶   normal

(a)

stage 1   stage 2   stage 3   stage 4   stage 5

reverse   ⟶   quadrupole   ⟶   normal

(b)

Figure 5.17 (a) Sequence of field configurations for R→N reversal with octupolar transition field (after Fuller *et al.*, 1979). (b) Sequence of field configurations for R→N reversal with quadrupolar transition field (after Fuller *et al.*, 1979).

is correct cannot be resolved by data from only one record of a particular reversal – at least two distant records are required.

In 1977 Hoffman hypothesized that for R→N reversals observed at mid-latitudes in the northern hemisphere, the poles tend to lie predominantly in the hemisphere centred about the site meridian. In 1978 Hoffman and Fuller obtained data for the Matuyama–Brunhes reversal from five sites in the northern hemisphere which gave some support to this suggestion. However, Williams and Fuller (1981a) obtained an unambiguous *far-sided* VGP path for an R→N reversal recorded in the Agno batholith (a 14–15 Ma-old quartz diorite) in Luzon

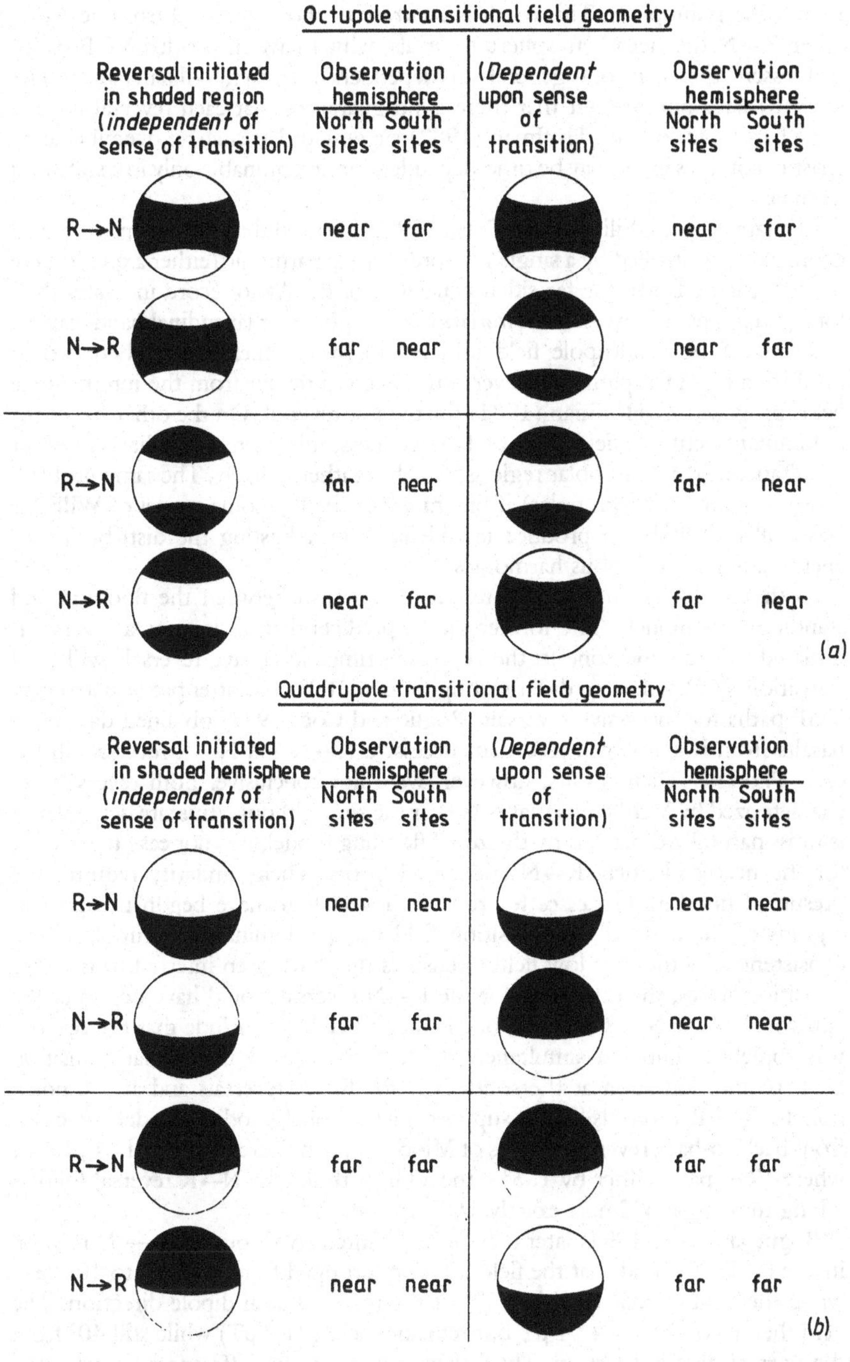

Octupole transitional field geometry

Reversal initiated in shaded region (*independent* of sense of transition)	Observation hemisphere		(*Dependent* upon sense of transition)	Observation hemisphere	
	North sites	South sites		North sites	South sites
R→N	near	far		near	far
N→R	far	near		near	far
R→N	far	near		far	near
N→R	near	far		far	near

(a)

Quadrupole transitional field geometry

Reversal initiated in shaded hemisphere (*independent* of sense of transition)	Observation hemisphere		(*Dependent* upon sense of transition)	Observation hemisphere	
	North sites	South sites		North sites	South sites
R→N	near	near		near	near
N→R	far	far		near	near
R→N	far	far		far	far
N→R	near	near		far	far

(b)

Figure 5.18 (a) Predicted VGP paths for octupole transition fields (after Fuller *et al.*, 1979). (b) Predicted VGP paths for quadrupole transition fields (after Fuller *et al.*, 1979).

Island, the Philippines. This result is contrary to those obtained from nearly all other R→N northern hemisphere reversals, which have near-sided VGP paths, and lessens the claim of a general site dependence. In fact, Williams and Fuller (1981b) later pointed out that there is no guarantee that each reversal has the same harmonic content. Hoffman (1982) suggested that such an overall characteristic, if it does exist, may be time-dependent or determinable only in a statistical manner.

In a later paper, Williams and Fuller (1982) showed that if the transition field geometry is controlled by a single low-order zonal harmonic (either a quadrupole or an octupole field), the far-sided behaviour of the Agno record indicates that, for a given sense of reversal, the initiation hemisphere, or latitudinal band, cannot be constant. A quadrupole field initiated in the northern hemisphere would produce a far-sided path. However, this is very different from the intermediate fields proposed by Hoffman (1981) for the last reversal. On the other hand, for a dominant octupole field to cause far-sidedness, initiation would have to occur simultaneously in both polar regions, which is rather unlikely. The same problem arises (initiation at both poles) if one modifies the harmonic model of Williams and Fuller (1981b) to produce far-sidedness by adjusting the distribution of energy among the various harmonics.

Hoffman (1981) suggested a test to distinguish between the flooding and standing field models. The former model predicts that, as long as a reversal is initiated in the same zone in the core each time, successive reversals will have transition VGP paths that differ in longitude by 180°. The latter predicts identical VGP paths for successive reversals. Bogue and Coe (1982) obtained data from basalts from Kauai which recorded successive R→N and N→R reversals in the early Pliocene. Their results, however, were not conclusive. Both reversals are characterized by VGP paths that pass near the site. Although this site dependence is most naturally explained by the zonal flooding model, it is not easy to account for the nearly identical R→N and N→R paths. Their similarity requires the presumed flooding processes for the two reversals to have begun in opposite regions of the core. If the transition field was predominantly octupolar, as is consistent with the very low field intensities that have been inferred from other transition zones, then flooding for the R→N reversal would have begun at the equator. For the N→R reversal, one is then forced to conclude that the reversal was somehow initiated simultaneously at both poles. Quadrupolar transition fields require initiation at the south pole for R→N reversals and at the north pole for N→R reversals. More support for the zonal flooding model has come from back-to-back reversal records of Miocene age in Crete (Valet and Laj, 1981) where VGP paths differ by 135° – the VGP path for the N→R reversal follows a longitudinal path almost exactly 180° from the site.

Bogue and Coe (1984) later measured the intensity through a R → N reversal in Kauai. They found that the field intensity dropped from 43.1 μT to 10.1 μT, while the field remained within 30° of the reversed axial dipole direction. The field then recovered to 44.4 μT but remained low (21.7 μT) while still 40° from the normal dipole direction. Thus, there appears to be differences between the initial and final stages of the reversal. The standing field model cannot explain this asymmetric intensity pattern. Moreover, the results from Kauai fit neither the octupolar nor the quadrupolar flooding model, but exhibit some characteristics of both. Bogue and Coe speculate that the core system is unstable, ready to reverse, but with no strong preference for one flooding scheme over another. It is also possible to model the asymmetric intensity pattern by a flooding mechanism that is itself asymmetric.

Much more work has since been done on transition fields and is discussed in Chapter 6.

5.5 Olson's model

Olson (1983) has investigated the behaviour of the magnetic field during a polarity transition based on the α^2 dynamo model for magnetic field generation in a turbulent core. Figure 5.19 shows schematically the four accessible equilibrium configurations of a homogeneous, axial, dipolar α^2 dynamo. The circular regions represent the core, the field lines piercing the core surface represent the poloidal (dipolar) field component and the field lines connected within the core represent the toroidal field component. Both positive and negative helicity* are associated with two equilibrium states, corresponding to normal (N) and reversed (R) polarities. The upper left hand configuration is the present-day 'normal' configuration. Its field line topology can be abbreviated symbolically as NN for normal dipole, normal toroidal polarities. Proceeding clockwise around Figure 5.19, the other three configurations are denoted by NR, RN and RR, respectively.

In this notation, the four transitions

$$NN \rightarrow RR, \; RR \rightarrow NN, \; NR \rightarrow RN, \; RN \rightarrow NR$$

are generated by multiplying the field vector by -1. Olson called these transitions *full reversals*. Similarly, there are eight transitions which can be generated by multiplying one field component (either toroidal or poloidal) by -1. Olson called these *component reversals*. Four of these involve a reversal of the dipole field (poloidal reversals); the other four involve reversal of the toroidal field (toroidal reversals).

The governing conservation equations are invariant for transitions with constant helicity, i.e. for the four full reversals. On the other hand, the governing equations are not invariant for transitions in which the helicity changes sign. In this case all reversals are of the component type (see Figure 5.19). Thus, an α^2 dynamo will respond to a change in the sign of helicity with a component reversal. If the helicity changes sign twice in a short time interval, as might be the case in a transient fluctuation, both field components can reverse. A full reversal is necessarily a two-step process in which the toroidal and poloidal fields reverse sequentially rather than simultaneously. Olson showed that rapid reversals

---

* The helicity is a measure of the correlation between the turbulent velocity U and the vorticity $\omega = \nabla \times \mu$. Specifically the helicity is $U \cdot \omega$.

of the dipole field occur when the helicity changes sign, and derived a set of equations which govern the dipole field behaviour during a transition.

If the helicity changes sign once, simulating a long-term fluctuation in core conditions, the α effect will change sign once, which forces a component reversal. In response to the change in helicity, both the toroidal and poloidal field amplitudes decrease rapidly. As the amplitudes approach zero, a bifurcation occurs, at

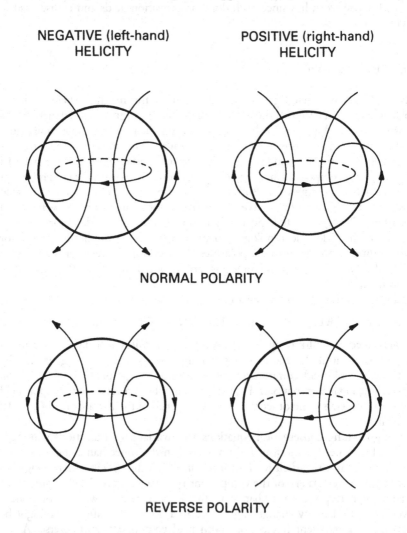

Figure 5.19 The four equilibrium configurations of a homogeneous, axial dipolar α^2-dynamo. The curves piercing the circular core boundary represent the dipole field; the curves closed within the core represent the toroidal field. Transitions between equilibrium configurations in which the sign of the helicity is conserved are called full reversals. Transitions between equilibrium configurations in which the sign of the helicity changes are called component reversals. After Olson (1983).

which point the dipole field continues through zero, while the toroidal field conserves its polarity and recovers without reversing (see Figure 5.20, which shows the response to a sudden change from negative to positive helicity). Beyond the bifurcation point the fields re-equilibrate in the RN configuration. Olson found that, in all cases, the reversing component is decided on the basis of relative strength near the bifurcation point – the stronger component retains its polarity, while the weaker one is forced to reverse. The time required to complete the reversal can be as short as 7500 a.

If the helicity changes sign twice within a short interval, the α effect will change sign twice and two component reversals may result. The poloidal and toroidal fields may both reverse, producing a full reversal. Alternatively, the transient fluctuation may be such that neither field component has time to reverse, and only an excursion in field amplitude is seen.

Whether or not a fluctuation in helicity produces a full reversal depends on its duration and severity. The effect of variable duration is illustrated in Figures 5.21(a) and 5.21(b). It is clear that the field response depends critically on the duration of the helicity. For reversals in helicity lasting $< \Delta t^* \approx 0.35$ (~ 2500 a) there is insufficient time for either field component to reverse sign, and the response is a transient reduction in amplitude only. For helicity reversals lasting longer than $\Delta t^* = 0.35$, full reversal occurs, with the weaker dipolar field reversing first.

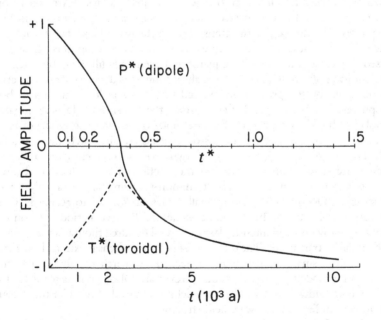

Figure 5.20 The response of dipole and toroidal field amplitudes of an α^2- dynamo to a sudden change from negative to positive helicity, applied at time $t^* = 0$. The field amplitudes and timescale are dimensionless. The component reversal illustrated here is NN→RN. After Olson (1983).

Olson offers as a possible explanation of the cause of reversals fluctuations in the core's net helicity in response to fluctuations in the level of turbulence from two competing energy sources – heat loss at the CMB and progressive solidification of iron at the ICB. He showed from symmetry considerations that these two energy sources produce helicity with opposite signs. The α effect generated by IC growth tends to oppose and destabilize the α effect generated by heat loss at the CMB. Over short time intervals, the contribution from growth of the IC may exceed the contribution from heat loss to the mantle, causing a transient reversal in the sign of the helicity. It should be pointed out that Olson's model does not take account of the non-dipole components which form part of the transition field. Schloessin and Jacobs (1980) had earlier suggested that reversals might be the result of competition between changing conditions at the ICB and CMB.

Olson and Hagee (1990) carried out numerical calculations of mean field dynamos to determine the long-term behaviour of azimuthally averaged magnetic fields in the Earth's OC maintained by transverse isotropic induction. (The α effect from convection in a rapidly rotating sphere is transversely isotropic with respect to the rotation axis). Olson and Hagee represented the inductive effect of core convection by a non-linear, anisotropic α effect with magnitude R_α concentrated near the ICB to simulate convection driven by growth of the IC. They showed that the α^2 dynamo is periodic but not sinusoidal, and is characterized by two timescales. The intensity of the dipole field decays slowly on a long timescale of $\sim 5 \times 10^5$ a as reversed flux accumulates at high latitudes in both hemispheres. The dipole polarity transition, on the other hand, takes place over a much shorter timescale ($< 5 \times 10^4$a). Reversals are accomplished as rings of reversed flux expand towards the equator and eventually fill the entire OC.

The addition of a large-scale toroidal shear flow ($\alpha^2\omega$ dynamo) causes reversals to be initiated by the growth of reversed flux zones at low latitudes in both hemispheres. For both α^2 and $\alpha^2\omega$ dynamos the transition field is dominantly octupolar within the OC, and they are examples of the flooding model of Hoffman (1977). VGP paths from successive reversals are alternately near- and far-sided. The long-term behaviour of α^2 dynamos is very sensitive to R_α, the magnetic Reynolds number based on the α effect. Large values of R_α, representing vigorous convection, result in non-reversing and relatively strong dipolar fields. On the other hand, small values of R_α lead to relatively weak, oscillatory dipolar fields. Intermediate values of R_α give periods of constant polarity of nearly uniform intensity lasting several hundred thousand years, separated by polarity transitions. The duration of polarity epochs is extremely sensitive to R_α – only a small change in R_α is needed to switch the dynamo from a non-reversing mode to one with frequent reversals. Olson and Hagee stress that their model considers only axisymmetric fields – removal of this constraint could result in very different large-scale field structure.

Olson and Hagee's calculations predict a positive correlation between the duration of a polarity interval and the strength of the dipole field. This supports the findings of Pal and Roberts (1988) that the dipole moment is appreciably greater during fixed polarity superchrons. Pal and Roberts based their conclusion on

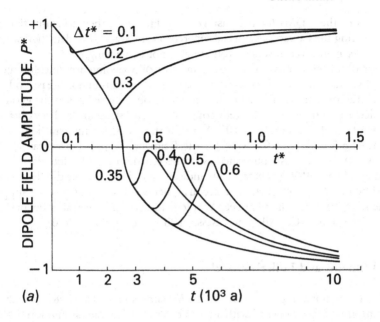

(a)

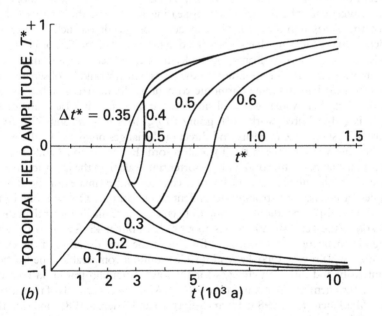

(b)

Figure 5.21 (a) The response of the dipolar field amplitude of an α^2-dynamo to transient reversals in helicity, commencing at $t^* = 0$ and having a duration Δt^*. The field amplitude and timescale are dimensionless. Full reversals of the type NN→RR occur for helicity fluctuations exceeding $\Delta t^* \simeq 0.35$ (2500 a). (b) Same as Figure 5.21(a) for toroidal field amplitudes. After Olson (1983).

estimates of the VDMs for the past 180 Ma. However, they do admit that the data covering the Cretaceous normal superchron are 'scant'. The data they present for the Permian reversal superchron are also not convincing.

Sherwood *et al.* (1993) carried out a number of palaeointensity measurements on mid-Cretaceous basaltic lavas from the Rajmahal Traps in north-eastern India and several formations in Israel in an attempt to obtain estimates of the geomagnetic field strength during the Cretaceous long normal superchron. They obtained a value of 4.7 ± 2.2 (1SD) $\times 10^{22}$ A m^2 for the VDM for the Israeli lavas and a value of $5.5 \pm 1.9 \times 10^{22}$ A m^2 for the Rajmahal Traps – both mean VDMs are lower than previously published estimates based only on Thellier data (Sherwood and Shaw, 1991). Their results indicate that the Mesozoic dipole low had ended before the onset of the Cretaceous quiet zone, and thus give no support to the suggestion of a link between dipole field strength, reversal frequency and true polar wander (Courtillot and Besse, 1987; Prévot *et al.*, 1990).

5.6 Williams and Fuller's model

Rather than simulating the VGP path, Williams and Fuller (1981b) analysed inclination records associated with a given reversal. The basic assumption of their model of the transition field is that the dipole field decays in a nearly exponential fashion, with part of the energy lost appearing in the first three terms (g_2^0, g_3^0, g_4^0) of the zonal non-dipole field. The decay of the dipole field energy was modelled on the intensity changes observed in the record of the Tatoosh intrusion (Dodson *et al.*, 1978). Williams and Fuller then calculated synthetic inclination and intensity records corresponding to the fields which would be observed if the transition field had the same harmonic content as the model. The last reversal (the Brunhes–Matuyama) was well simulated at most sites if 70% of the dipole energy is redistributed, with 20% going to g_2^0, 30% to $- g_3^0$ and 50% to g_4^0. Figure 5.22 shows the intensity and directional changes predicted by the model and the observed changes in the Tatoosh record. Both the model and observed data give a change in direction with a shorter timescale than the intensity change. However, both intensity and inclination changes and estimates of the time to complete a reversal are strongly dependent upon latitude. Observations in the northern hemisphere indicate that intensity changes take much longer than directional ones (the Tatoosh intrusion is about 47° N), but at 45° S there is a gradual change in inclination that continues for almost the total duration of the event. This may explain why Opdyke *et al.* (1973) found a comparable time for both the intensity and direction changes for the lower Jaramillo transition recorded in a southern hemisphere oceanic core (36° S). More recent records of transitional fields (Mankinen *et al.*, 1985; Herrero-Bervero and Theyer, 1986) indicate that additional terms (including non-zonal ones and perhaps more complex dipole variations) are necessary to describe transitional fields.

Clement and Kent (1984) analysed a detailed record of the lower Jaramillo (R→N) polarity transition from a southern hemisphere, deep-sea sediment core. Over 850 samples were taken across 140 cm of section – the transition itself

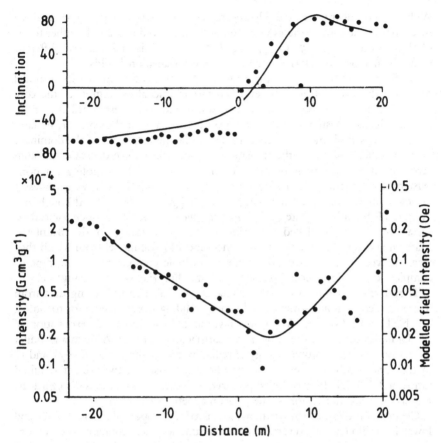

Figure 5.22 Comparison of inclination and intensity records from the Tatoosh intrusion, Washington (Dodson *et al.*, 1978), with the model for the Brunhes–Matuyama boundary. The site location is 46.8° N, 121.8° W. This record is a R→N reversal dated at 17 Ma recorded in a granodioritic intrusion. The modelled intensity is shown in Oe and the Tatoosh intensity data are for a mean of samples demagnetized to 100 Oe. After Williams and Fuller (1981b).

spanned ~ 70 cm. The inclinations shallow gradually early in the reversal and pass through very steep negative values (−80°) late in the transition. The declinations show little variation until the inclinations have moved through the near-vertical. The intensity drops to less than 15% of pre-transitional values, its fluctuations lasting much longer than the directional changes. Depending on the criteria adopted, the duration of the directional change is estimated to be 11.2–4.5 ka, and of the intensity change 20–15 ka. Clement and Kent tried to model the reversal, using the technique developed by Williams and Fuller (1981b) for the Brunhes–Matuyama reversal where the dipole energy is redistributed to the g_2^0, g_3^0 and g_4^0 terms. No agreement was found when the model parameters of

Williams and Fuller were used. However, a good fit to the data could be obtained by changing the sign of the g_2^0 term and allowing it to dominate the other terms (70% going to $-g_2^0$, 10% to $-g_3^0$ and 20% to g_4^0). This indicates that different reversals may be characterized by very different transitional fields.

Hoffman (1979) simulated the Brunhes–Matuyama transition with a model based on the assumption that reversals start in a localized region of the core which extends or 'floods', both north–south and east–west, until the entire core is affected. Assuming the reversal process to start at the equator, he later (1981) analysed his model solution to determine the behaviour of the dominant Gauss coefficients during the transition. As the axial dipole decreases through zero, the most dominant non-dipole terms are a zonal octupole g_3^0, a non-axisymmetric quadrupole having strength $[(g_2^1)^2 + (h_2^1)^2]^{1/2}$ and, to a far lesser extent, a higher-order term having strength $[(g_4^1)^2 + (h_4^1)^2]^{1/2}$. Although not unique, Hoffman's estimate of the most significant Gauss components and their variation for the modelled transition field corresponding to the Brunhes–Matuyama reversal is not tied to the phenomenological model from which they were derived, e.g. the strength of the non-dipole components at the time of complete dipole decay could result from a standing (i.e. non-reversing) field.

Theyer *et al.* (1985) tested the Williams and Fuller model, using data from five piston cores along a meridional band extending from the equator to almost 40° N. For the Brunhes–Matuyama reversal and the onset and termination of the Jaramillo subchron they obtained a satisfactory fit to the William and Fuller model in which the dipole energy is redistributed to the terms g_2^0, $-g_3^0$ and g_4^0 in the ratio 2:3:5. However, for the Olduvai transition, the best fit required ratios of 1.1:5:3.9. It must be emphasized that such zonal models disregard the important 'third dimension' – the palaeomagnetic declination.

Clement (1987) has shown that records of the upper Olduvai (N→R) and lower Jaramillo (R→N) reversals obtained from deep-sea sediment cores in both the northern and southern hemispheres are consistent with Olson's (1983) model, in which helicity fluctuations can cause reversals of an axial dipole field generated by a homogeneous α^2 dynamo. In Olson's model the field remains dipolar during the reversal and so does not account for non-dipolar transitional fields – records of the transition field show that the field does not remain dipolar during the reversal and that higher-degree terms must become important. Because of the difficulty in including higher-order terms in Olson's model, Clement adopts the approach of Williams and Fuller in which non-dipole geometries are specified as the dipole field decays and builds up in the opposite direction, the dipole intensity varying according to Olson's model. He found that good fits to both the upper Olduvai and lower Jaramillo reversals could be obtained by partitioning the energy lost by the dipole field, with 40% going to $-g_2^0$, and 55% to $+g_4^0$ and maintaining a stationary equatorial dipole of 5%. Both reversals are modelled as full (two-component) reversals, the only difference between the two models being the duration of the helicity fluctuation. The palaeomagnetic record of the upper Olduvai reversal at the northern hemisphere site together with synthetic records produced by the above model are shown in Figure 5.23. Although the predicted transitions are in reasonable agreement with the observed transition records, there are some discrepancies. All of the synthetic intensity records drop to low values coinciding with the directional change. However, unlike the observed records, the model predicts that the directions begin to change at the same time as the intensities begin to drop. The modelled intensity values also recover too slowly. Some sort of time-varying equatorial dipole term, rather than the assumption of a stationary equatorial dipole, seems to be needed to produce better fits to the declination records. Clement's model also gives a very good fit to the Steens Mountain R→N transition (see also Section 6.7) if the signs of the non-dipole terms are reversed, i.e. partitioning the energy lost by the dipole term, with 40% going to $+g_2^0$ and 55% to $-g_4^0$. It is remarkable that the same field symmetry that is used to model the relatively young, deep-sea sediment transitions can also provide a good fit to the 15 Ma Steen's Mountain volcanic record. Evidence of persistent transitional fields over a short time has been found from

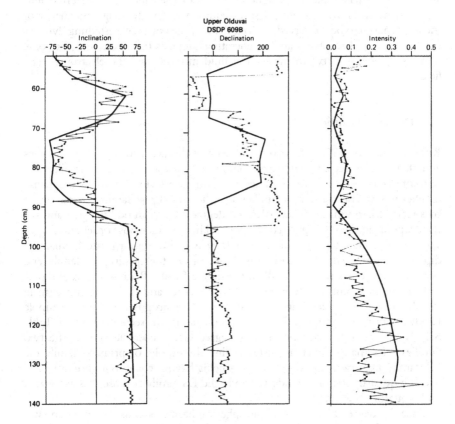

Figure 5.23 Palaeomagnetic record of the upper Olduvai reversal from HPC hole 609B presented as inclination, declination and magnetization intensity ($\times$ 10^{-3} A/m) plotted against depth. Synthetic records produced using the model discussed in the text are superimposed (heavy solid curves) on the observed record. Depths are measured from the top of core 14, section 1. After Clement (1987).

back-to-back reversals from Kauai (Bogue and Coe, 1982, 1984) and from Crete (Valet and Laj, 1984). The sequence of reversals from Crete, however, also shows that transitional fields may vary significantly between sequential reversals after remaining similar for several previous reversals. Additional evidence for persistent transitional field characteristics has been given by Hoffman (1986) in records from lavas from Australia and New Zealand (see Section 6.6). Williams *et al.* (1988) later showed that if low-order zonal harmonics are combined with a drifting non-dipole field, most of the features of the transition field can be reproduced, including the evolution of loops and hang-ups in the VGP path. These are the consequence of the phase relationships at various sites between the growth and decay of the zonals and the passage of features of the drifting non-dipole field.

Although some success has been achieved based on the suggestion of Williams and Fuller (1981b) that the magnetic energy of the dipole is transferred into higher harmonics, there is no physical requirement for the magnetic energy of the geodynamo to be conserved. In fact, the very idea of a self-sustaining dynamo that is capable of reversing and regenerating suggests that there must be some transfer of energy between mechanical fluid motions and the electromagnetic field.

5.7 Dynamo families

Roberts and Stix (1972) showed that dynamo fields are made up of two families and that, under certain symmetry conditions* in the core, these families are non-interactive – when these symmetry conditions are violated, the two dynamo families will interact. The two families are called the dipole family (characterized by spherical harmonics whose order and degree sum to an odd integer) and the quadrupole family (characterized by spherical harmonics whose order and degree sum to an even integer). This terminology is a little unfortunate in that the dipole has terms (the equatorial dipole g_1^1, h_1^1) in the quadrupole family and the quadrupole has terms (g_2^1, h_2^1) in the dipole family. Because the axial dipole term g_1^0 plays a central role in geomagnetic studies and is the leading term in the dipole family, Merrill and McFadden (1990) proposed calling the dipole family the primary family, and the quadrupole family the secondary family. Table 5.1 shows the separation of the low-degree Gauss coefficients into the two families. Hoffman (1991) suggested calling the dipole, or primary, family the EA family because it is associated with fields having equatorial antisymmetry, and the quadrupole, or secondary, family the ES family because it is associated with equatorial symmetry.

Since the degree plus order of any spherical harmonic must be either an even or an odd integer, the primary and secondary families taken together always provide a complete description of the magnetic field. Roberts and Stix (1972)

* Specifically, the mean velocity field must be symmetric and the α effect antisymmetric with respect to the equator.

Table 5.1. Primary and secondary families

	Primary family	Secondary family
Dipole	g_1^0	g_1^1, h_1^1
		g_2^0
Quadrupole	g_2^1, h_2^1	g_2^2, h_2^2
Octupole	g_3^0	g_3^1, h_3^1
	g_3^2, h_3^2	g_3^3, h_3^3

showed that solutions of the dynamo equations under certain assumptions can lead to selection rules in terms of the field configuration that may be generated from pre-existing configurations – of particular interest are the selection rules for $\alpha\omega$ dynamos. A more detailed account of the theoretical basis for the primary and secondary families has been given by Lee and Lilley (1986). Under the symmetry conditions specified by Roberts and Stix, an initial seed magnetic field contained entirely within one family can only generate other terms in that family. Although the symmetry conditions are sufficient for the two dynamo families to be non-interacting in many kinematic models, they are not sufficient to separate the two families in all models. This is because both families are driven by the same velocity field, and so, if the magnetic field from one family affects the velocity field, this can affect members of the other family.

Merrill and McFadden (1988) have developed a class of models (called M^2 models) based on these ideas in an attempt to explain reversals of the geomagnetic field. They suggested that a reversal, or aborted reversal, will occur when some instability increases the coupling between the primary and secondary families at a time when the relative magnitude of the field of the secondary family is high with respect to the primary family contributions within a given polarity interval. Their preferred model is one in which a major disturbance in the velocity field breaks down the symmetry with respect to the equator which requires the two families to be decoupled. M^2 models appear to be supported by palaeomagnetic data, and Merrill and McFadden predicted that the relative contribution of the secondary family to the magnetic field should be smaller when the reversal rate is low than when it is high. In particular, the relative contribution of the secondary family should have been lower during the Cretaceous normal superchron than at other times during the past 180 Ma. McFadden *et al.* (1991) obtained support for this prediction from an analysis of the palaeosecular variation of lava flows, which they used to estimate the relative contributions of the two families back through time. They used the method of Merrill and McFadden (1988), in which the scatter of virtual geomagnetic pole positions can be separated into its contributions from the primary and secondary families. Theoretical models by Zhang and Busse (1988, 1989) suggest that an increase in the secondary family may cause a decrease in the symmetric shear velocity and so increase the probability of having a relatively large antisymmetric velocity component breaking the condition for the two dynamo families to be non-interacting. McFadden *et al.* (1991) also showed that there are conditions under which one family can dominate the other. A bifurcation process may occur such that the dynamic processes during polarity transitions and non-transition times are significantly different. Although there is support, from both palaeomagnetic data and theoretical considerations, for M^2 models McFadden *et al.* (1991) point out that there are still many unanswered questions, such as: Why does a particular critical interaction between the two dynamo families occur? How might such an interaction lead to a reversal? Gubbins and Zhang (1993) later obtained the full set of symmetry operations for the dynamo equations which form an Abelian Lie group. Reflections in the equatorial plane giving rise to the two families discussed above form a smaller subgroup.

Further ideas of McFadden and Merrill on the reversal process are given in the next section.

5.8 McFadden's and Merrill's models

McFadden *et al.* (1985) maintain that polarity transitions are not characterized by significant periods when convection in the OC ceases, but that dynamic processes continue to act throughout a reversal. Estimates of the free decay time of the Earth's magnetic field suggest that the dipole field will decay more slowly than the non-dipole field. This is precisely the opposite of what is observed to happen during a reversal. Hence, it would seem that reversals of the geomagnetic field are not associated with the free decay of the field followed subsequently by a build-up of the field, but rather that they are associated with dynamic processes throughout. In some modelling processes (e.g. Clement and Kent, 1984), some spherical harmonic terms grow, while others decline during polarity transitions. This would indicate that reversal models which call on the cessation of convection over a significant period of time to cause the initiation of a reversal cannot be valid – again consistent with the above speculation of McFadden *et al.* that dynamic processes occur throughout a reversal. Finally, estimates of the time required for directional changes in a polarity transition are of the order of 4000–5000 a, which is too short a time for reversals to occur as a consequence of the free decay of the magnetic field. This again indicates that a dynamic breakdown of the field is required in which there is an interaction of the velocity field with the magnetic field.

McFadden and McElhinny (1982) analysed all VDMs determined for the past 5 Ma which have associated VGPs with latitudes greater than 45° (to ensure that they relate to periods of stable polarity). Histograms of normal, reversed and combined polarity data show that, although the data show maximum frequency around some central mean value, there is a fairly sharp cut-off in the observed VDMs at the low VDM end and the data are skewed to the high end. McFadden and McElhinny showed that neither a Gaussian nor a lognormal distribution provides a satisfactory fit to the data. Their preferred model is one in which the TDMs (true dipole moments) have a truncated Gaussian distribution (which they call nested) and the field strength of the non-dipole components is linearly proportional to the TDM (see Figure 5.24) They found no support for a model in which the field strength of the non-dipole component is a constant ratio of the mean dipole field strength, nor for Cox's (1968) model of a cyclic variation of the dipole moment during stable polarity periods.

The grouping of the TDMs around a central mean suggests that the dipole moment is an inherently stable equilibrium in the region of the stable mean, and that it is disturbed from this value by fluctuations in the velocity field and then driven back towards this equilibrium should the field become too strong or too weak. McFadden *et al.* (1985) suggested that there is a critical region for the TDM (about one-third of its mean value) below which the field cannot maintain a stable polarity. This would be associated with a significant change in the velocity field. They further speculated that the Lorentz force dominates during stable

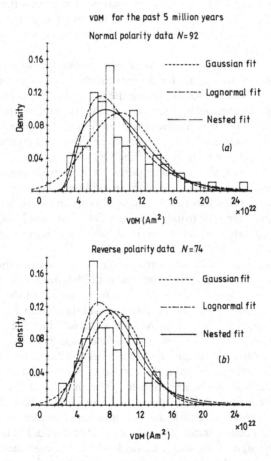

Figure 5.24 Nested distribution fit to VDM data with Gaussian and lognormal fits for comparison: (a) normal polarity, (b) reversed polarity. After McFadden and McElhinny, (1982).

polarity times and that the critical region is one in which the dominance of the Lorentz force is lost and in which there is a significant change in the character of the velocity field. Thus, during transitional times there is a tendency towards a geostrophic rather than a magnetostrophic balance of forces.

McFadden and Merrill (1986) showed, from an analysis of palaeogeomagnetic data, that strong constraints can be placed on classes of models of the geodynamo. In particular, a statistical analysis of the geomagnetic reversal sequence shows that the observed changes in the rate of reversals are not the result of changes in any inherent inhibition or encouragement in the reversal process, i.e. there must be some physical process that triggers reversals arising from an energy source independent of that which powers the main magnetic field. A statistical mechanical approach to reversals showed that times of increased mean polarity length should be associated with times of decreased temperature difference ΔT across the OC, implying that ΔT was a minimum during the long Cretaceous normal polarity superchron and has been increasing since that time. McFadden and Merrill (1986) further showed that, although changes in ΔT will affect the reversal rate, individual reversals (or attempted reversals) cannot have any significant effect on ΔT.

McFadden and Merrill suggested two models (called the hot-blob model and the cold-blob model) to account for the two energy sources – one of the sources being primarily responsible for producing the main magnetic field and the other for introducing disturbances leading, on occasion, to reversals. In the hot-blob model, core convection, which drives the main field, is due primarily to cooling at the CMB. The source of an instability is an occasional plume or hot-blob given off at the ICB due to freezing of the OC with growth of the IC. In the cold-blob model, convection is driven primarily by chemical convection associated with freezing of the OC at the ICB. Instabilities are generated by cooling at the CMB through heat loss into the mantle, cooler and more dense cold blobs sinking and destabilizing the main convection. Schloessin and Jacobs (1980) and Olson (1983) suggested models for reversals which are in some ways similar to those of McFadden and Merrill.

There does not seem to be any way to discriminate between the two models – McFadden and Merrill tend to favour the cold-blob model. In this model the disturbances triggering reversals originate near the surface of the OC, and will thus affect the outermost part of the core first and should cause a marked effect on the secular variation. Not until the blob sinks well into the core will it be able to destabilize the core's main velocity field enough to trigger a reversal. Precursors to a polarity transition should therefore be observable. In the hot-blob model, disturbances originate near the bottom of the OC and no precursors to a polarity transition should be seen. Smaller blobs, insufficient to cause a reversal, could lead to excursions of the magnetic field or pulses in the secular variation.

Merrill *et al.* (1990) also considered the question of asymmetry between normal and reversed polarity states. The velocity field of the fluid in the Earth's OC cannot sense the sign of the magnetic field. Thus, if any geomagnetic polarity asymmetries do exist, they must arise from boundary conditions or initial conditions (Merrill *et al.*, 1979). Although different lines of evidence have in the

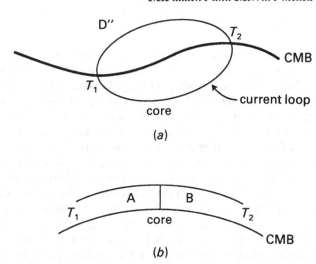

Figure 5.25 Sketch indicating two possible ways in which thermoelectric effects could produce a standing toroidal field. (a) Core–mantle bumps. The temperature T_1 would exceed T_2, the difference being the adiabatic temperature gradient multiplied by the amplitude of the bump. Negative charge (electrons) would flow from T_1 to T_2 in D″, and the return flow would occur in the outermost core, as depicted by the loop. (b) Variation in chemistry in D″. A and B represent different materials in D″. If there exists a horizontal temperature difference in D″, i.e. $T_1 \neq T_2$, then a current would flow in D″ with the circuit closure occurring in the outermost core. After Merrill *et al.* (1990).

past seemed to indicate the reality of such asymmetry, in most cases the evidence is not robust. Merrill *et al.* (1990) believe that the mean field inclination anomaly ΔI data (see Section 1.2) are the most reliable. Using a global data set (from lava flows and sediments) for the past 5 Ma, they found that $\Delta I \simeq -3.7°$ for normal and $-7.2°$ for reversed states at $I = 0$. Schneider and Kent (1988), using data from sediments only, obtained values of about $-2°$ and $-4°$, respectively. Both these studies indicate that to a first approximation g_2^0/g_1^0, in the reversed polarity state is roughly twice that in the normal state. Merrill *et al.* conclude that (at least for the past few Ma) polarity asymmetry seems to be a real feature of the magnetic field. They consider a number of possible mechanisms to account for this polarity asymmetry and conclude that thermal electric effects associated with core–mantle topography and/or chemical variations in the D″ layer (~ 200 km thick) at the bottom of the mantle are the most likely cause.

Figure 5.25 illustrates two possible ways in which thermal electric effects could produce a standing field, and rough estimates of the magnitude of the effect indicate that it is reasonable. It is difficult, however, to reach any definite conclusions, since values of some of the parameters in the calculations are not sufficiently well known and it seems likely that thermal electric effects would produce primarily toroidal magnetic fields which would be unobservable at the Earth's surface – their existence could only be seen after conversion to poloidal fields via the dynamo process.

5.9 Gubbins's model

Bloxham and Gubbins (1985) have produced maps of the radial component of the magnetic field at the CMB at selected epochs from 1715 to 1980. They identified a number of features, including static flux bundles (permanent regions of intense flux), static zero-flux patches (permanent regions of very low flux), rapidly drifting flux spots (observed in the southern hemisphere from around 90° E drifting westwards towards south America with changes in intensity) and localized field oscillations (such as that under Indonesia). A patch of flux of opposite sign to that expected for a dipole field occurs beneath southern Africa. A similar patch lies beneath South America. The above-mentioned maps of Bloxham and Gubbins show that the patch beneath southern Africa has intensified throughout the twentieth century – a result that they attribute to the expulsion of toroidal flux by fluid upwelling. Gubbins (1987) suggested that the location of these reverse-flux zones is related to lateral temperature variations at the CMB, and that their growth could provide a mechanism for polarity reversals. The idea of a constant region where reversals are initiated is similar to that of the flooding models of Hoffman (see Section 5.4).

Bloxham and Gubbins (1987) and Gubbins and Bloxham (1987) pointed out that the positions of the lobes of flux concentration in their maps are consistent with the concentration of magnetic flux by columnar convection rolls in the OC, which are aligned parallel to the Earth's rotation axis and tangential to the IC (Busse, 1975). The flux concentrations occur at the intersection of the columnar convection rolls with the CMB, and the zones of low flux observed in the maps at the geographical poles are the result of the dynamical effect of the IC.

Gubbins (1987) showed that the present decline in the dipole moment is directly related to the intensification and southward movement of these patches, and suggested that this fall can on occasion lead to a polarity reversal. The Earth may have a steady dynamo with superimposed waves driven by features at the CMB, such as topography beneath Indonesia or the hot region beneath southern Africa. Gubbins suggested that at times of large amplitude the waves could lead to abnormally large reverse-flux features and induce a polarity reversal by a mechanism similar to that of a periodic $\alpha\omega$ dynamo. He further suggested that the increase in frequency of polarity reversals since the Cretaceous normal superchron is associated with the growth of the hot region at present beneath southern Africa.

As discussed in Section 6.4, VGP paths for the last few million years tend to be concentrated in two longitudinal bands, one running through the Americas and the other through Asia and Australasia. These preferred longitude bands are those along which north–south flow is seen in core models and are also regions of fast seismic wave propagation (and, hence, low temperatures) in the lower mantle. Gubbins's model predicts that all reversals should be similar (i.e. have the same transitional field) within the lifetime of the pattern of mantle convection. Support for the model is seen in records from Crete (Valet and Laj, 1984) which show similar characteristics over several polarity intervals. However, negative evidence is seen in three successive ~ 5 Ma older but in the same geographical area (Laj *et al.*, 1988).

Clement (1992) found that the Cobb Mountain VGPs tend to fall along antipodal tracks, but centred over Africa and the central Pacific, which are very different from the preferred longitudinal bands for other recent reversals (Laj *et al.*, 1991). In fact, in the same maps used for the correlations of Laj *et al.*, the Cobb Mountain VGP paths correlate better with regions of slow seismic wave propagation! Because these reversals occurred over time intervals too short for significant changes to have occurred in the lower mantle, the differences in these transitional fields are difficult to reconcile with the hypothesis of lower-mantle control.

Chauvin *et al.* (1990) have also pointed out that the growing of one reverse-flux zone may not systematically lead to low intensities everywhere, a feature which is seen in all reversal records. However, Gubbins (1987) estimated that the formation of the two reverse-flux patches in the southern hemisphere is enough to reduce the dipole moment by about 30% – the timescale for each is ~ 700 a. Such features usually die away without growing large enough to threaten the main dipole, but may on occasion accumulate and produce a complete reversal. Gubbins estimated that another two reverse-flux patches would be sufficient to reduce the present dipole moment to zero. Even if Gubbins's model is not correct, it is highly probable that thermal and topographic variations at the CMB constrain magnetic field variations, particularly long-term changes. The question of possible mantle control on reversals is discussed in more detail in Section 8.5.

5.10 Statistical analyses

McFadden and Merrill (1986) pointed out that the fractional amount of time spent in a transition is very small and that geomagnetic reversals are fairly rare events. Thus, their times of occurrence appear to be probabilistically rather than deterministically controlled. Cox (1968, 1969) was the first to show that variations in the length τ of polarity intervals may be described by a probability function derived from the theory of Bernoulli trials. If p is the probability of a polarity change during one cycle of the dipole field, then the probability that a polarity reversal will occur on the xth cycle after $(x - 1)$ cycles of non-reversal is $p(1 - p)^{x-1}$. Thus, the probability function of τ is

$$P(\tau_p < \tau \leqslant \tau_p + \tau_D) = p \, (1-p)^{\tau_p/\tau_D} \tag{5.22}$$

where τ_D is the period of the dipole field, x is an integer and $\tau_p = (x - 1)\tau_D$. The mean value of τ for this distribution is

$$\mu = \tau_D/p \tag{5.23}$$

Cox (1968) used all available data for the past 10.6 Ma to plot the cumulative distribution (Figure 5.26) together with curves of the theoretical cumulative distribution,

$$P(\tau \leq \tau_c) = \sum_{i=1}^{n} p(1 - p)^{i-1} \tag{5.24}$$

where n is an integer and $\tau_c = n\tau_D$. Almost all the data lie between the curves for $p = 0.03$ and 0.06. If the mean observed interval length of 0.22 Ma is taken as the theoretical mean μ and τ_D as 10^4 a, the corresponding value of p is 0.045.

If p is small and x is large in equation (5.22), $P(\tau)$ may be approximated by the Poisson distribution.

$$P(\tau) = \lambda \exp(-\lambda\tau) \tag{5.25}$$

where the parameter λ characterizes the observed variations in the length of polarity intervals. The mean of this distribution is $\mu = 1/\lambda$, so that $\lambda = p/\tau_D$. For time intervals longer than some minimum time τ_{min}, Equation (5.25) will be

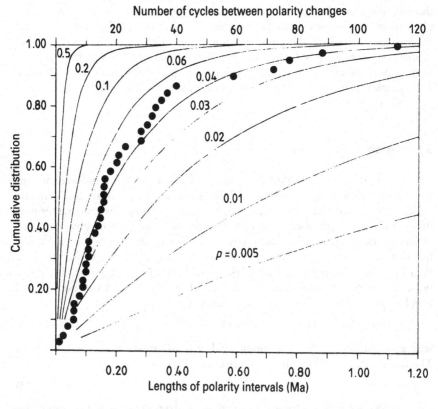

Figure 5.26 Cumulative distribution of polarity intervals. Solid circles are observational data. Curves are theoretical cumulative distributions obtained from Equation (5.24). After Cox (1968).

valid if and only if the probability of a reversal in any time interval $\tau > \tau_{min}$ is independent of whether a reversal occurred during any prior interval. Thus, if reversals are truly Poisson, the core has no memory longer than τ_{min}, which implies that the dynamo processes responsible for reversals have time constants no greater than τ_{min}.

τ_{min} is probably comparable with the longer time constants of the secular variation ($\sim 10^4$ a). Cox (1969) also found that the apparent average duration of polarity intervals was greater during the time $10.6 < t < 45$ Ma than during the past 10.6 Ma, while during the time $45 < t < 75$ Ma the average length was longer still. It appears that the average length of polarity intervals (and, hence, the value of p) has changed during the Earth's evolution, reflecting changes in the physical conditions in the Earth's core.

Similar statistical behaviour of the sequence of reversals is predicted by most models that have been proposed. In general, these models assume that reversals are due to instabilities in the MHD process and that an instability may take place at any time following a previous instability.

The exact physical nature of the instabilities varies with the model. They may result from too much symmetry in the fluid motions of the core (Nagata, 1969) or from some critical configuration of convection cells (Parker, 1969; Levy, 1972b, c). The models presume that the core retains no memory of past fluid configurations beyond the average lifetime of the MHD process. This condition defines a renewal process. Because the average time intervals between successive reversals are much longer than the time constants of the dynamo, further simplification can be made by assuming that the probability of a reversal per unit time is a constant independent of the time of the previous reversal. In this case, reversals describe a Poisson process (Cox, 1968, 1969, 1970; Nagata, 1969), and the time intervals between successive reversals show an exponential distribution (see Equation (5.25)). Although reversal timescales obtained from marine magnetic anomalies do show some properties of renewal processes (Phillips *et al.*, 1975; Phillips and Cox, 1976), they cannot be used in a direct test of the Poisson model. This is because the effects of short polarity events, which may account for nearly half the true number of reversals, are difficult to detect in marine magnetic anomalies. It is likely that a large number of these short events are missing from even the most recently determined timescales.

It is commonly assumed that the space between adjacent like-polarity points is occupied by the same polarity. In other words, no short-polarity zones have been missed in the space between the sampling points concerned. However, such conditions would also obtain if 2, 4, 6 . . . or any even number of reversals had taken place in the interval between the sampling points. On the other hand, when adjacent sampling points have different polarity, it is commonly assumed that one magnetic reversal has taken place exactly halfway in the space between. However, the data could also be explained if 3, 5, 7, . . . reversals or any odd number of reversals had taken place.

Lowrie and Kent (1983) discussed the effect of adding short events to the magnetic polarity timescale – this has a dramatic effect on the polarity interval distribution between 0 and 40 Ma. The number of polarity intervals increases from 126 to 226 and the mean length of a polarity interval is reduced from 0.314 Ma to 0.175 Ma. The exact duration of such short polarity events cannot be determined. If they are 20 ka long, the polarity interval length distribution differs significantly from a Poisson distribution – if they are about 30 ka long, there is no significant difference. The effect of additional short-polarity events also dramatically affects the distribution of normal and reversed polarity interval lengths. This change results from the unequal distribution of the short polarity events which are predominantly of positive polarity and concentrated in the late Cenozoic. The origin of such short-wavelength magnetic anomalies is vital for statistically modelling magnetic polarity reversals. It is not known how many of the 57 possible small events indicated by LaBrecque *et al.* (1977) for example are coherent. They have been interpreted as real polarity changes (Blakely and Cox, 1972; Blakely, 1974) or as magnetic field intensity fluctuations (Cande and LaBrecque, 1974). Schouten and Denham (1979) suggested that they may result from variability in the magnetized layer in areas of high spreading rate. Lowrie and Kent admit that it is not possible to decide whether these short-wavelength features represent real polarity changes, intensity fluctuations or source layer variations.

The observed sequence of reversals for the past 70 Ma satisfies a Poisson distribution reasonably well except that the observed number of short polarity intervals is somewhat smaller than predicted (Cox, 1969; Naidu, 1971). With the discovery of previously undetected short events (Blakely and Cox, 1972), the fit between the observed and predicted distributions has improved but still is not perfect. It is uncertain whether this is because of the failure to detect a significant number of short events in the palaeomagnetic record or because the underlying geomagnetic reversal process is not perfectly Poisson (Harrison, 1969; Cox, 1969; Aldridge and Jacobs, 1974; Tacier *et al.*, 1975)*. Even if reversals turn out to be non-Poisson, they may still be generated by any of several other classes of stochastic processes. For a Poisson process $p(t)$ is constant, whereas for a general renewal process $p(t)$ varies so that the probability of a reversal increases or decreases with increasing time t; however, $p(t)$ depends *only* on the time since the previous reversal and not on the length of prior polarity intervals. Thus, if geomagnetic reversals are generated by a renewal process, the lengths of polarity intervals will be independent. If normal and reversed polarity intervals have different mean lengths, the lengths of polarity intervals will not be independent.

---

* Aldridge and Jacobs (1974) constructed mortality curves for each normal and reversed phase of the Earth's magnetic field over the past 45 Ma which revealed a departure from the simple exponential distribution for the lifetime of these phases. The significance of this departure was demonstrated by comparing simulated distributions of an exponential random variable with actual distributions of the normal and reversed phases. Selective removal of shorter phases from the sequence of exponentially distributed random variables distorted the mortality curves for these generated phases so that they resembled the mortality curves of the observed phases. This demonstration provides evidence that there are probably undetected short phases throughout the past 45 Ma.

Naidu (1971) suggested that the probability density function for reversals is better described by a gamma distribution.*

$$P(\tau) = \lambda(\lambda\tau)^{k-1} \exp{(-\lambda\tau)}/\Gamma(k) \tag{5.26}$$

for which the mean value $\mu = k/\lambda$. This reduces to Equation (5.25) if the lengths of polarity intervals are Poisson distributed ($k = 1$). The parameter k *measures* both the proximity of the observed distribution to an exponential distribution and the dispersion of the intervals about their mean. Naidu's original purpose (1974) was to see whether polarity intervals were independent. He came to the conclusion that for the period 0–48 Ma they were not independent and that harmonic components with a fundamental frequency of 0.75 cycles/Ma were present in the data. He offered no physical explanation for this. He extended his work in a later paper (1975) and claimed that reversal intervals for the period 0–76 Ma were not independent – implying that the geomagnetic dynamo possesses a memory. It should be noted that the lengths of polarity intervals cannot be analysed as a simple stochastic process unless the intervals are independent of each other. Further work by Phillips *et al.* (1975), Ulrych and Clayton (1976) and Laj *et al.* (1979) indicates that there is no statistical reason for not accepting that reversals are independent. Naidu later (1976) agreed with them but still believes that for the period 48–72 Ma reversal intervals are not independent. Phillips and Cox (1976) have also examined the spectrum of reversal timescales, obtaining a generalized expression for the power density spectrum of a random telegraph wave with gamma-distributed polarity intervals. They were unable to reproduce Naidu's expression for the theoretical spectrum, nor could they find any periodicity.

If the Earth's field is not generated by a Poisson process, the probability of a reversal per unit time drops to zero immediately after a reversal and subsequently rises slowly to a steady state value. Figure 5.27 shows three theoretical curves, all with the same mean frequency of reversals (6 Ma^{-1}) but with varying values of k. For a Poisson reversal process ($k = 1$), the instantaneous probability of a reversal per unit time rises immediately to the steady state value of $6 \times 10^{-6} \, a^{-1}$. For $k = 2$, the 50% rise time is about $2 \times 10^5 \, a$ and for $k = 4$, it is about $8 \times 10^5 \, a$. The existence of a non-zero rise time should not be too surprising. The Earth's magnetic field would not be driven to reverse unless the new polarity represented a temporarily lower and thus more stable energy state. Because the core fluid does retain some coherency over periods of at least $10^3 \, a$, we should expect this increased stability to last a while.

Phillips and Cox (1976) carried out maximum entropy spectral estimates for various reversal timescales – all the spectra they obtained resemble the theoretical spectrum of a Poisson reversal process. From studies on synthetic data they found that the behaviour is consistent with a gamma process having a k value between

* McFadden and Merrill (1986) point out some of the problems encountered in a statistical analysis of the reversal record. In particular, they develop certain aspects of the gamma distribution and show how the parameters involved can be used to interpret physical processes acting in the Earth's core.

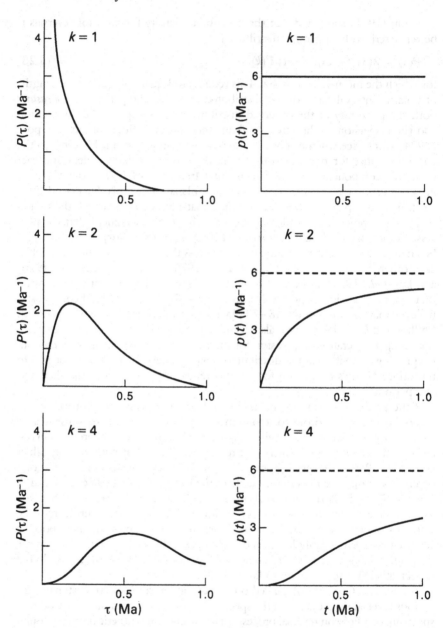

Figure 5.27 Three examples of renewal processes all with the same average frequency of reversals of 6×10^{-6} a $^{-1}$. Curves on the left are probability density functions for the lengths of polarity intervals. Curves on the right show the instantaneous probability that a reversal will occur per unit time. The upper two curves describe a Poisson renewal process ($k=1$). The lower four curves describe renewal processes in which the instantaneous probability that a reversal will occur increases with the passage of time after a reversal. After Cox (1975b).

1 and 2. In a later paper Phillips (1977) found that the parameter k undergoes a slow variation between 1.2 and 2.6. He found no evidence for discontinuities in k, the average value for the entire past 76 Ma being 1.72. If the reversal sequences during the Cenozoic and middle Mesozoic can be modelled with a gamma reversal process having a time-varying μ and a constant k, separate analyses of the reversed and normal polarity intervals should lead to the same conclusions. Phillips found that the average values of k during the Cenozoic for normal intervals was consistently larger (2.84) than those for reversed intervals (1.31). This difference, if true, would imply that normal and reversed polarity intervals have different distribution, and that the geodynamo is more stable after a transition to normal polarity than it is after a transition to reversed polarity. This problem has been addressed in later work by McFadden (1984) and McFadden and Merrill (1984), and will be discussed later in this section.

Although reversals occur very quickly on a geological timescale, they do require a finite time – typically around 5 ka for the directional change. Thus, the occurrence of a reversal must for at least a short time inhibit the future occurrence of another reversal, so the process cannot be truly Poisson. If the reversal rate increases, but the transition time remains constant, the time taken for a polarity transition will become a larger and larger proportion of the average interval between reversals, thereby causing greater inhibition. Thus, as McFadden and Merrill (1993) point out, the inhibition is driven by the increase in reversal rate, producing a positive, not negative, correlation between inhibition and reversal rate. McFadden and Merrill (1984, 1986) had earlier noted such a correlation and in their 1993 paper developed a model to account for it. Their model indicates that, after the $\sim$ 5 ka interval required for a successful reversal of the direction of the magnetic field during which the probability of a further reversal is zero, the probability for further reversals gradually recovers to its initial value over a period of $\sim$ 45 ka. Moreover, to first order the length of the total transition (50 ka) appears to be constant throughout the last 160 Ma.

In their 1984 paper McFadden and Merrill interpreted their observed positive correlation between estimated inhibition and reversal rates over the last 80 Ma to be the consequence of short intervals being missed from the reversal chronology. However, updated chronologies have not added short intervals into the past few million years and Cande and Kent (1992) have shown that 'tiny wiggles' in the marine magnetic records can be satisfactorily accounted for by variations in the intensity of the palaeofield. McFadden and Merrill (1993) therefore abandoned their earlier interpretation and considered two further models – boundary layer instability models and models involving internal instabilities. Both models appear possible, although the first class of models would seem to demand a chemical layer at the top or bottom of the OC. McFadden and Merrill prefer the second class of models, which involve internal instabilities. A key issue is that the total transition time appears relatively constant while other features of the geomagnetic field, such as the rate of reversals, have changed significantly. McFadden and Merrill suggest that this can be explained if the characteristic ohmic diffusion timescale (that associated with decay of the dipole field) can be correlated with the total transition interval.

Most mathematical descriptions of convection-driven dynamos are not asymmetric with respect to polarity. These models assume a dynamo that is spherical and axisymmetric. Kinematic solutions to the dynamo equations commonly assume symmetry about the equatorial plane as well. Slight departures from this geometry might conceivably affect the symmetry of normal and reversed states, but it is difficult to see how such departures could be maintained over the 150 Ma period of observations. All evidence suggests that reversals can be described by an alternating renewal process having a time-varying mean and gamma-distributed polarity intervals. As already pointed out, one of the difficulties in a statistical study of reversals is the extreme sensitivity of a histogram of polarity changes to the number of short polarity events. This sensitivity is due to the fact that the Earth's magnetic field has only two polarity states; thus, a short polarity event inserted in the middle of a long period of opposite polarity not only adds the short polarity event to the histogram but also erases a long period and adds two middle-length periods. Laj *et al.* (1979) analysed the series of geomagnetic reversals, using a correlation function of the telegraph signal obtained from the Heirtzler timescale by assigning ± 1 to intervals of normal/reversed polarity (Heirtzler *et al.*, 1968). Unlike the function used by Naidu (1971, 1975), their autocorrelation function is only very slightly sensitive to undetected short polarity intervals. Laj *et al.* also showed that other models, which do not need any triggering mechanism as in Cox's model, are compatible with the observed results. This is important, since, as already noted, it is very difficult to trigger a long period oscillator using short-period stochastic fluctuations of reasonable amplitude. Laj *et al.* concluded that, within statistical noise limits, successive polarity intervals are independent and distributed in time according to a Poisson process.

McFadden and McElhinny (1982) analysed all VDMs determined for the past 5 Ma which have associated VGPs with latitudes greater than 45° to see whether there were any differences between the statistical parameters of the normal and reversed polarity distributions. They showed that the data give no reason to reject the hypothesis of a common mean and variance for the (untruncated) Gaussian distribution of the normal and reversed polarity TDM, although a common truncation point can be rejected – the truncation point for normal polarity TDM being larger (see Figure 5.28). This indicates that there is a difference between the properties of the two polarity states. It could be interpreted as meaning that a stable polarity can be sustained for lower dipole moments in the reversed polarity state than in the normal polarity state, implying that the reversed state is inherently more stable than the normal state. Alternatively it could be interpreted as meaning that in the normal polarity state a stable polarity can be sustained for larger deviations of the VGP from the spin axis than in the reversed polarity state, implying that the normal state is inherently more stable than the reversed state.

McFadden (1984) has given a comprehensive and valuable discussion of the statistical analysis of geomagnetic reversal sequences. He gives a method which overcomes the problems of non-stationarity and the consequent dependence of the lengths of polarity intervals, and which also avoids the problems associated with the use of sliding windows. In particular, he showed that incomplete data from a Poisson process lead to a distribution of interval lengths that is essentially

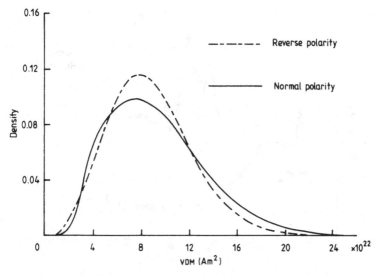

Figure 5.28 Comparison of nested distribution fits to normal and reversed polarity VDM data. After McFadden and McElhinny (1982).

indistinguishable from a gamma distribution. Using the methods developed in this paper, McFadden and Merrill (1984) analysed the statistical properties of marine magnetic anomaly timescales covering the period 165 Ma to the present. They showed that the data no longer support the hypothesis that during the last 86 Ma there were two essentially stationary sequences separated by an extremely rapid change in the mean length of polarity intervals at about 45 Ma (see also Section 3.4). Instead it seems that there has been a continuous and essentially smooth non-stationarity in the reversal process over the last 86 Ma.

Figure 5.29 shows the estimates of the parameter k for the LaBrecque *et al.* (1977) timescale using the method given by McFadden (1984). A visual inspection would seem to indicate that there is a difference in the stability of the normal and reverse polarity states. However, as McFadden pointed out, the parameter k contains very little information about the relative stabilities of the two states – k is effectively a ratio and the natural parameter to estimate is ln k, not k. Plotting estimates of k instead of ln k greatly exaggerates the apparent variability. This is shown by comparing Figures 5.29 and 5.30. The only difference between them is the addition of one normal polarity interval (from 34.00 to 34.01 Ma) in a total of 198 intervals!

McFadden and Merrill (1984) concluded that there is no difference in k for the two polarity states and, hence, no difference in their stabilities. They point out that it is the extreme sensitivity of the estimation procedure for k which led Phillips (1977) and Lowrie and Kent (1983) to conclude erroneously that reverse and normal polarity states have significant different stabilities. McFadden (1984) showed that if reverse and normal polarity rates of reversal are the same, the true

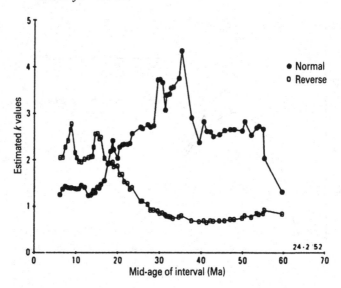

Figure 5.29 Estimates of *k* for both polarities for the LKC timescale. Sliding window covers 25 intervals of each polarity and shifts by one interval of each polarity each time. After McFadden *et al.* (1987).

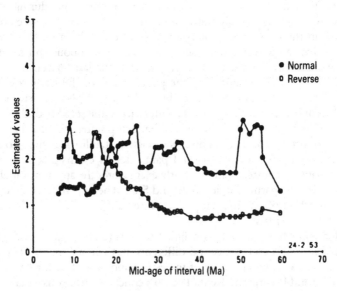

Figure 5.30 Estimates of *k* for both polarities for the LKC timescale with one short interval added at 34.00 to 34.01 Ma. Sliding window as for Figure 5.29. After McFadden *et al.* (1987).

values of k must also be the same. Any apparent differences in k are the result of using a sliding window and the statistical variation in its estimate.

Using the timescale of Harland *et al.* (1983) for the period 119–165 Ma, McFadden and Merrill concluded that the normal and reversed polarity sequences for this period also have the same statistical properties with no difference in their relative stabilities. They therefore suggest that the rate of reversals decreased in an approximately linear manner from 165 Ma until it reached zero at 106 Ma ($\lambda = 0$, i.e. $\mu = \infty$), implying that the process ceased. At 86.5 Ma the process of reversals began again, the rate of reversals increasing linearly with time. (There is a slight indication that the rate of reversals reached a maximum about 10 Ma ago and is again decreasing.) In their view the process causing instabilities leading to reversals on occasion gradually slows down until no further instabilities are produced, the field remaining in the polarity it happened to be in at that time. Thus, it is merely a matter of chance whether a long quiet interval happens to have normal polarity (such as the Cretaceous normal superchron) or reversal polarity (such as the Permo-Carboniferous reverse superchron).

McFadden *et al.* (1987) also investigated the effect that missed short polarity events would have on the properties of the two polarity states. Blakely and Cox (1972) and Blakely (1974) used a stacking procedure to enhance marine magnetic profiles and suggested that many short-wavelength anomalies previously unrecognized may be reversal events. It is not possible to identify all such 'tiny wiggles' as true reversals – they may also represent excursions, changes in the intensity of the Earth's magnetic field or changes in rock magnetic properties. McFadden *et al.* (1987) took the extreme case and assumed that all 57 of the tiny wiggles that have been seen in the Cenozoic do in fact represent short reversal events. They then investigated the effect of including these events in the reversal chronology record of LaBrecque *et al.* (1977), using the method of analysis developed by McFadden (1984), which takes into account both missed reversal events and unsuccessful reversals. Because of the disproportionately large number (46) of the 57 additional events identified as being of normal polarity, their analysis provides a severe test of the robustness of the method of analysis.

The duration of the additional short events cannot be determined from marine records, and equal durations of 20 ka were chosen arbitrarily for all the events. McFadden *et al.* showed that the data, either with or without the addition of the 57 short events, are compatible with the normal and reverse sequences having the same value of k and a common mean length or reversal rate. However, inclusion of these short events does change the structure of the non-stationarity in reversal rate, being no longer linear with time. This is mainly due to the fact that, of the 57 added events, only 7 occur prior to 38 Ma. The reversal rate still appears to change smoothly, approaching zero at the end of the Cretaceous normal superchron. Their conclusions were not changed if the durations of the short events were 30 or 40 ka instead of 20 ka. Marzocchi and Mulargia (1990) later analysed 11 different magnetic timescales and also showed that the data do not support any asymmetry between normal and reversed distributions.

Constable (1990) has given a plausible statistical model for reversals, based on characteristics of the secular variation inferred from the present field. In an earlier paper Constable and Parker (1988) had used such a model to describe the statistical distributions observed in field directions during times of 'normal' secular variation. Constable (1990) extended these ideas to include reversals by introducing an autoregressive time dependence for the statistical variation. Reversals are simulated by simply allowing the axial part of the dipole to decay and grow back with opposite sign while letting the rest of the secular variation continue as usual. Simulation of reversals at different sites and allowing the field to start from different initial configurations but with the same statistical properties showed that the model is capable of exhibiting many of the observed features of the reversal process (see Figure 5.31). Constable believes that a correlation time of 2000 a is a reasonable estimate of the timescales on which secular variation features persist and therefore the left-hand panel is likely to be the most representative. The total field intensity decays to about 10–20% of its initial value and large directional changes occur once the axial dipole has decreased to about one-half of its original intensity. Looping and loitering occur in the VGP paths, often near the pole positions, followed by a fairly rapid transition from one hemisphere to the other.

Gaffin (1989) has analysed the reversal record for the past 165 Ma for evidence of chaotic dynamics by searching for fractal scaling exponents in the statistical distribution of polarity intervals. He counted the number of polarity intervals $N(T)$ longer than a given cut-off size T and plotted $\ln N(T)$ against $\ln T$. The resulting graph suggests a straight line pattern (with an exponent ~ -1.5) over a fairly large cut-off range (0.5–10 Ma). A stationary Poisson process does not predict this behaviour – it would predict that $\ln N(T)$ is linear with T, not $\ln T$. Gaffin showed that a non-stationary Poisson model is an excellent fit to the data, and is a result of the trend towards longer intervals near the Mid-Cretaceous (see Figure 5.32). Thus, no special significance should be attached to the 3/2 scaling exponent. Gaffin suggested that the long-term trend or non-stationarities in reversal rate may make it difficult to demonstrate chaos in the reversal record. Lutz and Watson (1988) had earlier shown that such a long-term trend can seriously hamper the detection of short-term periodicities in the reversal rate (see Section 3.4).

Seki and Ito (1993) also examined the palaeomagnetic data for the past 165 Ma and claimed that the distribution of reversal intervals is better satisfied by a power law than by a gamma or exponential distribution. They interpret their finding as indicating that the geodynamo is in a critical state of marginal stability. To explain the power law distribution, they developed a model of the geodynamo in which turbulent eddies in the OC are assumed to behave as magnetic spins.

Finally, it must not be forgotten that the observed polarity sequence is not necessarily the same as the polarity sequence of the core itself. This is because observations are not perfect and some reversals actually recorded in the Earth's rocks will have been missed. Again, the process by which rocks acquire their magnetization acts as a filter to suppress the recording of short polarity intervals. Unfortunately, the rock magnetic filtering process is not amenable to direct deconvolution and in the past there have been several attempts at indirect methods for determining the rate at which actual core reversals have occurred.

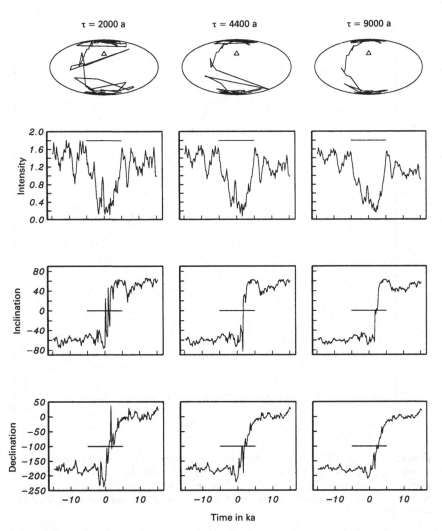

Figure 5.31 Simulated transition records for a reversal at 40° N and 180° E for a correlation time $\tau = 2000$ a (left panel), $\tau = 4400$ a (middle panel) and $\tau = 9000$ a (right panel). The horizontal bars indicate the duration of the reversal. Triangles represent the observations site. After Constable (1990).

There has been some speculation that some of the instabilities in the dynamo process are what Cox (1981) has called 'infertile', i.e. they do not lead to reversals. There is some evidence for the existence of infertile instabilities during normal polarity (but not during reverse polarity) over the past 80 Ma; but during the period of 120 Ma back to 170 Ma most of the instabilities were fertile.

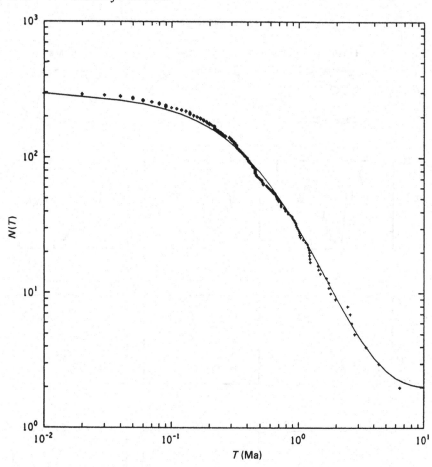

Figure 5.32 Empirical log-log plot of the number of polarity intervals $N(T)$ greater than a threshold length T for the Harland *et al.* reversal record (1983) leaving out the Cretaceous superchron. The smooth line is a plot of the theoretical $N(T)$, predicted by a non-stationary model for interval distribution using the smooth trend model of Lutz and Watson (1988). After Gaffin (1989).

6
Transition fields

6.1 Introduction

It is not easy to divide the subject matter of reversals of the Earth's magnetic field into separate chapters. Models for reversals, the subject of Chapter 5, must of necessity consider transition fields and they have been taken into account when evaluating the success of some models. This chapter deals specifically with transition fields, but on occasion also considers the relevance of certain models.

There have been many attempts to account for the structure of the geomagnetic field during reversals. The first records appeared to support models in which the transitional field was dominated by low-order zonal terms. Subsequent results, however, indicated that in most cases the transitional field cannot be purely axisymmetric. This question and other problems are discussed in this chapter. Our understanding of the morphology of the geomagnetic field during a polarity reversal is limited because it has been difficult to obtain well-documented transition records of the same reversal from widely distributed site locations – particularly from the southern hemisphere. The paucity of data from the southern hemisphere makes it difficult to constrain even the most fundamental symmetries of the transition field with respect to the equator.

The palaeomagnetic records seem to favour a simple model in which the reversing field consists of a reversing axial dipole g_1^0 superimposed on random or correlated noise, depending on the duration and mode of the recording process, which are different for lavas and for sediments with different deposition rates. The noise is essentially compatible with the secular variation of the non-dipole field observed over recent historical time. This is in agreement with the statistical model of Constable (1990) – see Section 5.9. Courtillot et al. (1992) therefore suggest that the Earth's field is the sum of two apparently independent processes, one a time-varying but dominant geocentric axial dipole with random fluctuations and reversals and a correlation length of a few thousand years, and the other, which has shorter timescales, a random non-axial dipole field with a white or pink spectrum at the CMB.

Van Hoof and Langereis (1991) have pointed out some of the dangers in interpreting palaeomagnetic records from relatively slowly deposited sediments. As an example they showed that the palaeomagnetic components in two sedimentary reversal records in marine marls have a delayed remanence acquisition and probably do not reflect the true geomagnetic field during the transition. The accuracy of some palaeomagnetic techniques to clearly isolate the primary remanent direction (if it still exists and has not been replaced by subsequent geochemical processes: Karlin *et al.*, 1987) is still one of the major problems.

The processes by which sediments acquire their magnetization are not well understood, and depend on a number of factors, including magnetic mineralogy, duration of the transitional field state and sediment deposition rate. Sediments are an imperfect recording medium for reversals and often smooth a transition on a timescale generally long with respect to the transition. Langereis *et al.* (1992) examined marls from the Mediterranean which recorded several reversals over the past 7 Ma. They measured pairs of (normal, reversed) VGPs immediately before and after the reversal which they call near-transitional. They used these poles (which are not antipodal) to simulate the VGP reversal paths which would be recorded when one VGP pole is gradually replaced over time by the VGP pole of the opposite polarity. They thus attempted to mimic the acquisition of remanence over a time interval that is long compared with the duration of the reversal. Langereis *et al.* also measured VGPs and simulated transitions from stable field states well away from the reversal, which they call a non-transitional model. The Mediterranean records show that there is good agreement with the non-transitional model, and even better correlation with the near-transitional model. In one record the observed VGPs initially follow the non-transitional path over North America, but then 'jump' to the near-transitional path over Australia. This switching behaviour between antipodal VGP paths is often seen in sedimentary records. Langereis *et al.* concluded that the longitudinal confinement of VGPs (see Section 6.5) can arise from the smoothing of the non-antipodal stable directions just before and after a geomagnetic reversal, because of the filtering effect of the remanence acquisition process in the sediments.

Weeks *et al.* (1992) question the suggestion of Langereis *et al.* (1992) that sedimentary VGP paths are an artefact due to smoothing which mixes non-antipodal pre-transitional and post-transitional directions. They point out that it is the timescale of the smoothing that is critical. Rochette (1990) had suggested that the acquisition time of the magnetization of intrusions and some sediments recording transitional fields may be long compared with the time taken for the field to reverse, in which case the directions recorded are simply combinations of differing proportions of normal and reversed components of magnetization, acquired before and after the reversal. To explain the records of sediments carrying DRM, or PDRM by Rochette's mechanism, the magnetization must be acquired over a sufficiently long time before and after the reversal for virtually all traces of the intermediate field to be lost ($\sim$ several tens of thousands of years). Weeks *et al.* point out that marine cores from the Mediterranean have been shown to record geomagnetic intensity variations on a timescale of less than 5 ka. In their model of a reversal, the decay of the dipole field causes the field intensity to decrease initially with little or no change in direction. The VGP then moves relatively rapidly to the opposite hemisphere, the intensity of the dipole field being close to its minimum value, followed by a more gradual movement to the new reversed-polarity location. The sequence of events is thus a very slow departure from stable polarity, a comparatively rapid equatorial crossing at weak fields and, finally, a slow recovery to the new polarity. Their general conclusion is that simple smoothing of pre-transitional and post-transitional directions, as suggested by Langereis *et al.* (1992) cannot explain the observed VGP paths during reversals, which represent the result of field behaviour. Smoothing over unrealistically long timescales would be required to generate intermediate directions of magnetization.

Hillhouse and Cox (1976) provided the first evidence that the geomagnetic field during a transition is not dipolar. They showed that the VGP paths for the Matuyama–Brunhes transition as recorded in sediments in Lake Tecopa, California, and in sediments from Boso Peninsula in Japan are completely different. They also showed that the dominant direction of the non-dipolar component was parallel to the present non-dipolar field. They interpreted those results in terms of a standing field model in which a substantial part of the non-dipole field during the main part of the transition is standing, while the axial dipole field decays to zero and then grows back in the opposite polarity. Using improved instrumentation and techniques, Valet *et al.* (1988b) resampled the Lake Tecopa section to increase the time resolution of the record. The most important finding of this new study is that AF demagnetization did not completely remove a very strong overprint. The transitional directions obtained from thermal demagnetization are thus quite different from those of Hillhouse and Cox. Valet *et al.* showed that the 'intermediate' directions seen by Hillhouse and Cox were caused by the superposition of different amounts of overprint. Figure 6.1 shows the complete record of the declination and inclination obtained with stepwise thermal demagnetization by Valet *et al.* Their VGP path is confined within a narrow band of longitude centred on 220° E which is significantly different (by 120°) from the path obtained by Hillhouse and Cox. However, it does not overlap with the path obtained by Niitsuma (1971) for the sediments of Boso Peninsula. Valet *et al.* therefore maintain that the conclusion of Hillhouse and Cox that the transitional field is not dipolar is not affected by their new results. Their VGP path is close to the site longitude, indicating that the transitional field was strongly axisymmetric.

Valet *et al.* (1983) found two R→N reversals at different stratigraphic levels of a single section of fine-grained Tortonian marine clays near the village of Skouloudhiana in western Crete. The VGP paths corresponding to these two transitions, which are separated in time by about 1 Ma, are almost identical. The two VGP paths are largely constrained in longitude along a mean great circle about 80° west of the site, and therefore cannot be classified as near- or far-sided. The same VGP path has also been observed for a R→N transition recorded in a nearby section at Potamida (Valet and Laj, 1981). However, the following N→R transition recorded at Potamida has a quite different VGP path, about 135° away from the R→N transition. This is not in accord with the standing field model which the coincidence of the Skouloudhiana paths would favour.

Valet and Laj (1984) later carried out a more extensive study of four reversals (three of which are successive) recorded in two upper Tortonian marine clays

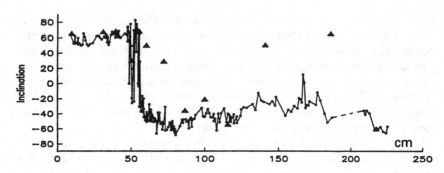

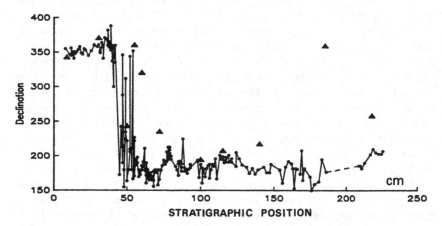

Figure 6.1 Complete record of declination and inclination obtained with stepwise thermal demagnetization of more than 200 specimens. The triangles represent what would have been obtained from AF stepwise demagnetization. The stratigraphic positions are measured from the base of the Bishop Tuff. After Valet *et al.* (1988b).

near Skouloudhiana and Potamida. They found the north VGP paths associated with the two R→N transitions and the south VGP path for the N→R transition in Skouloudhiana to be identical. On the other hand, the south VGP path for the upper N→R transition at Potamida is significantly different, being about 60° in longitude away from the other paths. These results indicate that the reversal process starting from either polarity remained unchanged for ~ 1.3 Ma, the time corresponding to the first three reversals. A change then occurred between the two upper reversals. This change must have been rather abrupt, since the time between them is less than 250 ka.

Valet *et al.* (1988a) later carried out a further, more detailed, study of these four reversals from three sections and confirmed the above main result, viz. that the reversal process starting from either polarity can remain unchanged for ~ 1.3 Ma. It should be noted, however, that the long persistence of the reversal process observed in Crete was not seen by Herrero-Bervera and Theyer (1986) in the Olduvai and Jaramillo reversals, where different transitional characteristics were observed for each transition. Their work is discussed in Section 6.3.

The data from Crete presented by Valet *et al.* are clearly inconsistent with the standing field model but can be described in generalized flooding field models (Hoffman, 1981). On the other hand, the rather inconclusive data of Clement and Kent (1985) for the upper Olduvai and lower Jaramillo transitions obtained from southern hemisphere deep-sea sediments is consistent with a standing field which persisted across the two reversals. In the records of Valet *et al.* the VGP paths are ~ 90° W or E away from the site longitude (depending on the sense of the transition), indicating that non-axisymmetric terms are present – they suggest that sectorial or tesseral harmonics are required to account for the significant deviations from the north–south axis.

Valet *et al.* (1986) carried out another detailed study of a reversal recorded in a sequence of Tortonian marine clays (~ 5.87 Ma) near the village of Skouloudhi-ana: 295 cores were obtained over a stratigraphic height of only 4.7 m. The high time resolution of the record, which is less than a few hundred years, allowed them to study in detail the characteristics of the transitional field. Two types of geomagnetic field variations were identified – very fast fluctuations and long-term variations which tend to be masked by the high-frequency fluctuation (Figure 6.2). Figure 6.2(a) shows a significant increase of the amplitude of the fluctuations during the transition. Such fluctuations have usually been assumed to be the result of random noise. Valet *et al.* (1986), however, suggest that these rapid fluctuations are real and are directly related to the reversal process and could be the result of an increase in turbulence in the Earth's core during the transition. A similar increase in turbulence has been suggested by Prévot *et al.* (1985a,b) to account for the directional jump seen in the volcanic record of the Steens Mountain reversal (see also Section 6.7).

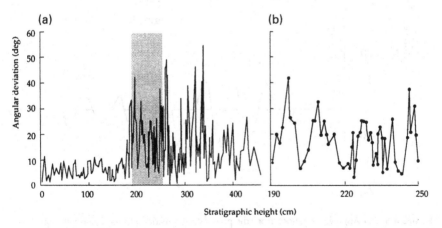

Figure 6.2 (a) Angular deviations between successive magnetization vectors as a function of stratigraphic height. The shaded part of the plot is enlarged in (b) to show that some of the fluctuations include several successive points. After Valet *et al.* (1986).

Fluctuations also occur on a considerably longer timescale, but are masked by the high-frequency fluctuations, so that a smoothing of the data is necessary to reveal them (see Figure 6.3A). Three well-defined oscillations occurred before the transition with the same wavelength, suggesting a periodicity between 2 and 4.5 ka (Figure 6.3A, curve a). This is in the same range as the periodicities reported for the non-dipole field variations obtained from high-deposition-rate lake records during the past 15 ka (Creer and Tucholka, 1982). Variations with similar time constants persist during the transition (Figure 6.3B), suggesting that some characteristics of the non-dipole field are not affected by the reversal of the main field. The observed continuity of the secular variation during the reversal lends experimental evidence in support of Le Mouel's (1984) theoretical contention that the non-dipole field cannot disappear during a reversal.

Clement and Kent (1991) also observed directional fluctuations in a record of

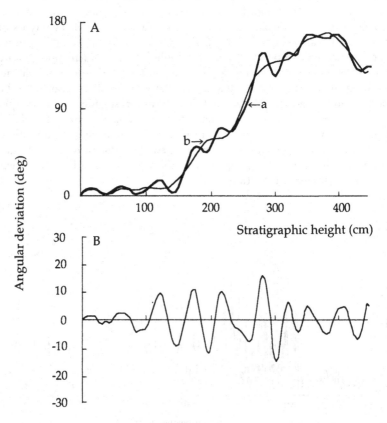

Figure 6.3 (A) Angular deviations from the pre-reversal direction plotted against stratigraphic height: (a) after smoothing by cubic splines with 27 knots; (b) after smoothing with 15 knots. (B) Periodic oscillations obtained by subtracting curve (b) from curve (a) in A. After Valet *et al.* (1986).

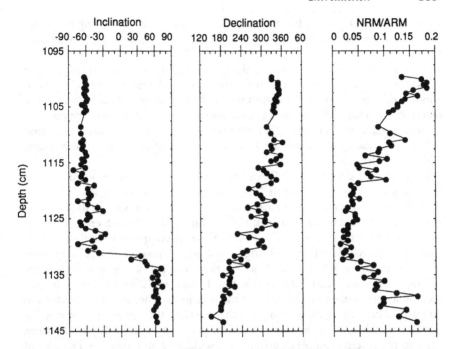

Figure 6.4 Unit vector mean directions and normalized intensity from each stratigraphic level plotted versus sub-bottom depth. Inclinations and declinations are plotted in degrees. NRM intensity after partial AF demagnetization at 20 mT normalized by ARM intensity after treatment at 20 mT represents the best estimate of relative geomagnetic intensity variations across this interval. After Clement and Kent (1991).

the Matuyama–Brunhes reversal from a southern hemisphere deep-sea sediment core. The inclinations begin to oscillate just before full normal polarity is reached – three swings occur before normal polarity is stabilized. The declinations also show oscillations – five occur within the transition zone, the amplitudes diminishing as full normal polarity values are approached (Figure 6.4). Keating and Ishimatsu (1992) also observed directional oscillations in a record of the Matuyama–Brunhes reversal in an oriented hydraulic piston core from the Australian–Antarctic basin. The transitional directions repeatedly oscillate between the stable reversed polarity direction of the Matuyama and the Brunhes normal direction. The directions appear to backtrack along the transitional path four times. Rapid oscillations in the declination before the main reversal have also been seen in the Olduvai subchron (see Section 6.3).

Valet *et al.* (1988b) resampled sediments in Lake Tecopa, California, that recorded the Matuyama–Brunhes transition. Despite the sharpness of the transition zone large fluctuations occur on a very short timescale – several are defined by successive directions and are not related to random noise (see Figure 6.1). Valet *et al.* admit that it is often difficult to determine to what extent these fluctuations truly represent large and sudden jumps of the geomagnetic field –

some could possibly result from incomplete cleaning of the very strong overprint. However, they believe that they do record real variations of the geomagnetic field.

A common feature of records of the magnetic field recorded in lavas is that similar directions can be seen in successive flows followed by large changes to another direction. This jerky behaviour of the transition field is in most cases probably the result of irregularities in the extrusion rate of the flows. The possibility that rapid changes in the field itself may be the cause is discussed in Section 6.7 for the case of the transition field recorded in Miocene lavas from Steens Mountain, Oregon.

Roperch and Chauvin (1987) studied the transition field recorded in lavas from volcanic islands in French Polynesia in the south-central Pacific Ocean. Such work is important, since most studies have been from sites in the northern hemisphere on sedimentary records. On Huahine Island they observed the transitional field for two sections of successive flows. A complete reversal (N→R) is seen in both records, the paths being very similar. However, in one section the field returns to intermediate directions – there is a gap in the sampling of the second section. In a later paper Roperch and Duncan (1990) combined the two records to construct a single path for the transition. They believe that this can be justified because of the proximity of the two sites, their essentially same K–Ar age (2.9–3.1 Ma) and similar features in the geomagnetic record. The directions close to the reversed dipole direction are associated with a very low intensity of magnetization, and they suggest that the fully reversed field is not observed and that the data are more consistent with an excursion (N→N). The data from Huahine further suggest that there is a significant difference in the geometry of the transitional field between the beginning of the reversal, which is controlled by axisymmetric terms, and the following more complex phase, characterized by intermediate fields with low intensity and large directional changes. Roperch and Duncan (1990) show that a decaying axial dipole and a growing axial quadrupole (having the same sign) or octupole (of opposite sign) can fit the data for the beginning of the transition – they prefer the quadrupole model.

Roperch and Chauvin (1987) and Chauvin et al. (1990) also sampled a volcanic sequence almost 700 m thick in the Punaruu Valley on the island of Tahiti. Four transition zones were recorded – the Cobb Mountain, the lower and upper Jaramillo, and the Matuyama–Brunhes. The data confirm their findings from Huahine that the beginning of a transition is consistent with a zonal field, but significant non-dipolar fields without any preferred axisymmetry seem to dominate the latter part of a transition. Hoffman (1982) had earlier suggested that the onset of a reversal should show fewer departures from axial symmetry than the following phases, and Hide (1982) has given some support for this conjecture.

Chauvin et al. (1990) obtained very low values for the palaeointensities of the transitional field – 3 to 8 µT. Large and sudden changes in direction are much easier to produce with such low intensities than with a normal field intensity. This could account for the large loops seen in their most detailed record (the upper Jaramillo). They speculated that it could be difficult for a sediment to adequately record a transition when the magnetic field strength needed to align grains becomes too weak. Thus, some sediments may not be able to record the middle part of a transition, but only the beginning and end when the field decays or grows towards the other polarity.

Valet and Meynadier (1993) analysed the records of the relative intensity of the geomagnetic field for the last 4 Ma obtained from marine sediments of the equatorial Pacific. The intensity pattern with reversals of the field is an approximate asymmetric saw-tooth curve, with a gradual decrease before the transition and a very rapid recovery following the direction changes. They regard the long-term decrease preceding a reversal as a continuous process initiated after the intensity maximum following the previous transition. Valet and Meynadier suggest that the rapid field regeneration after a reversal is due to field advection and that the following stable polarity state characterized by a long relaxation process is dominated by diffusion.

One exception to this behaviour is the period following the upper Olduvai, when there was no large field recovery. In this case, however, several successive short events (Gilsa, Ontong Java 1 and 2, and Cobb Mountain) were followed by a significant recovery of the field. Rapid field excursions associated with intensity lows do not follow the above pattern of major field reversals.

Bogue and Coe (1984) obtained palaeointensities from two transition zones in Kauai, Hawaii, which they interpreted as a back-to-back (R→N→R) reversal pair. The field directions were very similar, which gives support to the standing field model which predicts identical sequences of transitional field directions for repeated reversals from a single site. However, the field intensity in the R→N transition is asymmetric, whereas a standing field model predicts symmetric intensity patterns. Bogue and Coe attempt to explain their data by the zonal flooding model of Hoffman (1977, 1979). However, neither an octupolar nor a quadrupolar model fits the data, which show some characteristics of both models. They suggest that the observations could be explained if very different flooding schemes operate during a single reversal. The initial decrease and partial recovery of the intensity may represent an aborted attempt by the field to reverse by flooding from the equatorial zone. Following a return towards the initial reversed state, flooding may have begun again in a different region near the south pole of the core and continued until the reversal was complete. An alternative suggestion is that the flooding mechanism is itself asymmetric. Bogue and Coe give as one possibility flooding which proceeds both northward and southward, but at different rates, from an initiation point near the core's equator. The agreement of such a model with the observations is quite good.

6.2 The Matuyama–Brunhes transition

The Matuyama–Brunhes reversal has been observed in a number of records from widely spaced sites, although few from the southern hemisphere. The first attempts to analyse the harmonic content of the field during this reversal were mostly consistent with zonal axisymmetric components (see, e.g., Williams and

Fuller, 1981). Later work has shown that the transitional field cannot be purely axisymmetric and non-zonal terms have been incorporated into models.

Valet *et al.* (1989) obtained four records of the Matuyama–Brunhes reversal from ODP deep-sea cores in the Atlantic Ocean covering a latitude range from the equator to 50° N along a mean longitude of 20° W. One of the aims of their study was to test the hypothesis of axisymmetry. If quadrupolar terms dominate the transition, an equatorial site would show the largest variation in inclination (from 0° to either +90° or −90°). On the other hand, octupolar terms would show no change in inclination at the equator. In the model of Williams *et al.* (1988), which combines low-order zonal harmonics with a drifting non-dipole field of intensity and drift similar to those of the present non-dipole field, the directional changes are controlled by the phase relationship of the variation of strength of the zonal harmonics and the drifting non-dipole field. However, high inclinations should still be observed during the transition at an equatorial site, related to either the zonal terms or features of the non-dipole field over the site. Valet *et al.* (1989) found no marked changes in inclination, which remained close to zero in the transition record at their equatorial site. This would imply that low-order zonal harmonics are not dominant during the reversal. An octupolar family would require the absence of any other component (e.g. the declination should jump from 180° to 0°) and this is not observed.

Valet *et al.* (1989) also estimated the duration of the Matuyama–Brunhes transition at different latitudes. This depends to a certain extent on the criteria used to define the length of the transition, which could be crucial for records with low sedimentation rates. They did not find any dependence on site latitude – the mean duration was quite short (~ 2.3 ka). For a field dominated by octupolar terms, the duration of the reversal at equatorial latitudes should be significantly shorter than at mid-northern latitudes. This reinforces the hypothesis that zonal terms did not dominate during the transition. Moreover, the VGP paths for Hole 609B by Clement and Kent (1987) and for Hole 664D (Valet *et al.*, 1989) are very different, but are constrained along almost antipodal longitudes 90° W and 90° E from the sites meridian (see Figure 6.5). Since the two sites are located on the same longitude, similar VGP paths would be predicted if the transitional field were dominated by zonal terms.

Although the Matuyama–Brunhes transition has been observed at a number of widely separated site locations, there are few records from the southern hemisphere. Clement and Kent (1991) improved this situation with a detailed study of the transition recorded in a deep-sea sediment core V16–58 from the southwestern Indian Ocean. The VGP path is longitudinally constrained, centred over the 300° meridian tracking northward through the Americas. It is neither clearly near- nor far-sided, falling on average 120° W of the site location, suggesting a non-axisymmetric geometry. It is instructive to compare this path with those obtained from two other deep-sea sediment cores – that from DSDP site 609 in the North Atlantic and that from ODP site 664 in the equatorial Atlantic, which are essentially on the same longitude. If the Matuyama–Brunhes transitional field was dominated throughout by dipolar fields, the VGP paths from different locations should coincide. Figure 6.6 shows that the VGP path from the southern hemisphere mid-latitude site (V16–58) is very similar to that of the northern mid-latitude site (DSDP 609), both paths tracking northward through the Americas. However, these paths are very different from that obtained from the equatorial site (ODP-664), which is nearly antipodal, tracking northward through Asia. This difference between the VGP paths obtained from mid-latitude and equatorial sites reinforces earlier interpretations that dipolar fields were not dominant during the reversal (Valet *et al.*, 1988b, 1989). These records suggest that the transitional field was dominated by non-dipolar, non-zonal terms which are symmetric about the equator.

Clement and Kent (1991) suggested that the lowest-degree non-zonal, non-

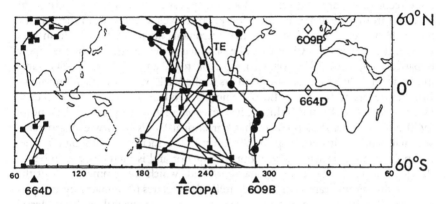

Figure 6.5 VGP paths relative to the records of the Matuyama–Brunhes reversal from ODP Hole 664D, Lake Tecopa and DSDP Hole 609B. The location of the sites is shown in the figure. After Valet *et al.* (1989).

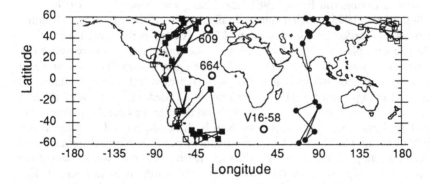

Figure 6.6 Matuyama–Brunhes VGP paths obtained from DSDP Site 609 (open square symbols), ODP Site 664 (solid circles) and piston-core V16–58 (solid squares). The site locations are indicated by the open circles. These records display a distinct symmetry with respect to the equator. The paths from mid-latitudes from both the northern and southern hemispheres track northward through the Americas, while the VGP path from the equatorial site is nearly antipodal to these, tracking northward through Asia. After Clement (1991).

dipolar term h_2^1 may account for several features of the transitional field seen in these three records. If an h_2^1 term is introduced in conjunction with a reversing axial dipole, using the method of Williams and Fuller (1981), the resulting synthetic VGP paths tend to cluster about the 90° or 270° E meridian, depending upon the site hemisphere and the sign of the h_2^1 term. However, a time-invariant h_2^1 term cannot account for the symmetry about the equator suggested by the above three records, since it produces antipodal synthetic VGP paths from sites in opposite hemispheres, unlike the observed VGP paths for DSDP 609 and V16–58, which are nearly coincident.

In a later paper Clement (1991) showed that an h_3^1 term provides the most simple geometry to explain the non-zonal symmetry about the equator seen in these records. A pure h_3^1 geometry does not appear to fit transitional VGP paths obtained from sites in the Pacific hemisphere as well as those from the Atlantic sector. However, Clement showed that reasonable fits to VGP paths from sites in both the Pacific and Atlantic can be obtained if the energy lost by the dipole is partitioned, with 10% going to g_2^0, 20% to g_3^0 and 70% to h_3^1. The same model also produces reasonably good fits to the VGP paths (for the Matuyama–Brunhes reversal) from the Boso Peninsula (Niitsuma, 1971), Lake Tecopa (Valet et al., 1988b) and Tahiti (Chauvin et al., 1990; Roperch and Duncan, 1990). This model thus predicts that transitional VGPs from widely separated sites will tend to cluster along the 90° E or 270° E meridian. Although h_3^1 is an octupolar term, it is symmetric about the equator and is therefore a member of the quadrupole family. The dominance of fields which are symmetric about the equator during the transition between full polarity states (dominated by antisymmetric or dipolar symmetries) gives some support to the reversal model of Merrill and McFadden (1988), in which reversals result from critical interactions between the dipole and quadrupole families (see Sections 5.7 and 5.8). The h_3^1 geometry is also similar to the geometries of the quadrupolar solutions of the dynamo models of Cuong and Busse (1981) and Zhang and Busse (1988, 1989).

Sun et al. (1993) carried out a detailed study of the Matuyama–Brunhes transition recorded in a loess section (L8) at Xifeng in the central part of the Chinese loess plateau. To reduce high-frequency noise, the data were smoothed, using a five-point filter (see Figure 6.7). The figure shows that the reversal of direction does not follow a single smooth transition but shows a number of repeated swings. The transition record can be divided into five stages, the first four of which represent a R→N→R cycle and the fifth the final R→N transition. In stage 1 the VGP path shows loops in both hemispheres during the R→N path, although the N→R path shows confinement to the North America/Pacific region. In stages 2 and 3 the field does not reach full N polarity. In stage 2 the VGP moves rapidly from the southern to the northern hemisphere, before returning more slowly to the reversed position, remaining throughout in a narrow band of longitude off the west coast of Africa. In stage 3 the VGP moves gradually to the northern hemisphere before returning rapidly to the reversed position, and shows less significant longitudinal confinement. In stage 4 the VGP moves abruptly from the southern to the northern hemisphere, passing through the Pacific, returning more gradually to the southern hemisphere along

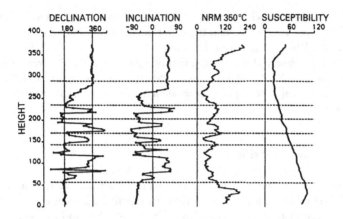

Figure 6.7 Smoothed curves of declination, inclination, NRM intensity ($\times$ 10^{-8} A m^2/kg) at 350 °C and susceptibility ($\times$ 10^{-8} m^3/kg). The five reversal stages are marked with horizontal dashed lines. The heights refer to the base of loess unit L8. After Sun *et al.* (1993).

a narrow band of longitude close to the site. The VGP path of the final reversal of direction (stage 5) shows a gradual northward progression across Africa. The path is $\sim 90°$ W of the site. The inconsistency of the VGP paths for the different stages of the Matuyama–Brunhes transition shows that the reversal cannot be adequately explained by models invoking preferred longitudinal bands over the Americas and their antipode (see Section 6.5), or by models in which the transitional field is dominated by time-dependent low-order zonal harmonies (see Section 5.6).

A decrease in field intensity occurred some time before the onset of major directional changes and the intensity recovered after the directional changes had finished, the intensity curve showing strong symmetry with respect to the transition zone. Within the transition zone the average intensity falls to $\sim 30\%$ of its pre-transitional value and shows strong fluctuations ranging from 10 to 50% of the pre-transition value. Sun *et al.* estimate a duration of 32 ka for the decay and recovery of the main dipole intensity and 20 ka for the period of major directional changes. These estimates were based on an accumulation rate in the section sampled of 100 a/cm. This figure was obtained by relating the accumulation rate to loess grain size and using the top loess unit L1 as a calibration. Thermoluminescence gives an average for the base of L1 of 100 ka. An alternative age estimate of 80 ka for the base of L1 has been proposed based on correlations between the loess susceptibility record and the deep-sea δ^{18}O record (see Section 8.2). This would reduce the above age estimates of the durations by 20%.

Zhu *et al.* (1993) also studied the Matuyama–Brunhes transition field as recorded in Chinese loess from Xifeng. They found that the transition field is characterized, not by a single reversal, but by at least three reversals and one aborted reversal. The chronology of the transition was determined from the SPECMAP δ^{18}O timescale (see Section 8.2). Zhu *et al.* estimated that the duration of the directional changes was ~ 3.6 ka and that of the intensity changes ~ 8 ka. These are much less than the duration obtained from the same section by Sun *et al.* (1993). The reason for this discrepancy is not clear. Some of the polarity changes observed were so fast that no intermediate VGPs were observed. What intermediate VGPs were obtained indicate that the transition field is not entirely non-dipolar, the dipole component being comparable with the non-dipole component.

6.3 The Olduvai transition

Herrero-Bervera and Theyer (1986) analysed records from north-central Pacific deep-sea cores of the onset of the Olduvai transition at a low-latitude (~ 30° N) site and its termination at a mid-latitude site (~ 37° N), and of the onset and termination of the Jaramillo transition at a site near Kauai (~ 19° N). A comparison of all four VGP paths shows that they are characterized by individual features peculiar to each transition. The data indicate that the reversal process is asymmetric, in that each reversal sense is associated with a different field configuration. All four records are unambiguously associated with significantly non-axisymmetric transition fields. Another common feature of the VGP paths is that they all tend to lie on the American continent, being clustered around the 90° W meridian (see Section 6.5). The standing field model cannot explain the diverging behaviour of these paths, because it predicts identical transitional field behaviour for repeated reversals when viewed from a single site. The generalized flooding model, on the other hand, can successfully explain the data, although indicating a degree of asymmetry in the transition field. None of the four transitions show smooth pole-to-pole paths – all display major loops before settling into the opposite polarity position.

Herrero-Bervera *et al.* (1987) continued their investigation of the onset of the Olduvai transition, using data from two azimuthally oriented deep-sea cores from two low-latitude sites in the north-central Pacific, separated by ~ 6° in latitude and ~ 8.5° in longitude. Both cores showed intermediate directions of magnetization, with transitional inclinations reaching values up to 70°. Moreover, up to three sequential 180° changes in declination were seen before the main reversal, two of which showed rapid oscillations of the magnetic field (see Figure 6.8). Liddicoat (1982) had reported earlier the presence of directional oscillations of the geomagnetic field before the Gauss–Matuyama reversal. On the basis of solutions of the non-linear dynamo equation that assume a time-delayed feedback, Yoshimura (1980) predicted that during a polarity transition a number of oscillations may take place. The number of such oscillations, whether odd or even, determines the subsequent polarity. The intensity dropped to 15% of its recovery value and the duration of the intensity changes may have been as long as 28 ka before the onset of the Olduvai reversal was complete. In contrast, the directional changes lasted ~ 5 ka. For one core the geometry of the field during the initial and middle stages of the transition appears to be controlled by non-axisymmetric terms. The final stage of the transition, on the other hand, was significantly

CORE K78019

SITE LAT: 8.95° N   LONG: 170.30° W

10 mT

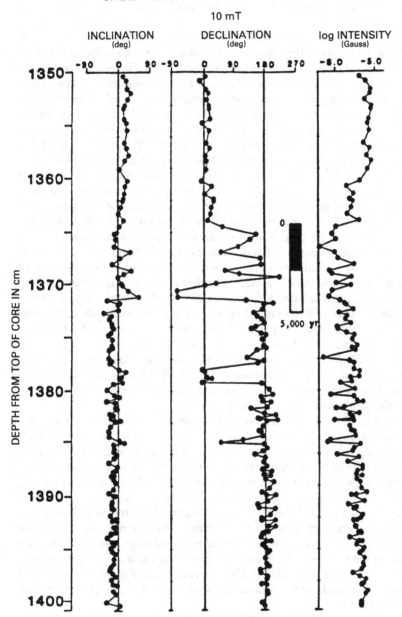

Figure 6.8 Magnetostratigraphic plot of inclinations, declinations and intensity of magnetization in the record of the onset of the Olduvai reversal in core K78019. The total length of the bar represents 5000 a. Demagnetization level at 10mT. After Herrero-Bervera *et al.* (1987).

controlled by zonal harmonic terms. The record from the second core suggests that the field geometry during most of the transition was dominated by non-axisymmetric terms, although dominance by zonal harmonics cannot be excluded during some of the stages.

Tric *et al.* (1991a) obtained a high-resolution record with more than 100 transitional directions of the upper Olduvai transition from sediments in the Po Valley, Italy. The VGP path consists of three stages (see Figure 6.9). In the first stage the VGP moves quickly from the north to an intermediate equatorial latitude along a path largely confined along a meridian over North America, ~ 90° W of the sampling site (Figure 6.10). This is then followed by a period of near-standstill, during which the VGP is confined to equatorial latitudes with directional scatter of ~ 10° between successive samples. Tric *et al.* estimated that this first stage represents 35% of the total duration of the reversal. During the second stage the VGP moves back to the north pole along nearly the same meridian and then executes one rapid oscillation from high northern latitudes to equatorial ones and then back to north again. Tric *et al.* estimated that this stage also represents ~ 35% of the duration of the transition. In the third stage the VGP reaches the south pole, following the same path as that observed in the first stage and continuing over South America. This stop-and-go behaviour has been seen in other detailed volcanic and sedimentary records. Other records of the upper Olduvai reversal obtained from different sites also show that the VGP path is confined to the same longitude band, e.g. that obtained by Herrero-Bervera and Theyer (1986) from a deep-sea north Pacific core at about the same latitude, but almost exactly opposite in longitude. A record of the transition obtained by Clement and Kent (1986) from a north Atlantic deep-sea core also indicates a VGP path which passes over the Americas. However, in this case, when the reversal appears to be completed, the VGP moves back to northern latitudes along a great circle almost antipodal to the Americas before moving back to the south pole with a sudden jump.

Tric *et al.* (1991a) compiled a list of 48 sedimentary records of reversals and found that in more than two-thirds of the cases the VGP paths are similarly confined along a meridian over the Americas or antipodal to them, irrespective of the sampling site and of the sense of the transition. This would indicate that to a large degree the transitions are controlled by a dipolar field. There are, however, some differences, e.g. the VGP path for the Upper Olduvai transition obtained by Clement and Kent (1985) from a southern Indian Ocean core is very complex. Again, the paths of several Upper Miocene reversals in Crete are confined to either one or the other of these preferred longitudes, depending on the sense of the transition (Valet and Laj, 1981; Valet *et al.*, 1988a). Moreover, the records of two of three Middle Miocene reversals in Zakinthos are confined to longitudes antipodal to the Americas, irrespective of the sense of the transition (Laj *et al.*, 1988).

Herrero-Bervera and Khan (1992) obtained a high-resolution record of the termination of the Olduvai (N→R) transition from a northern, mid-latitude deep-sea sediment piston core in the Pacific. The reversal was sampled at 3–5 mm intervals across 130 cm of section – at least ten intermediate directions

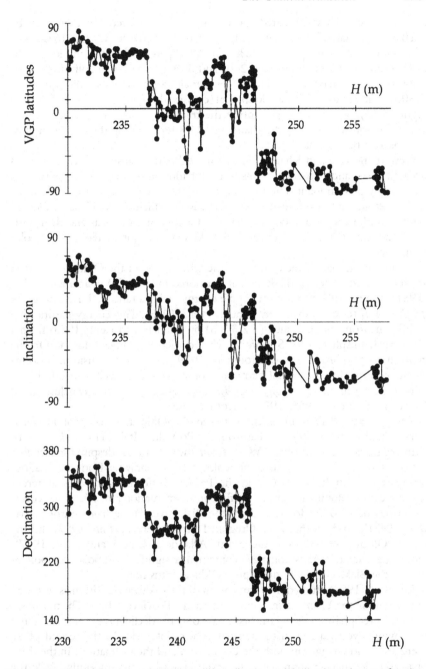

Figure 6.9 Plots of the observed VGP latitudes, inclination and declination values as a function of the stratigraphic height for the Upper Olduvai transition recorded from the Crostolo sediment. The transition presents periods of rapid change alternating with a period of near-standstill. After Tric *et al.* (1991a).

were recorded. They estimated the duration of the directional change to be ~ 10 ka and that of the intensity decrease to be between 12 and 16 ka. The minimum intensity drop was ~ 22%. The VGP path is far-sided and longitudinally confined to a band between 260° and 300° E. The initial half of the transition is characterized by two loops – one small loop that crosses the equator at ~ 300° E and another broad loop that crosses the equator at ~ 270° E (see Figure 6.11). The VGPs representing the transitional field recover very slowly in comparison with the rapidly changing VGPs representing the field right after the onset of the reversal.

Herrero-Bervera and Khan (1992) showed that the transition is characterized by both zonal and non-zonal terms, the contribution of non-zonal quadrupolar terms being predominant. This is in contrast to the earlier observation of Herrero-Bervera *et al.* based on four deep-sea records, in which the VGP paths showed general confinement to a longitudinal band across the Americas and their antipode, indicating that the transitional field may be dominated by a dipolar component.

If the transitional characteristics for the termination of the Olduvai subchron are interpreted in terms of the zonal harmonic model of Williams and Fuller (1981), the redistribution of the dipole energy yields a ratio of 1.1:5:3.9 to the g_2^0, g_3^0 and g_4^0 terms, respectively (Theyer *et al.*, 1985). This energy partitioning is different from that obtained ($-2:3:-5$) by Clement and Kent (1985) from a southern hemisphere (36° S) core which recorded the upper Olduvai. On the generalized flooding model of Hoffman (1979, 1981), the transition is dominated by a quadrupolar zonal harmonic component g_2^0 as well as quadrupolar, non-zonal harmonic terms g_2^1 and h_2^1. This agrees with the conclusions of Clement and Kent (1985, 1987) for other cores.

There is remarkable agreement between the Olduvai reocord of Herrero-Bervera and Khan (1992) and that from the Po Valley, Italy (Tric *et al.*, 1991a), with regard to the path of the VGPs (over the Americas), despite the fact that the two records come from different geological environments, from site localities far apart (one in Italy and one in the Pacific), different lithologies, different magnetic mineralogies and entirely different sedimentation rates. A similar correlation has been found for records of the Blake event (one from Europe: Tric *et al.*, 1991b; and another from Oregon: Herrero-Bervera *et al.*, 1989) and the Upper Olduvai record from the North Atlantic (Clement and Kent, 1987), indicating that the transition field contained a significant dipolar component during the Blake event and the Upper Olduvai transition.

Liu *et al.* (1993) studied the termination of the Olduvai transition as recorded in red loam in the Yushe basin, south-east Shanxi Province, China. The transition exhibits three phases. During the first phase only the declination changes, making two swings westward from the normal axial dipole field. In the second phase there are rapid changes in both the declination and the inclination. In the third phase the declination approaches the reversed axial dipole smoothly, while the inclination exhibits a fairly slow up-and-down swing, passing through zero twice, before finally reaching the reversed axial dipole direction. The estimated duration

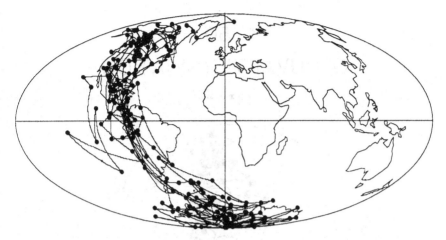

Figure 6.10 Plot of the VGP path during the reversal shown in Figure 6.9. The large north–south oscillations are confined to a longitude over North America. After Tric *et al.* (1991b).

of these three phases are ~ 8.9, 5.3 and 6.6 ka. All the VGPs, except three within the rapid phase, lie within 60° W of the site.

Rolph (1993) has carried out a detailed study of the Matuyama–Jaramillo transition as recorded in a loess section near Lanzhou, China. The duration of directional changes was estimated to be ~ 5 ka, while the field strength was reduced for at least 9 ka. Fluctuations in the intensity are linked to large directional swings which in some cases, both before and after the dipole reverses, are sufficiently large to cause an almost full reversal of the local field. Fourier analysis of the record shows a periodicity of ~ 750 a in the intensity and periodicities in the horizontal and vertical field components which can be correlated with the more recent secular variation. By smoothing the data with a low-order polynomial fit, Rolph showed that the inclination data are consistent with the reversal being dominated by a zonal quadrupole of negative sign ($-g_2^0$). Clement and Kent (1985), in their study of the lower Jaramillo reversal recorded in a deep-sea sediment core from the southern hemisphere, modelled the reversal by partitioning the energy lost by the dipole, with 70% going to a $-g_2^0$ term, 20% to $-g_4^0$ and 10% to g_3^0. During the reversal the VGP path is restricted in longitude, initially following the coast of Africa until it reaches the equator. It then swings west to the Caribbean before moving up through North America to complete the reversal. The second part of this path lies within one of the preferred longitude bands seen by Laj *et al.* (1991) – see Section 6.5. This study shows that although the quadrupole term plays a dominant part before and after the transition while the dipole is weak, the geometry of the transitional field is also strongly influenced by non-zonal terms, with the initial part of the transition occurring within a longitude band 90° W of the site, midway between the preferred longitude bands of Laj *et al.*

K7501
OLDUVAI TERMINATION
N➤➤R

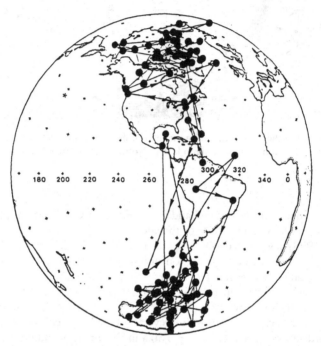

Figure 6.11 Plot of the VGP path for the termination of the Olduvai (N→R) reversal in core K7501. Site locality is shown by the asterisk in the Pacific Ocean. Latitude and longitude grid is shown at 20° intervals. After Herrero-Bervera and Khan (1992).

6.4 The Cobb Mountain transition

Clement and Martinson (1992) compared two records of the Cobb Mountain event from North Atlantic deep-sea sediments – ODP site 647 on the southern Labrador Sea and DSDP site 609, approximately 1300 km away (see Figure 6.12). In order to quantify the similarities and differences between the two records, they removed the effects of differences in sedimentation rate at the two sites by aligning the records, using the correlation technique of Martinson *et al.* (1982). The method involves comparing the agreement between the best-fitting alignment of each of the vector components. A change in sedimentation rate would affect all the vector components in the same way, and therefore any differences in the mapping functions describing the alignments would indicate

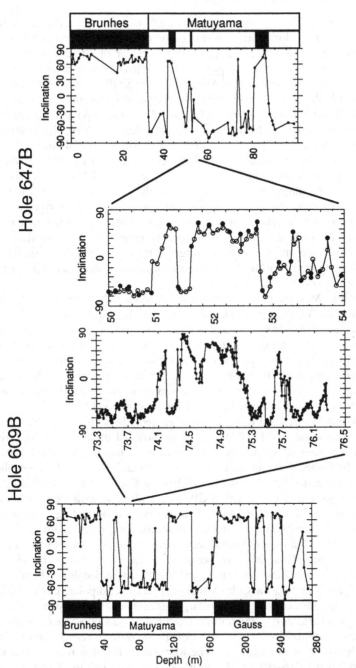

Figure 6.12 Magnetostratigraphy from DSDP hole 609B and ODP hole 647B with expanded intervals showing the inclination records of the Cobb Mountain subchron from both locations. The degree of similarity between these two records suggests that these deep-sea sediments provide accurate records of geomagnetic field behaviour. After Clement and Martinson (1992).

that differences exist in the records which cannot be explained by changes in the rate of sedimentation alone. Any remaining differences must be the result of local factors affecting the remanence acquisition process or of site-dependent field behaviour. To test whether site-dependent field behaviour may account for any remaining differences, Clement and Martinson align VGPs instead of directions.

Both records show an interval of full normal polarity directions bounded by transitions characterized by polarity rebounds. The most striking similarities in the two records are seen in the inclination (Figures 6.13 and 6.14). Large fluctuations occur in the two records both below and above the normal polarity interval. The most notable feature of both declination records is the lack of fluctuations associated with the large inclination rebounds. Clement and Martinson showed that, although the alignment between each of the vector components is very good, there are differences in the mapping functions for each of the components throughout the entire length of the records, indicating that relative sedimentation rate changes between the two records cannot account for these differences. Although inclination and declination changes at the two sites are very similar in slope, the timing of the inclination changes relative to the declination is not the same in both records.

Clement and Martinson also found good agreement between the mapping functions for the VGP alignments, indicating that the sequences of VGP positions observed at the two sites and the relative rates of movement are very similar. Intervals of very rapid field changes are seen in both records. The VGP paths of the upper polarity rebound in both records show a clockwise loop with a small counterclockwise loop superimposed, although the magnitudes of the loops are different. The observation of these loops in two independent records argues that the looping is a real feature of the transition field.

Clement and Martinson then compare their records of the Cobb Mountain event with the sequence of polarity reversals recorded in lava flows in Tahiti which, on the basis of K–Ar age determinations correlate with the reversals bounding the Cobb Mountain and Jaramillo events and the base of the Brunhes chron (Chauvin *et al.*, 1990). They found that the Tahitian lavas recorded fields of the Cobb Mountain event with VGPs nearly identical with those obtained from sediments in the North Atlantic. However, because of the large gaps in the lava record, it is not possible to determine whether the relative timing of the VGPs recorded at Tahiti compares directly with that of the VGPs recorded in the North Atlantic. Thus, it is not possible to determine whether the similar VGPs were recorded at the same time, which is crucial in deciding whether the transition fields bounding the Cobb Mountain event were dipolar. The similarity between the records certainly suggests large-scale, if not dipolar, symmetries in the geomagnetic field during the transition bounding the Cobb Mountain event, as has been suggested by Tric *et al.* (1991a) for other recent reversals. The situation, however, is not clear-cut. The Matuyama–Brunhes record obtained from piston core V16–58 in the south-west Indian Ocean (Clement and Kent, 1991) gives a VGP path which is clearly identical with that obtained from DSDP site 609 in the North Atlantic (Clement and Kent, 1987). However, the record from ODP site 664 from the equatorial Atlantic gives a VGP path

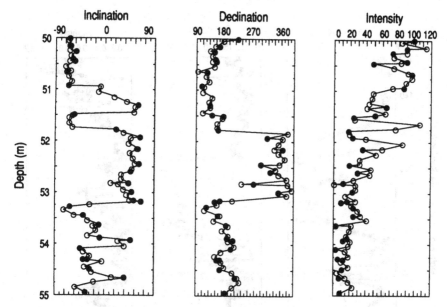

Figure 6.13 The record of the Cobb Mountain subchron obtained from ODP site 646 in the Labrador Sea, presented as inclination, declination (both in degrees) and NRM (after treatment at 9 mT) intensity (in mA/m). The results of the shipboard pass-through cryogenic measurements after treatment at 9 mT are plotted as open circles, and the ChRMs defined by progressive AF demagnetization of discrete samples are plotted as solid circles. The agreement between both sets of measurements is very good. The Cobb Mountain is defined in this record as a full, normal-polarity interval spanning nearly 1.3 m of section. After Clement and Martinson (1992).

which is nearly antipodal to these other two (Valet *et al.*, 1989), which would indicate that the field was not dipolar during the Matuyama–Brunhes reversal. Whatever the configuration of the geomagnetic field during recent reversals, however, it appears that large-scale symmetries were present in many of them.

The sequence of reversals recorded at DSDP site 609 differs from other reversal sequences in that the transition field reverts back to a similar configuration after a different configuration has persisted through two or more intervening reversals. Thus, the Upper Olduvai transition shows important similarities to the Matuyama–Brunhes transition, while differing significantly from the intervening Cobb Mountain event and lower Jaramillo event, which are themselves similar.

In a later paper Clement (1992) presented two new records of the Cobb Mountain event from the western Pacific – from ODP sites 767 in the Celibes and Sulu Seas. These records show sequences of VGP positions that are very similar to those observed in the North Atlantic and Tahiti, further strengthening the evidence of dipolar transition fields during the Cobb Mountain reversal. These five transitions were obtained from different types of palaeomagnetic

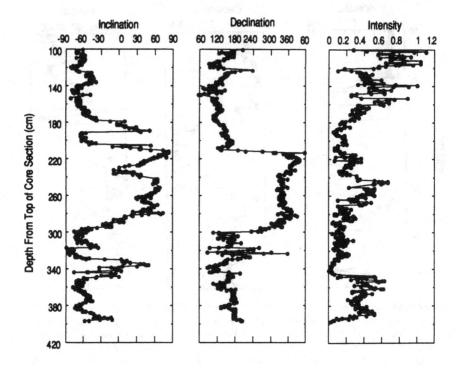

Figure 6.14 The record of the Cobb Mountain subchron obtained from DSDP site 609 in the North Atlantic Ocean (Clement and Kent, 1987). The similarities in the sequences of directions recorded at these two sites indicate that these sediments provide accurate records of geomagnetic field behaviour during the Cobb Mountain subchron. Units are the same as in Figure 6.12. The depths are in centimetres from the top of the core section. This record spans the interval from 73.30 to 76.26 mbsf. After Clement and Martinson (1992).

records (lavas and deep-sea sediments) from widely separated locations, making it difficult to ascribe the similarities in the records to artefacts of the recording processes.

6.5 Longitudinal confinement of transitional VGPs

The concentration of transitional VGPs into two longitudinal bands suggests that there exists a strong geographical control of the reversal process within the Earth's outer core. Bloxham and Gubbins (1985) showed that the main features of the historical geomagnetic field originate from four major regions of concentrated magnetic flux, the concentrations of flux forming two pairs of flux lobes (Figure 6.15a,b) separated by ~ 120° of longitude (see Figure 6.15c). The longitudinal concentrations of VGPs (Figure 6.15a,b) nearly coincide with the two longitudinal bands containing the two pairs of flux lobes. This suggests that the processes in the core which produce the two pairs of flux lobes observed by Bloxham and Gubbins in the historical geomagnetic field may have been present during the Matuyama–Brunhes reversal, over 0.78 Ma ago. There are also some suggestions that such longitudinal confinement of VGPs may occur in other reversals. Multipole northern hemisphere records of the upper Olduvai reversal from Italy (Tric *et al.*, 1991a) and from deep-sea sediments from both the Pacific (Helsley *et al.*, 1989) and the North Atlantic (Clement and Kent, 1986) yield VGP paths which track southward through the Americas. In fact, Tric *et al.* noted that VGP paths from more than two-thirds of all recently obtained transition records from different reversals are confined to meridians over the Americas or Asia – suggesting that this longitudinal grouping is a persistent feature of the reversing field (Valet and Laj, 1984).

Laj *et al.* (1991) also pointed out that the longitude bands preferred by the VGP paths are those along which north–south flow is seen in core models (Bloxham and Jackson, 1991). The preferred longitudes also represent regions of fast seismic-wave propagation (and, hence, low temperature) in the lower mantle (Olson *et al.*, 1990). The VGP paths appear to link the morphology of the transitional fields to features of the present-day field and to temperature anomalies in the mantle. Since the time constants of mantle convection are many orders of magnitude longer than those of core motions, the persistence of these VGP bands over ten million years or more (if confirmed) would suggest that fluid flow in the core is controlled, or at least modulated, by the temperature at the core–mantle boundary, which is determined by the pattern of mantle convection (Gubbins and Bloxham, 1987). It is not at all clear, however, how inhomogeneities in the lower mantle affect flow in the outer core, and how a particular flow pattern translates into specific features of the transition magnetic field.

Valet *et al.* (1992) question the suggestion that multiple records of reversals during the last 12 Ma indicate that VGPs have a strong tendency to follow one of two paths during the reversal. They use a single weighted mean VGP longitude for each path to avoid bias towards transitions with a large number of VGPs. Using standard χ^2 tests, they concluded that different records of the same reversal do not have a common VGP transition path and are thus not compatible with a dipolar transition field. Moreover, they claim that there is no robust evidence for a preferred longitude sector for VGP paths from different reversals. They also found no systematic relationship between VGP paths and site longitude (nor with the antipodal meridian), as predicted by models involving zonal non-dipole terms. In fact, for the VGP paths confined to the longitudinal band across the Americas, the sites tend to lie 90° away, either in Europe or in the Pacific. Finally, Valet *et al.* found no age-related pattern in the distribution of VGP paths, which gives no support to the suggestion of a persistent component of the transition field. Their general conclusion is that the available data are not inconsistent with a uniform random choice of transitional paths. These could be generated by drifting and diffusing non-dipolar features.

Valet *et al.* (1992) also considered records of reversals recorded in lavas, which are far fewer. The data of ten reversals from five sequences show a wide

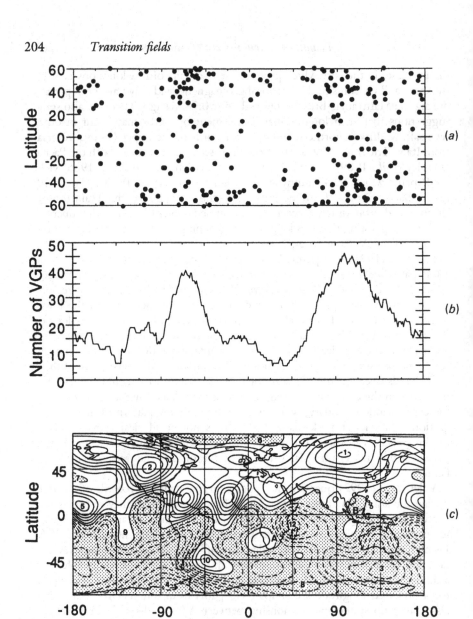

Figure 6.15 (a) The geographical distribution of Matuyama–Brunhes transitional VGPs from a number of site locations. (b) The longitudinal distribution of transitional VGPs plotted as the number of transitional VGPs in a sliding window of width thirty degrees longitude. The clustering apparent in (a) appears as two peaks of the frequency distribution. (c) Map of the radial component of the magnetic field at the surface of the outer core (after Bloxham and Gubbins, 1985). The concentrations of flux labelled 1, 2, 3 and 4 form two pairs of flux lobes in longitudinal bands which lie very close to the concentrations of transitional VGPs. After Clement (1991).

longitudinal scatter of the VGPs. The studies of the Steens Mountain, Oregon, reversal (see Section 6.7) and the Upper Jaramillo from Polynesia (Chauvin *et al.*, 1990) are the two best-documented and most complete volcanic records. Both VGP paths display large longitudinal loops similar to smaller-amplitude variations seen in one of the most detailed sedimentary records from Crete (Valet *et al.*, 1986). Such features are compatible with drifting sources of the non-dipole field and can be modelled with low-order zonal and drifting non-dipole harmonics (Weeks *et al.*, 1988; Williams *et al.*, 1988). The question of reversals recorded in lavas will be returned to later in this section.

Laj *et al.* (1992a) dispute the statistical analysis used by Valet *et al.* (1992). Valet *et al.* obtained a mean VGP longitude with a statistical weighting of the record. They then divided the data into longitudinal bins and used the χ^2 test with 18 sectors of 20° width to test for a preferred longitude in the distribution of VGP paths. Laj *et al.* point out that the χ^2 test can only be used if the expected frequencies in each bin are at least equal to a minimum (about 5). In the calculations of Valet *et al.* the expected frequences are well below the acceptable limit for the χ^2 test to be valid. Laj *et al.* believe that it is more appropriate to use the methods of circular statistics, and in a later paper (Laj *et al.*, 1992b) applied a series of tests based on such methods. They showed that the preponderance of transitional VGPs over the Americas and its antipode is not consistent with the statistical fluctuations of a random probability distribution. They used essentially the same database as that used by Valet *et al.* (1992) and included the records of the Blake, Cobb and Gilsa events. They did not consider records from intrusions or lavas, because of the possibility of substantial field averaging due to the long acquisition times of magnetization and because it is difficult to define VGP paths for volcanic records of reversals, since in most cases they give only scattered spot readings of the field. The methods of circular statistics are specifically designed for the analysis of data distributed on a circle and do not require any *a priori* partitioning of the data into longitudinal bins, thus avoiding subjective choices of the width and position of intervals on the circle. The methods also make it possible to judge whether a given non-random circular distribution is unimodal or bimodal. This is done by doubling the angles – when all the angles are doubled and the multiples reduced modulo 360°, a bimodal axial distribution is transformed into a unimodal circular sample to which the usual circular tests can be applied. A uniform distribution, on the other hand, remains uniform upon doubling the angles. The results of the analysis of Laj *et al.* (1992b) clearly indicate the observed bimodality. The mean angles for the bimodal distribution are 120 ± 28° and 300 ± 28° – these bands of longitude track across the Americas and its antipode.

Quidelleur and Valet (1994) have looked again at the sedimentary and volcanic records of the Mono Lake, Laschamp, Blake and Cobb Mountain geomagnetic excursions/reversals. They found that the VGP paths for excursions do not fall in any preferred longitudinal bands. The data for reversals alone (or reversals and excursions) fall into two classes: some VGP paths are longitudinally confined to sectors over the Americas and their antipode and are significantly 90° away from their site longitude; the other VGP paths are scattered all over the world with no relationship to the longitude of the sampling site. They conclude that the confinement of those VGP paths within two preferred bands of longitude could be due to artefacts linked to the acquisition of magnetization in sediments. (This will be discussed in more detail later in this section.) The scattered records could in addition be due to the non-dipole field.

Constable (1992) has analysed the records of stable polarity direction measurements recorded in lavas over the past 5 Ma. Earlier analyses, while showing that the VGPs were both far-sided (i.e. beyond the geographical pole with respect to the observation site) and right-handed (i.e. having positive declinations), concluded that the data gave VGPs which were equally scattered in longitude around the geographical North Pole. In contrast, Constable showed that individual VGPs (there were more than 2000 in her analysis) were biased towards two longitudes – the same preferred longitudes seen in the reversal records. Her conclusion appears to be insensitive to both the distribution of sampling sites and the effects of recent plate motions. One of the longitudinal bands (that over the Americas) coincides with that expected from the reversal of a non-axial dipole field exactly like that present today – the other requires only a change in sign in the non-axial dipole terms of today's field. This is a direct consequence of the present VGP lying in the longitude over the Americas, which is principally due to the orientation of the equatorial dipole. As the axial dipole decreases, the location of the VGP is increasingly controlled by the equatorial dipole which maintains the path over the Americas – as is observed. Constable further showed that a non-zonal bias, similar to that observed in the reversal data, is also evident in data in the secular variation of the field over the past 5 Ma. Figure 6.16 shows a histogram of the longitudinal distribution of VGP positions for palaeosecular variation data for the past 5 Ma. It can be seen that there is a concentration in two antipodal bands, similar to those seen in the Matuyama–Brunhes and other reversal data.

McFadden *et al.* (1993) devised a new statistical test specifically designed to test the hypothesis of two preferred antipodal longitudinal bands, using all reversal records for the past 12 Ma. Their analysis shows that the records, taken as a group of independent observations, do show an overall preference for two antipodal longitudinal bands. They also found, as have others (Laj *et al.*, 1992b; Valet *et al.*, 1992) that there is a tendency for the sites at which reversals are sampled to lie 90° either side from their associated VGP paths. Egbert (1992) has investigated for a wide range of conditions the effect of the transformation in spherical co-ordinates used to find VGPs. He showed that, except in the special case where the field is entirely dipolar, the distribution of longitudes is always peaked at ± 90° in longitude away from the sampling site. Constable (1992) has used Egbert's calculations to compute the expected distribution of pole longitudes for any given site distribution, assuming reasonable statistics for the variability of the magnetic field with time. She showed that the bias expected as a result of site distribution is small compared with the overall signal, although it does peak in approximately the same longitudes. The bias to American longitudes is still present when all sites exhibiting any tendency to ± 90° are removed, although the antipodal band is not observed. There is also no longer a correlation

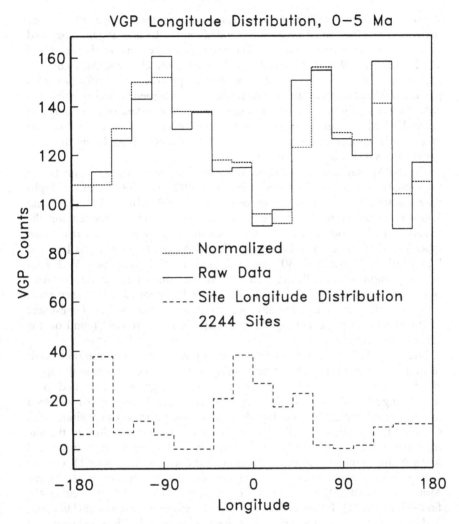

Figure 6.16 Histogram of VGP longitudes for globally distributed palaeosecular variation data from lava flows spanning the past 5 Ma. Solid line is for the raw data; dashed line is normalized for site distributions. After Constable (1992).

between the observed peaks and those predicted from the site distribution.

Van Hoof and Langereis (1992a) carried out a detailed palaeomagnetic study of the upper Kaena polarity reversal from late Pliocene marine marls in southern Sicily. The (R→N) reversal takes ∼ 3.3 ka and is followed by a stable normal interval of ∼ 3.3 ka. This is followed by an excursion lasting 3.6 ka. Relative intensities were lower for ∼ 5.6 ka with a minimum of 10% of the non-transitional values, but, during the excursion, there were no apparent intensity changes. Van Hoof and Langereis concluded, on the basis of the lithology and rock magnetic properties and of earlier results from the same marls (van Hoof and Langereis, 1991), that the features observed in the Upper Kaena record are largely determined by the remanence acquisition process in the sediments. This process is complicated by authigenic formation of magnetic minerals and cyclically fluctuating palaeoredox conditions, resulting in considerable smoothing and delayed NRM acquisition in certain sedimentary intervals. They believe that the excursion is not due to the behaviour of the geomagnetic field during the transition, but is a sedimentary artefact.

The NRM was found to have two separate characteristic components – a low-temperature (LT) component between 200 and 500 °C and a high-temperature (HT) component between 500 and 580–610 °C. Van Hoof and Langereis calculated the VGPs of both components. The HT component initially follows a path ∼ 60° east of the site, loitering south-west of Australia, then quickly crosses the equator between India and Australia and goes via Japan to high northern latitudes ∼ 180° east of the site. The LT component VGP path is more complicated, oscillating between two meridians – one passing over Australia and the other over the Americas. No VGPs intermediate between these two meridians were found. Initially most VGPs are on the southern hemisphere of the Australian path. During the final stage VGPs were only found on the (North) America path.

Linssen (1988, 1991) reported the results of five successive transition records from the lower Pliocene in Calabria, ranging from the upper Thvera to the upper Nunivak. The durations of the major directional changes were estimated to be in the range 0.3–3.0 ka. The VGP path of the upper Thvera record follows a great circle 90° west of the site meridian, indicating a non-zonal transition field. On the other hand, the VGP paths of the other records are close to the site meridian, indicating a mainly zonal transition field – the lower Sidufjall and lower Nunivak records are nearly identical. Rebounds in the Upper Thvera and Upper Sidufjall are similar to the stop-and-go movements seen in the Steens Mountain record (Mankinen *et al.*, 1985; Prévot *et al.*, 1985a). Neither the flooding model (Hoffman 1977, 1979) nor the standing field model (Hillhouse and Cox, 1976) fits the sequence of the five reversal records. The flooding model does not fit the results, since the Upper Thvera reversal seems to be dominated by non-axisymmetric terms. The standing field model only fits the results if it changed from zonal to non-zonal between the upper Thvera and the lower Sidufjall. Linssen further showed that the zonal harmonic model of the Matuyama–Brunhes reversal (Williams and Fuller, 1981) fits the lower Sidufjall and lower Nunivak records.

Van Hoof and Langereis (1992b) have since carried out a much more detailed study of the upper and lower Thvera transition records from marine marls in southern Sicily. The lower Thvera transition shows a smooth change from reversed to normal directions. The upper Thvera transition, on the other hand, is preceded by two 'excursions' and followed by another. Van Hoof and Langereis believe the two excursions preceding the transition to be sedimentary artefacts as a result of early diagenetic processes, the directions of the excursions being caused by the post-transitional reversed geomagnetic field. They probably acquired post-transitional directions due to the formation of secondary magnetite, while outside the excursions the sediments had already acquired the pre-transitional directions. The last excursion is probably the result of a recent overprint due to weathering.

The VGP paths of both the upper and lower Thvera transitions are strongly confined to the meridian passing over the Americas. Van Hoof and Langereis (1992b) believe that this reflects the smoothing of the non-antipodal stable directions before and after the transitions. They also compared the upper Thvera transition from Sicily with that obtained by Linssen (1988, 1991) from Calabria, about 250 km away. There are both similarities and differences between the two records. Both transitions have identical VGP paths, but in Calabria the transition is recorded lower in the sediment, the first and third excursions have different characteristics, and the second one is absent. This is a further indication that the upper Thvera record is unlikely to be a true registration of the transitional geomagnetic field.

Van Hoof *et al.* (1993a) later carried out a detailed study of the upper and lower Nunivak transitions as recorded in Pliocene marine marls in southern Sicily. This follows earlier work by van Hoof and Langereis (1991) in which only the lower Nunivak transition was studied and only declination and inclination records were available. The transitions are seen in two components – a low-temperature and a high-temperature component. The two components do not reverse simultaneously, nor are their transitional characteristics the same. The changes in these components take place at lithological boundaries and so cannot give a true representation of the geomagnetic field. Moreover, there are significant directional changes between these records and those of the same transition seen in Calabria. Magnetites are the main carriers of the remanence, and van Hoof *et al.* explain the directional changes by diagenetic formation of magnetite in which, shortly after burial, the remanence carried by newly formed secondary magnetites is superimposed on the initial remanence carried by primary magnetite. The VGP paths of the LT components of the two transitions are confined to the Americas. (The path of the HT component of the Upper Nunivak transition lies more in the Atlantic – no path for the HT component of the lower Nunivak transition could be calculated, since this component showed no intermediate directions.) The longitudinal confinement of the VGP paths is probably caused by smoothing of the stable directions before and after the transitions, as suggested earlier by Langereis *et al.* (1992).

Van Hoof *et al.* (1993b) have continued their work on the marine marls of southern Sicily with a detailed palaeomagnetic and geochemical study of the upper Cochiti (N→R) reversal. The marls consist mainly of carbonates and a mixture of clay minerals that show a pronounced rhythmic bedding which is characteristic for this formation in Sicily. Small-scale sedimentary cycles are quadripartite and show a distinct grey–white–beige–white colour layering. Van Hoof *et al.* found two consecutive and very rapid transitions (R→N and N→R) that coincide with distinct lithological boundaries. They explain their observations with a 'Fe-migration model' in which magnetite is formed under different diagenetic conditions. During deposition of the anoxic grey layer primary magnetite is formed. During post-oxic diagenesis upon burial by the white and beige layers, secondary magnetite is formed by migration of Fe^{2+}. Primary magnetite causes the geomagnetic field to be recorded during or very shortly after the formation of this magnetite, by growing through its critical blocking diameter and causing acquisition of CRM. Upon burial of the grey layer, secondary magnetite is formed in the white layers on either side of the grey layer. If at that moment a polarity reversal occurs, it will be recorded by the secondary magnetite. The presence of the already formed primary magnetite competes with the secondary magnetite, and, if the secondary magnetite is dominant, the resulting direction will reflect the new polarity.

The situation is further complicated by the presence of both a low-temperature and a high-temperature component. There is a difference in acquisition between the two components, the HT component showing a delay with respect to the LT component. Because of the very rapid change in the magnetic remanence, van Hoof *et al.* were unable to record any intermediate direction and, hence, gave no VGP paths.

Gubbins (1987) suggested a mechanism to explain reversals of the geomagnetic field (see Section 5.10), and he and Coe later (1993) extended his theory and showed that the field can be strong at the core surface during a transition and that VGP paths can be confined within relatively narrow longitudinal bands even if the transition field has a substantial non-dipolar structure. Thus, although the longitudinal bias of VGP paths is definite evidence for core–mantle interaction, the VGP paths are not evidence of near-dipolar transition fields. Gubbins and Coe further showed that blocking at the Pacific rim, as observed in the secular variation, can give rise to VGP paths through the two longitude bands preferred by much of the data. For N→R transitions the VGP follows the Asian path when the reversal is initiated in the southern hemisphere, the Americas path when the reversal is initiated in the northern hemisphere.

Runcorn (1992) has put forward an alternative explanation for the longitudinal confinement of geomagnetic reversal paths. He points out that the preferred pole paths bound the Pacific, where the secular variation and non-dipole field are smaller than elsewhere. He had suggested as long ago as 1956 that this may be the result of the varying core dynamo field being screened by a highly conducting layer in the lower mantle*. Runcorn now (1992) proposes that this layer has metallic conductivity under the Pacific hemisphere and is insulating elsewhere. As the dipole begins to reverse along an arbitrary meridian, it induces currents in the hemispherical conducting shell and the torque rotates the core relative to the mantle until the reversing dipole path coincides with the two boundary meridians in the Americas and eastern Asia. Thus, in his model the field reverses, not by disappearing or becoming non-dipolar, but through rotation of the dipole axis.

---

* The structure of the lower 200–300 km layer at the bottom of the mantle (the D″ layer) is discussed briefly in Section 8.5.

The controversy over the longitudinal confinement of intermediate VGP paths continues. Prévot and Camps (1993) obtained about 400 intermediate VGPs from 121 volcanic records of excursions and reversals less than 16 Ma old and found, in agreement with Valet *et al.* (1992), no evidence that they were confined to two preferred, antipodal longitudinal bands. They suggest, as have others (e.g. Rochette, 1990; van Hoof and Langereis, 1991; Langereis *et al.*, 1992) that the preferred longitudinal confinement of VGPs found from analyses of sedimentary records is an artefact caused by distortion and smoothing of the geomagnetic signal by the sediments. Prévot and Camps thus conclude that there is no evidence for control of transitional fields by temperature variations at the CMB.

This continuing controversy raises two issues. First, is the correct statistical analysis being used and is the analysis sufficiently rigorous to come to definite conclusions? Second, is the fidelity of the palaeomagnetic records. To help rectify this second problem, Athanassopoulos *et al.* (1993) are compiling an atlas of the raw data upon which the records are based, viz. description of sites, rock type, age, magnetization direction and intensity and demagnetization behaviour, and the analysis used in preparing the final record. The atlas will contain a standardized representation of all the raw data, Zijderveld plots (if available), declination, inclination, intensity plots and VGP plots.

6.6 Clustering of VGPs

Hoffman (1986) examined the behaviour of the transition field recorded in lavas from three mid-southern-latitude Cenozoic shield volcanos in Australia and New Zealand. The data indicate that a potentially long-lived intermediate field is developed early in the transition process, from which both successful and unsuccessful reversal attempts are made. In particular, the palaeofield recorded in lavas from the Liverpool volcano, New South Wales shows three departures of the field vector from full polarity before the actual change in polarity occurs, each departure being dominated by essentially the same intermediate field geometry. Another instance of a recurring intermediate field geometry was reported earlier by Watkins and Nougier (1973) in Brunhes-age lavas from Amsterdam Island, situated in mid-southern latitudes. Hoffman speculated that the core process responsible for these intermediate field events has possessed similar characteristics throughout the Cenozoic and that at least some unsuccessful reversal attempts have occurred during the Brunhes normal chron (the Amsterdam Island event was probably not successful). The Miocene R→N record from mid-northern-latitude basalts from Steens Mountain, Oregon, also show a stop-and-go behaviour (see Section 6.7). In this reversal, following a large and regular change in direction at the onset, the field vector takes up an intermediate orientation which remains essentially unchanged during the extrusion of several flows. This is followed by a rapid change to full normal polarity and then, before the completion of the reversal, a 'rebound' to the same intermediate field direction (Prévot *et al.*, 1985a,b). Stronger supporting evidence comes from intrusions, because of their ability to record in a more continuous fashion than lava sequences.

Records of two Miocene R→N transitions obtained from intrusions in the Pacific north-west of the United States, differing in age by several million years, display very similar field behaviour (Dodson *et al.*, 1978). Both records show a rapid change in inclination at the onset of the transition before going to a loosely stationary intermediate position for some time before the completion of the reversal. The record of the Matuyama–Brunhes transition from Boso Peninsula in Japan (Niitsuma, 1971) and the Gauss–Matuyama transition from Searles Valley in California (Liddicoat, 1982) both indicate that the reversal onset is characterized by a movement to an intermediate state which remains essentially stationary (cf. Shaw, 1975, who suggested that the Earth's magnetic field may have a third metastable state – see Section 3.2). This state seems to act as a kind of springboard from which reversal attempts, both successful and unsuccessful, are made. Such relatively stable intermediate field directions, or 'hang-up' points, are in fact quite common and suggest that certain field geometries may offer energy states that are preferred by the geodynamo during transitions. Hoffman (1986) suggested that these results may arise from a succession of nearly identical disturbances at the same place in the core over a relatively short time interval or from fluctuations in the intensity of a long-lived disturbance.

Hoffman (1991), using data from Plio-Pleistocene lavas from the Hawaii and Society Islands, suggested that the process of polarity reversals involves the development of specific quasi-stationary transitional field configurations which recur during successive polarity reversals spanning several million years. Because the observation sites are situated almost symmetrically about the equator, he based his analysis on the division of the dynamo field into the primary (EA) and secondary (ES) families (see Section 5.7).

Palaeomagnetic directions corresponding to the Olduvai–Matuyama (N→R) transition recorded on Molokai, Hawaii, are shown in Figure 2.3. It can be seen that few flows give directions deviating by more than 30° from that of an axial dipole. The intermediate directions that are found define for the most part a relatively tight cluster – the corresponding VGPs being located in the Atlantic off the coast of South America. Similar clustering has been found for other Hawaian lavas. The Matuyama–Brunhes (R→N) transition observed at Maui begins with a tight cluster defined by 22 successive lavas, indicating a VGP within South America, relatively close to the position found from the Molokai record (Coe *et al.*, 1985). Another example has been found from lavas of Pliocene age on the island of Kauai (Bogue and Coe, 1982, 1984). This transition exhibits a tight clustering of more than 20 flow directions, with the corresponding VGPs being south of the tip of South America. Similar results can be seen from a sequence of Plio-Pleistocene reversals recorded in lavas in the Society Islands (Chauvin *et al.*, 1990; Roperch and Duncan, 1990). Records from Tahitian lavas of the R→N Matuyama–Brunhes transition, the N→R Jaramillo–Matuyama transition and the closely spaced pair of reversals bounding the Cobb Mountain normal subchron (Chauvin *et al.*, 1990) all contain a cluster of transitional palaeo-directions which are similar to one another. However, the corresponding clusters of VGPs are near south-west Australia, nearly 180° away in longitude from the cluster of VGPs from Hawaii (see Figure 6.17). Further results were obtained by Roperch and Duncan (1990) from a transitional event (~ 3.0 Ma ago) from lavas on Huahine Island. Although they found many intermediate directions, several sites recorded a particular palaeofield orientation corresponding to a VGP near north-west Australia. All three Hawaiian VGPs lie within a 60° band of longitude centred near 310° E which overlaps the longitude band claimed by many to contain a large number of transitional VGP paths obtained from sediments (see, e.g., Clement, 1991; Laj *et al.*, 1991; Tric *et al.*, 1991a). The sediment studies also indicate a second preferred longitude band for VGP paths, antipodal to the first, centred along a meridian near western Australia – the VGP clusters recorded at the Society Islands lie within this band.

Hoffman (1991) has attempted to explain these results in terms of the two dynano families – the family associated with fields having equatorial antisymmetry (EA) and the family associated with equatorial symmetry (ES). Transition fields may be analysed in this manner if synchronous data are available from sites essentially along the same line of longitude and symmetrically situated about the equator. Hoffman suggested two models. In his preferred model the recorded site-dependent cluster localities correspond to two transitional field states separated in time. In this model the clustered transition data from the Hawaiian and Society Islands can be explained by a core process in which the EA family field (and, in particular, the axial dipole) vanishes independently of the ES family field. The model is also compatible with the contention that long-lived transitional field states may be significantly dipolar. Hoffman further suggested that the EA family dominates in the Earth's deep core – the axial dipole term (g_1^0) is much stronger than higher-degree harmonics at the CMB. Furthermore, since the ES family is largely associated with the westward drift (Merrill and McFadden, 1988), it seems reasonable to suppose that it is likely to be concentrated in the upper parts of the core. If this were the case, polarity reversals might result from the

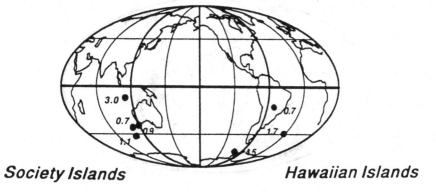

Society Islands   **Hawaiian Islands**

Figure 6.17 VGPs representing palaeodirectional cluster means recorded at the Hawaii hotspot (19° N, 205° E) and the Society Islands hotspot (18° S 209° E), for the Plio–Pleistocene transitions indicated by approximate age in Ma. The plot is centred on longitude 207° E. After Hoffman (1991).

occasional destruction of the deep-core EA field, leaving the intermediate field geometry dominated by the upper-core ES field.

Hoffman (1992), using data from lavas from five sites well distributed over the Earth and spanning ~ 10 Ma, found, not preferential longitudinal confinement of the path of the VGP, but two localized regions in the southern hemisphere, defined by clusters of sequential VGPs. The two regions, however, do lie mainly within the longitudinal bands deduced from sediment data (see Figure 6.18). These new data confirm Hoffman's earlier suggestion (1991) that the reversal process is dominated by long-lived and recurring transitional field states and that the two VGP cluster localities are associated with standing transitional orientations of the true geomagnetic pole during field reversal, i.e. transition fields are characterized by particular inclined dipolar states. Additional preliminary results from a transitional record obtained from lavas on Moorea (erupted at the Society Islands hot spot) near the end of the Olduvai subchron contain first a transitional VGP within the western Australia cluster patch and then a series of three VGPs within the other cluster patch in the south Atlantic. Hoffman believes that the lavas have recorded instants before and after a transformation in field geometry from one inclined dipolar state to the other.

Gubbins and Bloxham (1987) have obtained maps of the radial component of the Earth's magnetic field at the CMB at various times since 1715. The southern hemisphere is characterized by two flux patches. These flux patches seem to have strengthened over historic times, consistent with progressive weakening of the axial dipole, and Gubbins (1987) has suggested that continued growth could ultimately lead to a reversal. However, it should be noted that these flux patches on the CMB appear to drift westward, whereas the VGP cluster patches found by Hoffman suggest little movement over the past 10 Ma.

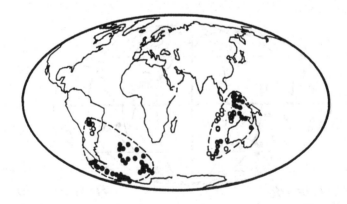

Figure 6.18 Clustered virtual geomagnetic poles (VGPs) contained in six single-cluster records (●) and four more-complex records (○) forming the indicated dual-patch distribution. These data were obtained from late Cenozoic lavas from five widely scattered sites about the globe. These transitional data strongly suggest that reversing fields are dominated by particular long-lived, inclined dipolar field states. After Hoffman (1992).

Hoffman (1992) has obtained a map of the vertical component of today's magnetic field at the Earth's surface stripped of its axial dipole – the g_1^0 term (see Figure 6.19). In the southern hemisphere there is an intense downward-directed field below the south Atlantic off the coast of Argentina and an intense upward-directed field over western Australia. These features show a strong correlation with the locations of the two palaeomagnetic transition VGP cluster patches (see Figure 6.18). Hoffman points out, however, that the features of the vertical component of today's stripped magnetic field are largely responsible for the non-dipole part of the field, whereas the palaeomagnetic data suggest that the cluster patches are associated with the dipole-dominated transitional field.

Dziewonski and Woodhouse (1987) in their analysis of normal-mode data found regions of anomalously fast seismic P waves in the lower mantle. There are two regional maxima – one in the southern hemisphere and one in the northern hemisphere. Hoffman plotted four maxima in these regions together with four localities of near-radial flux centres associated with today's surface field stripped of its axial dipole (Figure 6.20). The correlation of these features suggests to Hoffman that the particular recurring inclined dipolar states during field reversal arise from localized, long-standing thermodynamic features at the CMB. Two records from Australia (Hoffman, 1986) – a Miocene N→R reversal and the other ~ 34 Ma in age – both show clustering close to the northern hemisphere vertical field features shown in Figure 6.19. These older lava records suggest that each vertical feature seen in to-day's axial dipole stripped field, regardless of the

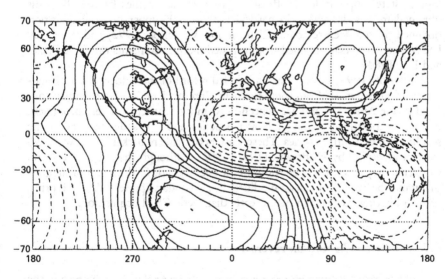

Figure 6.19 Contours of the vertical component of the 1975 International Geomagnetic Reference Field at the Earth's surface for the case in which the axial dipole term is removed. Downwardly directed and upwardly directed field are denoted by solid and dashed contour lines, respectively. Note the correspondence of the two Southern Hemisphere features with the VGP cluster patches (Figure 6.18). After Hoffman (1992).

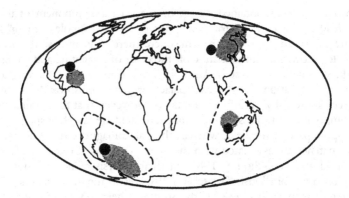

Figure 6.20 Regions of fastest P-wave propagation in the lower mantle (shaded) (from Laj *et al.*, 1991; Dziewonski and Woodhouse, 1987), localities of near-radial flux centres associated with today's surface field stripped of its axial dipole (●), and the transitional field VGP cluster patches derived from lavas. Although seismic velocity models determined from normal-mode data must have antipodal symmetry, the close pairing of velocity maxima with geomagnetic features provides credibility to the interpretation. That these paired observables are found well within the Southern Hemisphere VGP cluster patches suggests strongly that the sources of particular recurring inclined dipolar states during field reversal are tied to localized, long-standing dynamical features of the core–mantle system. After Hoffman (1992).

hemisphere in which it is found, may be associated with a transitional inclined dipolar state. Support for Hoffman's hypothesis also comes from VGP paths obtained from sediments – in particular, the N→R upper Olduvai transition obtained from a deep-sea core (Clement and Martinson, 1992) and the R→N late Miocene reversal obtained from exposed marine sediments on Crete (Valet and Laj, 1984).

Van Hoof (1993) has obtained a detailed record of the Gilbert–Gauss boundary in deep-sea marls from southern Sicily. Intermediate VGPs are largely confined to the Americas, and are concentrated in two clusters – a very prominent one at the north-east coast of North America and a less distinct one at the southern tip of South America. These clusters coincide remarkably well with two clusters identified by Hoffmann (1991, 1992). Although the confinement and clustering may reflect true behaviour of the geomagnetic field, van Hoof suggests that it may be caused, or at least modified, by a secondary (present-day) overprint, or by a filtering mechanism due to early diagenetic CRM acquisition.

6.7 The Steens Mountain reversal

One of the best-documented records of a polarity reversal is that observed in the 15.5 ± 0.3 Ma Miocene reverse to normal transition recorded in lava flows from Steens Mountain, south-eastern Oregon. The transition was first discovered by Watkins (1965a,b; 1969) and investigated in great detail by Mankinen *et al.*

(1985) and Prévot *et al.* (1985b). A shorter account of their work has been given by Prévot *et al.* (1985b). The record begins with an estimated several thousand years of reversed polarity with an average intensity of 31.5 ± 8.5 μT, which is only about two-thirds of the expected Miocene intensity. The polarity transition consists of two phases (see Figure 6.21). In the first phase, which lasted ~ 550 ± 150 a, the field temporarily reached normal polarity, the intensity increasing to pre-transitional values. The field then remained in this polarity for a brief period (100–300 a). Prévot *et al.* (1985a) suggest that the dipolar structure of the field could have been briefly regenerated in an aborted attempt to re-establish a stationary field. The second phase of the transition is characterized by very low intensities, with an average of 10.9 ± 4.9 μT for field directions more than 45° away from the dipole field and a minimum of ~ 5 μT. During this time, changes in direction described a long counterclockwise loop, in contrast to the changes in the first phase, which were confined near the local north–south vertical plane (see Figure 6.22). The second phase lasted ~ 2900 ± 300 a and both normal directions and intensities were attained at the same time. The record closes with an estimated several thousand years of normal polarity with an average intensity of 46.7 ± 20.1 μT, in agreement with the expected Miocene value. Large and apparently rapid intensity fluctuations are observed following the end of the transition, which may reflect some instability of the newly re-established dipole. Both directional and intensity changes appear to have occurred in a very irregular manner during the transition. Periods of regular, though probably fairly rapid, movements of the field were recorded in successive lava flows. There were also periods of sudden movement in which changes in direction were recorded during the cooling history of individual flows. There were at least two, and possibly three, large swings at astonishingly high rates (see Figure 6.22). Each of these transitional magnetic 'impulses' occurred when the field intensity was low (< 10 μT). Finally, there are periods (600 ± 200 a) of directional stasis during which the field showed little or no movement while several flows erupted. During this time the magnitude of the field increased greatly.

For the best-documented magnetic impulse the angular difference between the magnetization of near-surface samples from one flow and that recorded by interior samples from the same flow was 32 ± 14°. Mankinen *et al.* (1985) determined the rate at which this change occurred by estimating cooling times for the Steens Mountain lava flows by comparison with the cooling history of lava lakes in Hawaii. They obtained the extremely rapid angular change of the magnetic field of 58 ± 21°/a. The intensity change during this period was 6700 ± 2700 nT/a, which is about 15–50 times larger than the maximum rate of change of the non-dipole field observed in historical records. Changes in intensity during the transition sometimes occurred at the same time as changes in directions, but at other times they occurred independently. Mankinen *et al.* (1985) estimated that the total duration of directional changes during the transition was 3600 ± 700 a, while the total duration of the period of diminished intensity was 4400 ± 900 a.

It is obvious that the Steens Mountain transition cannot be explained by simple flooding or standing non-dipole field models. Figure 6.22 shows qualitatively

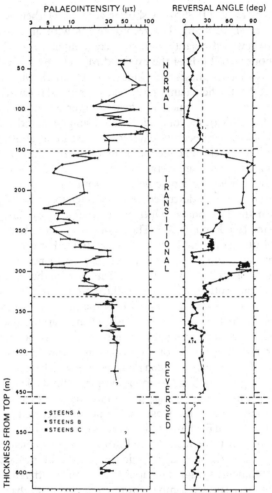

Figure 6.21 Palaeointensity and reversal angle (as measured from the dipole field direction, regardless of polarity) of individual lava flows versus thickness from top of section A, measured from the middle of flows. Correlations between Steens A, B and C from Mankinen *et al.* (1985). Reference section for calculating thickness is Steens A from the top to 318 m (A66, B66), Steens B from 318 m to 568 m (B102) and Steens C (Adel) below. Where necessary, minor adjustments in thickness have been made in order to have the data from the three sections represented in their assumed chronological order. Palaeointensity diagram (logarithmic scale): symbols (diamonds, circles, squares) correspond to the weighted mean for the flow; error bars are root mean square of the deviations about the unweighted average (see text); off-centring of symbol within error bar, when observed, is mainly a consequence of weighting. The absence of error bar indicates a single palaeointensity determination. Reversal angle diagram: symbols are the same as for the palaeointensity diagram; uncertainties about this angle, which are not represented for more clarity, are usually less than 5° (Mankinen *et al.*, 1985); the vertical dashed line is the semiangle of the cone which contains 95% of the individual flow direction during Miocene reversed and normal periods, as calculated from Columbia River Basalt Group lavas, Oregon and Washington. After Prévot *et al.* (1985a).

the importance of non-zonal components during the Steens Mountain reversal. Prévot *et al.* (1985b) suggest that the entire directional path observed during the Steens Mountain transition can be interpreted as corresponding to a rapidly damped oscillation of the field, as proposed by Braginskiy (1964). The absence of a purely zonal morphology for transition fields is compatible with the suggestion of Gubbins and Roberts (1983) that the frozen-flux approximation is valid during reversals. Under the frozen-flux approximation the transitional magnetic impulses seen on the Steens Mountain record can be accounted for by a large increase in fluid velocity at the core surface (Prévot *et al.*, 1985b). This would imply an increase in the level of turbulence within the liquid core during reversals. This increase in kinetic energy could result from a decrease in magnetic energy or could be of thermal origin.

Valet *et al.* (1985a) take issue with the interpretation of Prévot *et al.* (1985b) that the Steens Mountain transition consists of two different phases. They suggest that the second phase occurred at a significantly later time than the first phase, and is an excursion unrelated to the preceding reversal. Grommé *et al.* (1985) defend the interpretation of Prévot *et al.*, pointing out that all available field observations and geochemical and geochronological data contradict the suggestion of Valet *et al.* that the two phases represent two different geomagnetic events. Valet *et al.* (1985b) remain unconvinced and point out that, in some of the reversal records they have obtained from sedimentary sections in Greece, they have observed excursions occurring either before or after a transition.

Roberts and Fuller (1990) obtained new palaeomagnetic data from the Santa Rosa Mountains which are of similar age to, and about 200 km away from, Steens Mountain. The similarity between the directional variation of the magnetic field recorded by lavas at these two sites convinced Roberts and Fuller that the same field reversal was observed – as had been suggested earlier by Larson *et al.* (1971). The lavas of the Santa Rosa formation show an initial switching of the field from reversed to normal polarity followed by a large counterclockwise loop, as is also seen in the Steens Mountain record. Roberts and Fuller conclude, as did Mankinen *et al.* (1985), that the reversal occurred as a two-stage process. In a study of Cenozoic lava sequences in New Zealand, Hoffman (1986) found evidence for a two-stage process of field reversal – the field occasionally taking up an intermediate direction from which it sometimes goes on to complete the reversal and on other occasions reverts back to its original polarity.

Coe and Prévot later (1989) returned to Steens Mountain to investigate in more detail another flow (B51) from the region where their earlier study had indicated a possible transitional impulse in the geomagnetic field. The results of their further study were even more surprising. Figure 6.23 shows a group of nine flows with similar directions (the last being B52) which precedes a 90° jump to a normal direction (flow B50) followed by a group of six flows which are indistinguishable from one another. Coe and Prévot suggest three possibilities to account for the extraordinary unusual distribution of the directions of remanent magnetization within flow B51 – baking by the overlying flow, incompletely removed VRM, and rapid variation of the geomagnetic field direction during cooling and acquisition of primary remanence. However, secondary TRM caused

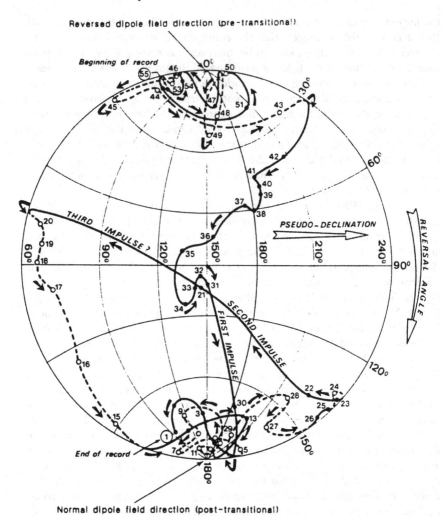

Figure 6.22 The Steens Mountain directional record. Stereographic projection (true angle) of the rotated field direction of the successive directional groups. Directions are rotated by 28.5° about the east–west horizontal axis to bring the dipole field directions coincident with the poles of the projection sphere. The reversal angle varies from 0° (reversed dipole direction) to 180° (normal dipole direction), and corresponds to lines of equal 'latitude' on the projection sphere. Lines of equal 'longitude' correspond to pseudo-declination, which is measured in the 'equatorial' plane from the projection of the north-seeking direction into this plane. A purely axisymmetrical field would have a pseudo-declination equal to 0 or 180° which corresponds to near-sided or far-sided virtual geomagnetic pole paths, respectively. Directional group numbers are indicated near most of the circles representing the successive average field directions. With few exceptions, 95% confidence semiangles are only a few degrees and could not be drawn on this diagram. Full circles and solid path are on the hemisphere of the projection sphere which contains the pole of projection (reversal angle 90°, pseudo-declination 150°); empty circles and broken path are on the opposite hemisphere. After Prévot *et al.* (1985b).

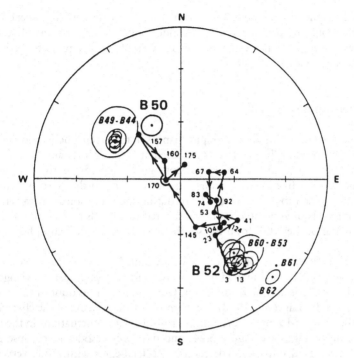

Figure 6.23 The first large directional jump during the reversal, as recorded at Steens Mountain section B. Flow-mean directions with their associated 95% confidence limits are shown by small dots. The forward direction of time in the lava flow pile corresponds to decreasing numbers, and the directional jump occurs between flows B52 and B50 (heavy α_{95} circles). Large dots are directions of remanence of samples from the intervening flow B51 after demagnetization to approximately 50 °C, with heights (in cm) of samples above the base of this 190-cm-thick flow indicated alongside and lines with arrows connecting the samples in ascending order. Note that these directions tend to fill in the gap in direction between flows B52 and B50, and that there is a strong serial correlation between vertical position in flow B51 and these directions. After Coe and Prévot (1989).

by reheating by the overlying flow B50 could only explain the variation in direction of the first few samples near the top of flow B51. Thermal demagnetization studies show that incomplete removal of VRM could not explain the observed record. They also found no evidence of any chemical changes throughout flow B51 that could influence the acquired TRM. Assuming that flow 51 really did record a change in the geomagnetic field as the flow cooled by conduction during the reversal, Coe and Prévot estimated that the field must have changed at the incredibly high rate of at least 3° and 300 µT/day. The implication of such a rapid change in the geomagnetic field would imply fluid velocities near the CMB of at least 1 km/h (analyses of the secular variation by Whaler, 1982, suggest fluid velocities at the CMB could at most be a few tens of km/a). Moreover, for such rapid magnetic field variations not to be electromagnetically screened by the mantle would require the electrical conductivity of the lower mantle to be 4–5 times lower than that estimated by analyses of the geomagnetic 'jerk' of 1969–70 (Courtillot *et al.*, 1984).

6.8 Other rapid reversals

There have been other instances of rapid changes in the geomagnetic field during a reversal, but nothing like as fast as that seen at Steens Mountain. Okada and Niitsuma (1989) obtained detailed palaeomagnetic records of the Brunhes–Matuyama reversal from marine sediments outcropping in the Boso peninsula, central Japan. The movement of the VGP during the transition is characterized by short intervals of very rapid motion separated by periods of relative stability. The last swing of the VGP from the southern to the northern hemisphere took only 38 a.

Laj *et al.* (1987) have carried out a very detailed study of three successive polarity reversals recorded in Middle Miocene marine clays – near the village of Macherado on the island of Zakinthos in Greece. A fuller account of this work has been given by Laj *et al.* (1988). These records exhibit many of the characteristics of the Steens Mountain reversal. They display large fluctuations in the rate of directional change (see Figure 6.24) and different episodes of collapse and recovery of the field intensity. The reversal ZM01 begins with a full, very fast N–S episode (see Figure 6.25). The intermediate directions suggest an ill-defined path east of the site's longitude. After a period of standstill at the south pole, the VGP moves back to the north pole. The path is not well defined – it is situated approximately in a longitudinal band 90° W of the site, indicating that large non-zonal terms are present at this stage of the transition. The VGPs cluster around the north pole for a time and then start moving again towards mid-latitudes more or less along the site meridian. In contrast to the rapidity of the first N–S–N cycle, the VGPs linger at north mid-latitudes for a relatively long time, with quite a large scatter of the field directions both longitudinally and latitudinally. After this period of standstill, the VGPs start moving again, first to the west, then southward along a rather narrow longitudinal band centred 90° W of the site. The south pole is reached after a final oscillation that brings the VGP back to equatorial latitudes. Thus, large changes in the structure of the transition field occur during this N–S segment, zonal terms probably dominating the onset, whereas non-zonal terms are apparent during the second half.

There is a sharp drop in the intensity of the field slightly before the beginning of the very fast N–S directional change. A significant increase in field intensity then occurs after the first N–S directional change, suggesting a temporary re-establishment of the usual (R) dipole field. The intensity then drops again while the VGP returns to the north pole and does not recover when it reaches a normal dipole polarity. The time during which intermediate directions occur (VGPs lingering at mid-latitudes) is also marked by a progressive increase in the field intensity. Finally, the field intensity drops again during the last part

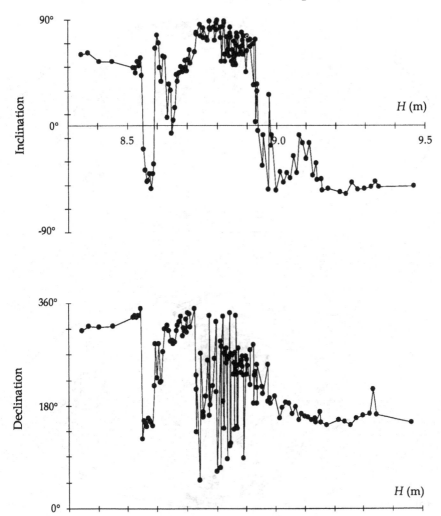

Figure 6.24 Plots of the observed inclination and declination values as a function of stratigraphic height for three successive reversals near Macherado, Greece. The transition is characterized by a very fast onset. The fluctuations present on the declination record between 8.7 and 8.9 m correspond to a period of very high inclinations. Their amplitude is thus amplified by the projection. After Laj *et al.* (1987).

of the record, ultimately reaching non-transitional values after completion of the reversal.

The Macherado record (ZM01) is characterized by periods of very rapid change alternating with periods of near-stationarity. The first N–S directional

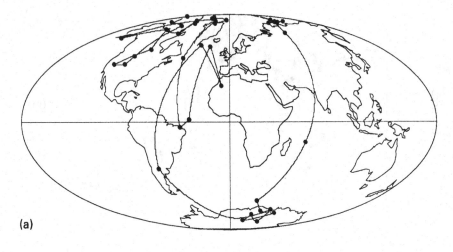

(a)

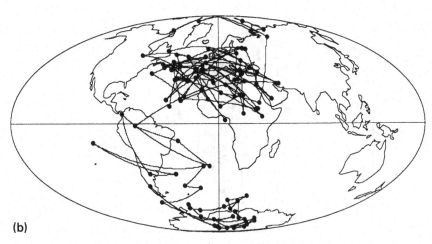

(b)

Figure 6.25 Plot of the VGP path during the reversal ZM01. The path of the first N–S–N cycle (a) has been separated from that of the final N–S change (b). This separation is arbitrary and only intended to increase the clarity of the plot. There are very few points on the first N–S–N cycle in spite of the fact that 5 mm samples from the same core were used, indicating a very fast rate of change of the geomagnetic field vector. This diagram is centred on the Greenwich meridian. After Laj *et al.* (1987).

change took 500–1000 a. The recovery and successive decay of the field intensity at the end of the first N–S phase, when the field direction is due south, is also very rapid – the entire process lasting only ~ 2000 a. In contrast, no recovery of the field intensity is observed during the subsequent period, lasting about

3000–4000 a, characterized by normal (northerly) directions. Thus, in spite of perfectly dipolar directions, this configuration cannot be considered as a re-establishment of a normal dipolar configuration, but is a transitional one. A more stable configuration is observed in the rather long period (30% of the total duration of the reversal) characterized by intermediate directions and a progress-ive increase of field intensity. The Macherado reversal thus provides additional evidence for the existence of transient intermediate states of the geomagnetic field characterized by non-negligible intensity values and non-dipolar directions. These intermediate states have been previously observed only in lava sequences. The occurrence of episodes of very rapid changes of the geomagnetic field sup-ports those models for reversals that involve large changes in the level of turbu-lence in the core under frozen-flux constraints (Prévot *et al.*, 1985a). The Macherado record (the very fast N–S onset) seems incompatible with the time constants for flux diffusion and may be entirely controlled by a frozen-flux-dominated reversal mechanism.

The characteristics of this record are different from those of the records of four sequential reversals obtained in Crete from upper Tortonian sediments of similar sedimentary and magnetic mineralogy (Valet and Laj, 1984; Valet *et al.*, 1988a). The Macherado records are not confined to narrow longitudinal bands as are the Cretan ones. Moreover, the mean paths corresponding to ZM01 and ZM02 are approximately antipodal, while those for ZM02 and ZM03 virtually coincide in spite of the fact that these last two reversals are of opposite sense. This suggests that large changes in transitional fields may occur over a period of ~ 6 Ma – the time between the Macherado and Cretan records. If confirmed, this would not support the model of reversals of Gubbins (1987), which requires that all records from the same geographical area be similar within the lifetime of mantle convection patterns.

7
Magnetostratigraphy

7.1 Introduction

Magnetostratigraphy organizes strata according to their magnetic properties acquired at the time of deposition. Because the polarity of the Earth's magnetic field has reversed repeatedly in the geological past and because polarity transitions, while lasting only a few thousand years, are synchronous over the entire globe, their record in marine or land-based sediments provides isochrons applicable to worldwide correlation. Khramov (1957) recognized the value of palaeomagnetism in the subdivision and correlation of sedimentary sequences in the Soviet Union. However, it was not until it was used in deep-sea sedimentary sequences many years later that the value of magnetic stratigraphy was fully appreciated. Magnetic reversals are distinct from most other stratigraphic criteria, which are characteristically diachronous. There are other synchronous events such as tephra layers which may be used for correlation but which are too restricted geographically for wide-scale use.

Magnetostratigraphy for the last 5 Ma, based on radiometric dating of sequences of terrestrial lava flows, has provided information with which to establish age relationships of sediment and fossil sequences. This chronology in turn has provided a basis for determining rates of geological change with fair accuracy. It must be stressed, however, that a magnetostratigraphic sequence by itself does not provide unequivocal dates for geological events preserved in sediments, since magnetic reversals are repetitive events and do not, in general, possess singular properties. Independent criteria are therefore required to verify the age of a sequence. However, examination of microfossils at only a few stratigraphic levels may enable a biostratigrapher to establish the age of a magnetic polarity sequence.

The development of a timescale for young sequences of volcanic rocks has been described by a number of people (see, e.g., McDougall, 1979, for a good review of the earlier work). Until the middle 1970s the polarity timescale was confined to rocks younger than about 4.5 Ma. A major difficulty in extending it to older volcanic and sedimentary sequences is the decreasing resolution of the K–Ar method of dating older rocks. The first use of combined magnetic polarity and K–Ar data was by Rutten (1959), who concluded from studies on lavas from volcanoes near Rome that the present normal magnetic interval has lasted for at least 0.47 Ma, and that an earlier interval of normal polarity possibly existed about 2.4 Ma ago. In the early 1960s a large amount of additional data on the same or closely related lavas was published, leading to the development of a polarity timescale for the last 4 Ma. Cox *et al.* (1964) collected all the available data and noted that during the last 3.5 Ma changes in polarity had taken place at irregular intervals, and that, within intervals of predominantly one polarity (called epochs), shorter intervals of opposite polarity (called events), about 0.1 Ma in duration, sometimes occurred*. During 1966 many papers incorporating new data and giving compilations of previous results were published. These results confirmed the general outline of the polarity timescale and provided evidence for additional short events.

In compiling a local magnetic polarity stratigraphy from a sequence of superposed rocks, of necessity only a finite number of samples can be taken. Thus, it is quite likely that some polarity changes will have been missed between adjacent samples. It is usually assumed that the space between adjacent points of like polarity has the same polarity, i.e. that no short polarity zones have been missed. However, the data could equally well be explained if an even number of reversals had taken place in the interval between the two sampling points. On the other hand, if adjacent sampling points have different polarity, it is usually assumed that one magnetic reversal has occurred midway in the space between them. It is clear that the same record would have been obtained if an odd number of reversals had occurred in this interval.

In 1966 there also took place the successful and accurate correlation by Opdyke *et al.* and Ninkovich *et al.* of the reversal history recorded on land with magnetic polarity measurements in deep-sea sedimentary cores. The first workers to record reversals in deep-sea sediments were Harrison and Funnell (1964), who demonstrated that the last reversal of the Earth's magnetic field occurred within the Quaternary. Later (1966) Harrison obtained an isochronous correlation in sections more than 3000 km apart, and made the first attempt to correlate these reversals with the polarity timescale. In the same year Ninkovich *et al.* examined a set of deep-sea cores from the North Pacific and demonstrated quite clearly the presence of the Jaramillo event in deep-sea sediments. This work was also among the first to estimate the duration of the Jaramillo and Olduvai events and to make estimates of the rates of sediment deposition. It was also the first to intercalibrate correlations, using both ash layers and palaeomagnetic reversals. In 1967 Hays and Opdyke extended the magnetostratigraphy of Antarctic sediments to about 3.4 Ma, using three long piston cores, and obtained strong evidence for the existence of three short polarity events within the Gilbert reversed epoch. A radiolarian biostratigraphy, based on the upward sequential disappearance of several radiolarian species, was dated by use of the polarity scale and evidence put forward for a possible connection between reversals of the Earth's magnetic field and radiolarian extinctions (see also Section 8.4).

One of the attractions of magnetostratigraphy is that correlation over long

---

* The terminology of polarity intervals is discussed in Section 7.2. For the moment the terms 'epoch' and 'event' will be retained.

distances using fossils is always difficult, even when planktonic groups are used. This is because different water masses within the oceans are marked by distinct planktonic assemblages, and even species that do range over wide areas often exhibit stratigraphic ranges in different oceanic regions. The value of magneto-stratigraphy was further demonstrated by Hays *et al.* (1969) when they dated and correlated a large number of microfossil events in eastern equatorial Pacific deep-sea cores of Quaternary and Pliocene age. They showed that changes in the fossils used for the differentiation of the Miocene–Pliocene boundary were only 4.5–5 Ma old and occurred near the base of the Gilbert epoch. Up until this time, little and often contradictory radiometric evidence had suggested an age of 9 Ma for this boundary. They were also able to date climatic cycles in the Quaternary and found a number of instances of the extinction of microfossils at the time of reversals (see also Section 8.4).

Johnson and McGee (1983) developed statistical models which can be used to test the interpretation of a set of magnetic polarity stratigraphy data or to estimate the age information contained in magnetic polarity data. Their models are based on the assumption that the occurrence of magnetic reversals in the geological past is given by Equation (5.25) and that each palaeomagnetic sampling site recovers the true magnetic polarity for its stratigraphic level (in time).

They estimated the mean time span for polarity intervals in the late Neogene to be 120 ka, in good agreement with a large body of magnetic polarity stratigraphy data. Once the mean time span for polarity intervals has been found, an estimate of the time spanned by the stratigraphic section may be made – again they obtained good results for a variety of magnetic polarity sections in the Neogene. Johnson and McGee further estimated the mean time span for polarity intervals in the early Palaeogene to be 327 ka. It must be stressed that their statistical models are based on the spacing of samples in time, not stratigraphic units. In a typical stratigraphic sequence, unconformities, unsuitable lithologies and differential sedimentation rates will tend to randomize the time spacing. Thus, although a time-uniform sampling programme was shown to be the most efficient for establishing a local magnetic polarity stratigraphy, a random sampling model is probably more applicable.

7.2 Magnetostratigraphic polarity units

The purpose of magnetostratigraphic classification is to organize rock strata systematically into identifiable units based on stratigraphic variations in their magnetic characteristics. In considering the chronostratigraphic value of polarity horizons and units, it must be stressed that, although they may be very useful guides to isochronous position, they have relatively little individuality (one reversal being very much the same as another) and therefore can usually only be unequivocally identified by the use of supporting evidence, such as palaeontological or radiometric data. A magnetostratigraphic polarity unit is an objective unit ideally based on a directly determinable property of the rocks – their magnetic polarity. The presence of the unit can be strictly assured only where this property can be identified. In these respects it is more similar to a lithostratigraphic or a biostratigraphic unit than to a chronostratigraphic unit. However, lithostratigraphic and biostratigraphic units are usually geographically quite restricted, whereas a polarity unit is potentially worldwide and in this respect more similar to a chronostratigraphic unit. Again it must be emphasized that they are not chronostratigraphic units, since they are defined primarily, not by time, but by a specific physical character – the polarity of remanent magnetism.

The best sequential record of reversals of the Earth's magnetic field for the past 150 Ma is preserved in the pattern of sea-floor spreading anomalies and has been dated by extrapolation and interpolation from radiometric and palaeontological evidence. However, because of the nature of these linear magnetic intensity anomalies from the ocean floor, it is not possible to designate any satisfactory type intervals or type boundaries. Instead, the standards of reference for the marine magnetic anomalies must remain profiles such as those described by Heirtzler *et al.* (1968), the boundaries of the units being determined by model fitting.

Ideally, the standard for the definition and recognition of a magnetostratigraphic polarity unit should be a clearly designated stratotype in a continuous sequence of rock strata – a specific section showing the polarity pattern of the unit throughout and clearly defining its upper and lower limits by means of boundary stratotypes. Alvarez *et al.* (1977) have attempted to establish a stratotype section of this sort in the Gubbio section of Italy, where upper Cretaceous polarity units have been identified geologically and geographically and related to lithostratigraphic and biostratigraphic data from this section.

Rock magnetic polarity units have been established in two different ways: (1) through a combination of radiometric or biostratigraphic age data with magnetic polarity determinations on outcropping or cored volcanic and sedimentary rocks, and (2) through the use of magnetometer profiles from ocean surveys to identify and correlate linear magnetic anomalies that are interpreted as reversals of the Earth's magnetic field recorded in the lavas of the sea-floor during sea-floor spreading, usually in the complete absence of any direct radiometric data. Any system should be flexible enough to allow for polarity intervals discovered at a later date to be conveniently and unambiguously inserted into the existing system. The first four periods of dominantly one polarity were called epochs, and named after past workers in geomagnetism (shorter intervals of one polarity that occurred during the above epochs were named after the locations where they were first observed).* In 1972 a subcommission on a Magnetic Polarity Time Scale (SMPTS) was established as part of the International Commission on Stratigraphy, which in turn is a part of the International Union of Geological Sciences. After a series of meetings this commission produced a set of recommendations in an attempt to establish an unambiguous nomenclature in magnetostratigraphy, which is as consistent as possible with conventional stratigraphic terminology. Recommendations by the subcommission, which were

---

* These names should be preserved, but should be considered to be primarily polarity zones and not formal chronostratigraphic and geochronological units.

rather general, were published in 1973 and guidelines were given by Watkins (1976).

The report, however, failed to satisfy fully the International Subcommission on Stratigraphic Classification (ISSC), who in turn produced a document on magnetostratigraphic polarity units (1978), which was then reviewed by members of the SMPTS. The resulting report was published as a supplementary chapter of the International Stratigraphic Guide (Anon., 1979) and is the prepared statement of both the ISSC and the SMPTS.

The terms recommended for describing subdivisions of time based on geomagnetic polarity are 'polarity subchrons', 'polarity chrons', and 'polarity superchrons'. Polarity chrons describe the main subdivisions and replace the older term 'epoch'. The term 'subchron' describes very short ($\leq$ 0.1 Ma) polarity intervals occurring within a chron and replaces the older term 'event'. Superchrons are used to describe very long periods of one polarity. The four most recent chrons were originally named after eminent researchers in geomagnetism (Brunhes, Matuyama, Gauss and Gilbert). This system of nomenclature is obviously impractical for earlier chrons because of their large number. Older chrons were defined as being predominantly normally or predominantly reversely magnetized intervals in conventional piston cores of deep-sea sediments. Hays and Opdyke (1967) numbered such intervals sequentially from 1 (the Brunhes), odd numbers referring to predominantly normal polarity intervals and even numbers to reversed polarity intervals. Opdyke *et al.* (1974) and Theyer and Hammond (1974) later extended this scheme to chron 23 near the base of the Neogene.

About the same time a second numbering scheme was set up by Le Pichon and Heirtzler (1968) based on the positive marine magnetic anomalies that are seen in magnetic profiles over ocean basins. The numbers go from 1 at currently spreading mid-ocean ridges to 32 (Pitman *et al.*, 1968). However, the initial set of 32 numbered anomalies could only account for a fraction of the later known chrons. Additional chrons representing smaller positive anomalies are labelled by adding letters, decimals and primes to the original set of 32 numbers (LaBrecque *et al.*, 1977; Ness *et al.*, 1980). Harland *et al.* (1982) identified unlabelled chrons by adding letters to the next youngest numbered chron, e.g. 5A and 5B follow 5, and 5AA and 5AB follow 5A. Subdivisions of currently numbered chrons such as 5A are described by additional decimal numbers, e.g. 5A.1, 5A.2. Subchrons are labelled with the number of the chron in which they occur followed by -1, -2 . . . , the numbers increasing in order of increasing age (see Figure 7.1).

To avoid any confusion between the chron numbering scheme based on the magnetic properties of Neogene sediment cores and ocean floor spreading chrons, LaBrecque and Hsu (1983) suggested prefixing ocean floor spreading chrons by C. In the *Geologic Time Scale* (1982) of Harland *et al.* previously unlabelled reversed chrons were given the number of the next-youngest normal chron with the letter r appended. LaBrecque and Hsu (1983) suggested referring to normally magnetized ocean floor chrons with the suffix N and reversely magnetized chrons with the suffix R. Harland *et al.* (1990) adopted these suggestions in their updated

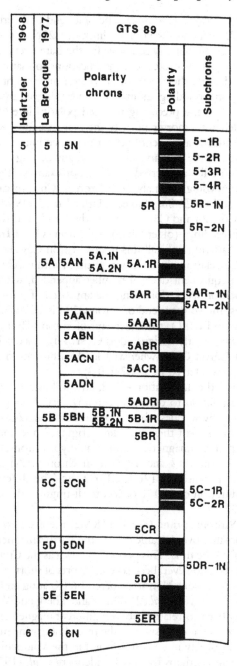

Figure 7.1 Numerical scheme for numbering chrons and subchrons derived from numbered marine magnetic anomalies. After Harland *et al.* (1990): redrawn from Figure 1 of Berggren *et al.* (1985a).

version of their 1982 *Geologic Time Scale*. The older end of a chron, e.g. C29N, is referred to as C29N(o), and the younger limit of C29N as C29N(y).

In the latest geomagnetic polarity timescale for the Late Cretaceous and Cenozoic, Cande and Kent (1992) make some modifications to that of Harland *et al.* (1990). Like Harland *et al.*, they refer to the longest intervals of predominantly one polarity by the corresponding anomaly number followed by the suffix n for normal polarity, or r for the preceding reversed polarity interval. When these chrons are subdivided into shorter polarity intervals, Cande and Kent (1992) refer to them as subchrons and identify them by appending, from youngest to oldest, a 0.1, 0.2, etc., to the primary chron name, and adding an n for a normal polarity interval or an r for a reversed interval. For example, the three normal polarity intervals composing anomaly 6C (chron C6Cn) are called subchrons C6Cn.ln, C6Cn.2n and C6Cn.3n, whereas Harland *et al.* (1990) refer to them as chrons C6C.1n, C6C.2n and C6C.3n. Similarly, Cande and Kent refer to the reversed interval preceding (older than) subchron C6Cn.1n as subchron C6Cn.1r, whereas Harland *et al.* call this interval chron C6C.1r.

For more precise correlation, the fractional position within a chron or subchron is referred to by the equivalent decimal number appended, within parentheses, to the chron or subchron name, following the approach of LaBrecque and Hsu (1983). As examples, the younger end of chron C29n is C29n(0.0) or C29(0.0) (= C29n(y) in Harland *et al.*), the older end of chron C29n (= C29n(o) in Harland *et al*). is conveniently designated as C29r(0.0), since it is equivalent to the younger end of Chron C29r, whereas a level within chron C29r and 3/10 from its younger end is referred to as C29r(0.3).

Cande and Kent use the designation -1, -2, etc., following the primary chron or the subchron designation to denote apparently very short polarity intervals corresponding to the tiny wiggles which, upon calibration, convert to durations of less than 30 ka. In view of their uncertain origin, Cande and Kent refer to these globally mapped geomagnetic features as cryptochrons. Thus, the tiny wiggles between anomalies 12 and 13 (within chron C12r) are called from youngest to oldest cryptochrons C12r-1, C12r-2, etc. This differs from Harland *et al.*, who used a duration of 100 ka or less to distinguish the class of shortest polarity intervals.

From Aptian to Santonian time (~ 83–118 Ma) there are no globally recognized magnetic anomalies over oceanic crust – the Earth's magnetic field appears to have been normal all the time. This period is known as the Creteceous Normal Superchron. There have, however, been several reports of short-duration reversed intervals within the Cretaceous Normal Superchron (Keating and Helsley, 1978; Hailwood *et al.*, 1979; Lowrie *et al.*, 1980b; Vandenberg and Wonders, 1980), but they have not yet been correlated between different magnetostratigraphic sections and have no consistent expression in the record of marine magnetic anomalies. A brief reversed polarity interval was found in the Valdovbia section of the Umbrian Apennines in northern Italy by Vandenberg *et al.* (1978) and Lowrie *et al.* (1980a), and has since been shown to correlate on a worldwide scale (Tarduno *et al.*, 1989; Tarduno, 1990). Several intervals of reversed polarity have now been found in the Contessa section (Tarduno *et al.*, 1992) – see Figure 7.2. Some of the

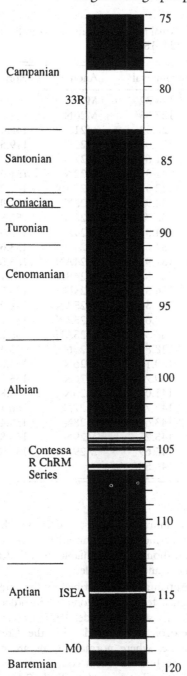

Figure 7.2 Possible mid-Cretaceous polarity timescale. Absolute ages after Harland *et al.* (1982). ISEA after Tarduno (1990). Contessa reversed ChRM series after Tarduno *et al.* (1992).

Table 7.1. Revised ages of normal polarity chrons for ocean-floor magnetic anomalies in the time range 124–158 Ma

Anomaly	Normal polarity interval		Anomaly	Normal polarity interval	
C34N	(83.00)	124.32	M20N	146.93	147.17
M1N	124.88	127.35	M20N	147.22	147.75
M2	127.70	128.32	M21N	148.43	149.30
M4	130.17	131.05	M22N	149.67	150.92
M6N	131.51	131.64	M22N	150.96	151.00
M7N	131.74	131.89	M22N	151.04	151.10
M8N	132.25	132.53	M22AN	151.77	151.87
M9N	132.75	132.99	M23N	152.01	152.31
M10N[a]	133.41	133.72	M23-1	152.53	152.56
M10NN[a]	134.01	134.31	M24N	153.06	153.32
M10NN	134.36	134.65	M24-1	153.61	153.64
M10NN	134.67	134.94	M24AN	153.80	153.90
M11N	135.17	135.87	M24BN	154.11	154.40
M11-1N	136.27	136.31	M25N	154.53	154.76
M11AN	136.64	137.30	M25AN	154.96	155.08
M12N	137.37	137.63	M25AN	155.15	155.23
M12-2N	138.28	138.36	M25AN	155.29	155.40
M12AN	138.53	138.82	M26N	155.48	155.56
M13N	138.92	139.14	M26N	155.62	155.69
M14N	139.50	139.73	M26N	155.74	155.81
M15N	140.46	141.02	M26N	155.85	156.00
M16N	141.47	142.76	M27N	156.12	156.28
M17N	143.28	143.61	M28N	156.41	156.64
M18N	144.80	145.25	M29N	156.81	157.55
M19N	145.58	145.69	M29R(o)	157.98	
M19N	145.75	146.56			

[a]For explanation see text.
After Berggren *et al.* (1985a).

reversed intervals are of very short duration (< 100 ka), which would account for their lack of resolution in marine magnetic anomaly records. However, two of the intervals are longer and should be identifiable in both detailed stratigraphic sections and marine magnetic anomaly profiles.

Polarity chrons of pre-Aptian age are described by the letter M running from M0 to M29 and are assigned to marine anomalies in order of increasing age (Larson and Pitman, 1972; Larson and Hilde, 1975; Cande *et al.*, 1978). The prefix M stands for Mesozoic. Compared with the Cretaceous–Tertiary– Quaternary reversal sequence, where normal chrons are numbered reversed chrons are numbered in the Jurassic–Cretaceous reversal sequence, although chrons M2 and M4 have normal polarity (see Table 7.1). An account of the magnetostratigraphic timescale has been given in a Special Report of the Geological Society by Hailwood (1989).

7.3 The polarity timescale

Because polarity reversals are potentially recorded simultaneously in rocks all over the world, magnestostratigraphic divisions, unlike lithostratigraphic and biostratigraphic, are not time-transgressive. However, the age of magnetization might not be the same as other geological events in the history of the rock. In igneous rocks the magnetization is acquired after the crystallization of the rock and before the setting of the K–Ar clock. In chemically altered rocks, on the other hand, the magnetization is generally acquired at the time of chemical alteration.

The degree of magnetochronologic resolution depends on the frequency of geomagnetic reversals and the availability of a well-defined record of the polarity sequence. For the last 83 Ma (the late Cretaceous and Cenozoic) and during the late Jurassic and early Cretaceous (~118–160 Ma), there were frequent reversals and a fairly precise magnetochronological timescale has been constructed. On the other hand, between about 83 and 118 Ma and between about 160–170 Ma, the magnetic field was of predominantly one polarity (normal). These superchrons correspond to the oceanic Cretaceous and Jurassic 'Quiet' zones, respectively. Although frequent reversals from about 170–200 Ma have been reported in magnetostratigraphic land sections (mainly from the Mediterranean region), no oceanic crust of this age has been found and a detailed sequence of reversals for this time interval is not well established. The geomagnetic field has only two stable polarity states, and a transition from one to another has in itself no diagnostic character. Successful magnetostratigraphic correlation depends on the identification of a sequence of polarity intervals of different durations which form a recognizable pattern, or characteristic 'fingerprint'.

The marine anomaly reversal timescale must be calibrated – by direct isotopic dating or by indirect biostratigraphic correlation. Two methods are used. The first is to drill through the ocean floor beneath a well-defined magnetic anomaly and determine the K–Ar age of the basalt layer or the biostratigraphic age of the sediment immediately above the basalt layer. The difference between the time of extrusion of the basalt and the first sedimentation on the ocean floor is not known, but is unlikely to exceed a few Ma. Sedimentary hiatuses and changes in sedimentation rates can upset the pattern of polarity states. The second method depends on the lengths of the polarity zones. These lengths vary widely and randomly from one zone to the next. Thus, a sequence of some 4–6 intervals gives a pattern which can be correlated with corresponding signatures elsewhere. In spite of the great difference between the widths of polarity zones in the oceanic crust and the thickness of the zones in sedimentary sections, the ratios of polarity zones will be about the same, provided that the rates of sea-floor spreading and sediment deposition do not change much over periods of about 10 Ma. Over appreciable time intervals, however, sea-floor spreading rates are more likely to be constant than sedimentation rates.

There are two steps in setting up a marine magnetic anomaly timescale. First one must construct a magnetic anomaly sequence from ocean-floor profiles which will give the relative spacing of the anomalies. The sequence must then be calibrated in time at particular points, or by particular line segments, and the ages of the anomalies interpolated between the calibration points or on each line segment.

7.4 Isotopic dating of magnetic anomalies

K–Ar isotopic dating of volcanic rocks has been the major method of calibrating Plio-Pleistocene short polarity intervals, since such intervals are not easy to resolve on marine magnetic profiles. The main chron boundary in the first million years is the Brunhes–Matuyama. This boundary has become important for the interpolation of ages in deep-sea sedimentary cores, particularly for the detailed interpretation of palaeoclimates from oxygen isotope data (Shackleton and Opdyke, 1973). In their revised polarity timescale for the past 5 Ma, Mankinen and Dalrymple (1979) gave an age of 0.73 Ma for the Brunhes–Matuyama boundary.

Berger and Loutre (1988) have calculated variations in the orbital parameters of the Earth for the past 10 Ma, and Imbrie and Imbrie (1980) have modelled the climate response to such variations (see Section 8.2). Several attempts have been made at using the geological response to climatic change to calibrate the geomagnetic reversal timescale. Johnson (1982) used rather low-resolution oxygen isotope ($\delta^{18}O$) data (a proxy for ice volume changes) and suggested that timescales based on radioactive decay were lower than those based on astronomical calculations – he estimated an age of about 0.79 Ma for the Brunhes–Matuyama boundary. A later study by Ruddiman et al. (1989), based on higher-resolution $\delta^{18}O$ data combined with orbital 'tuning' techniques, concluded that there was no discrepancy between the timescales based on radioactive decay and those based on astronomical calculations.

Shackleton et al. (1990) found that the effect of changes in the obliquity of the ecliptic and of axial precession together with its modulation by eccentricity variations are clearly present in the Quaternary and late Pleistocene oxygen isotope records from ODP site 677. They developed a reversal timescale for the lower Pleistocene on the assumption that there has been a constant phase relationship between the astronomical forcing and the climatic response over the past 2.6 Ma. Their proposed modification to the timescale would imply that the currently adopted radiometric dates, not only for the Brunhes–Matuyama boundary, but also for the Jaramillo and Olduvai subchrons and the Gauss–Matuyama boundary, underestimate their true astronomical age by between 5 and 7%. They placed the Brunhes–Matuyama boundary at 0.78 Ma, supporting the earlier estimate of Johnson (1982).

Hilgen and Langereis (1989) and Hilgen (1991a), using sapropels in the Mediterranean, had also concluded that K–Ar ages are systematically too young.[*] Tauxe et al. (1992) obtained a ^{40}Ar–^{39}Ar age of 0.746 Ma for sediments immedi-

---

[*] Hilgen (1991a,b) used inferred phase relationships between orbital cycles and sapropel cycles. A detailed account of his work on the astronomically calibrated polarity timescale is given in Section 8.2.

ately overlying the Brunhes–Matuyama boundary from samples taken from the Olorgesailie Formation in the Central Kenya rift, and Baksi *et al.* (1992) an age of 0.783 from a series of lavas from Maui. Spell and McDougall (1992) obtained new $^{40}Ar-^{39}Ar$ ages for Valles Caldera rhyolites, New Mexico, based on laser fusion dating of individual sanidine phenocrysts. They obtained a value of 0.78 ± 0.01 Ma for the Brunhes–Matuyama boundary compared with that of 0.73 Ma obtained earlier by Mankinen and Dalrymple (1979), who used samples from the same area. There appears to be no discrepancy now between the astronomical and the most recent isotopic data for the age of this boundary. Recent analyses have also shown that the Jaramillo subchron is older than the original estimate of 0.90–0.97 Ma by Mankinen and Dalrymple (1979). New $^{40}Ar-^{39}Ar$ ages by Spell and McDougall (1992) suggest that it extended from 0.915 to 1.010 Ma, while Tauxe *et al.* (1992) obtained an age of 0.992 Ma for its upper boundary. Astronomical estimates by Shackleton *et al.* (1990) suggest that it extended from 0.99 to 1.07 Ma.

Liddicoat (1993) has examined the Brunhes–Matuyama transitions recorded in exposed lake sediments near Bishop, California. He found numerous, rapid changes in the polarity in the central part of the transition – he does not believe that they represent the true behaviour of the field, but are the result of inaccurate recording. However, he did find a zone of normal polarity in the upper quarter of the transition. The zone is as much as 30 cm thick and is overlain by at least 10 cm of silt that records reverse polarity. A similar zone of normal polarity occurring in the Matuyama chron about 600 years before the beginning of the Brunhes chron has been reported in Japan by Okada and Niitsuma (1989).

Insufficient reliable polarity data and K–Ar age determinations led to uncertainty in the Pliocene and early Pleistocene geomagnetic timescale. In 1979 Mankinen and Dalrymple published a revised polarity timescale for the past 5 Ma, using a significant amount of new data and new values for the atomic abundance and decay constants of ^{40}K (Steiger and Jäger, 1977): their timescale has been widely used since. There has been considerable controversy as to whether the Olduvai and Gilsa normal events in the Matuyama chron are separate or one and the same – it seems more likely that there was only one change of polarity: Mankinen and Dalrymple placed the Olduvai subchron at 1.67–1.87 Ma. The number and duration of the Réunion events are also uncertain, although most evidence suggests that there probably two. Mankinen and Dalrymple placed the two Réunion events at 2.01–204 and 2.12–2.16 Ma. Two reversed subchrons have been found in the Gauss normal polarity chron, the Mammoth and the Kaena. Mankinen and Dalrymple placed the Kaena subchron at between 2.92 and 3.01 Ma and the Mammoth between 3.05 and 3.15 Ma.

Renne *et al.* (1993) have presented new $^{40}Ar-^{39}Ar$ dating and magnetostratigraphic data from the Hadar Formation, Ethiopia, and obtained an estimated age range of 3.21–3.30 Ma for the Mammoth subchron. This is in excellent agreement with the astronomically calibrated geomagnetic polarity timescale prediction of 3.22–3.33 Ma and reinforces findings that earlier estimates of ages in the Pliocene and early Pleistocene are too young. Walter (1994) has done further work on the Hadar Formation, using single-crystal laser-fusion $^{40}Ar-^{39}Ar$

analyses, and confirmed that the ages of both the Mammoth and Kaena subchrons need to be increased by as much as 4–5%.

Zijderveld *et al.* (1991) carried out a detailed magnetostratigraphic and biostratigraphic study of late Pliocene to early Pleistocene marine marl sequences in Calabria, Italy. In the Crotone area the upper and lower boundaries of the Olduvai subchron were clearly identified. The records showed some interesting results near the upper boundary. A relatively long normal polarity interval representing the main Olduvai subchron and lasting 115 ka was followed by a short (30 ka) reversed interval and then a short (15 ka) normal interval. Zijderveld *et al.* suggest that the complete N→R→N polarity succession with a total duration of 160 ka be considered to represent the Olduvai subchron, which thus has a short reversed interval in its upper part. The Réunion event was recognized in sequences from the Monte Singa area. Astronomical calibration gives ages older than conventional ages based on radiometric dating: for the Olduvai subchron 1.79–1.95 Ma as against 1.66–1.88 Ma.

Liddicoat *et al.* (1980) examined the palaeomagnetic polarity in a 930 m core from Searles Valley, California (see Figure 7.3). For 320 m below a depth of 185 m, the inclination is predominantly negative, and Liddicoat *et al.* assume it to represent the Matuyama chron. Below that the inclination is predominantly positive, which they assume to be the upper part of the Gauss chron. Assuming that the Brunhes–Matuyama and Matuyama–Gauss boundaries have been correctly recognized, it is possible to plot age against depth to identify the zones of positive inclination within the Matuyama. In this way Liddicoat *et al.* identified the Jaramillo, Olduvai and two Réunion events. Three other short segments of positive inclination, not included in the Mankinen and Dalrymple timescale (1979), are also present in the record (N_1, N_2, N_3 in Figure 7.3). Liddicoat *et al.* speculate that N_3 might be the same as the Cobb Mountain event (see Section 6.4). The N_2 event may correlate with the controversial Gilsa event. Finally, the N_1 event might correspond to the X event (Heirtzler *et al.*, 1968) seen in some magnetic anomaly data or to that observed in Icelandic lavas very close to the Matuyama–Gauss boundary (Kristjansson *et al.*, 1978).

In the Gauss chron Liddicoat *et al.* identified the Kaena and Mammoth subchrons and, in addition, found two other short segments of negative inclination (R_1, R_2 in Figure 7.3). Some reversely magnetized K–Ar dated basalt flows in the age range of R_1, R_2 have been reported – two from Oaha in Hawaii (Doell and Dalrymple, 1973), one in west Iceland (McDougall *et al.*, 1977) and one in St Vincent (Briden *et al.*, 1979). The advantage of using lacustrine deposits such as those in Searles Valley to look for short events is that the sedimentation rate is at least 50 times greater than that in most deep-sea sediments and is much less discontinuous than the eruption of lava flows.

Cox and Dalrymple (1967) identified two normal polarity intervals within the Gilbert reversed chron – these were named the Cochiti event, dated at 3.7 Ma and the Nunivak event, dated at 4.1 Ma. Two older events have also been found in marine sediments and called C_1 and C_2 by Foster and Opdyke (1970), with mean ages of about 4.37 and 4.56 Ma (Opdyke, 1972). Both these older intervals of normal polarity in the Gilbert chron have subsequently been identified in a

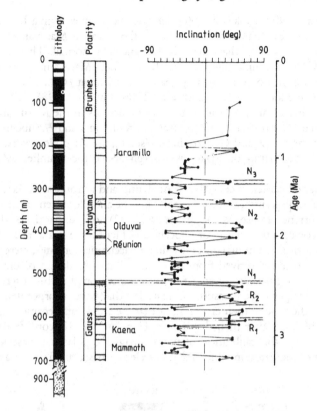

Figure 7.3 Palaeomagnetic polarity in Searles Valley, California. Shaded region highlights the unnamed zones of inclination – three in the Matuyama chron and two in the Gauss chron. In the lithological column: unshaded area, salines; black area, mud; spotted area, arkose. In the polarity column: hatched area, normal polarity; unshaded area, reversed polarity. Age is extrapolated from polarity boundaries in the timescale of Mankinen and Dalrymple (1979), and by assuming a uniform sedimentation rate. Note that the timescale is not linear. After Liddicoat *et al.* (1980).

continuous sequence of lava flows in western Iceland by McDougall *et al.* (1977) and named the Sidufjall and Thvera events. The ages of the boundaries of the Sidufjall event were estimated to be 4.33 and 4.45 Ma and those of the Thvera event 4.60 and 4.80 Ma. Mankinen and Dalrymple (1979) place these two events at 4.32–4.47 Ma and 4.85–5.00 Ma.

Champion *et al.* (1988) believe that there are probably ten short-lived subchrons within the Brunhes chron and the upper part of the Matuyama chron, in addition to the Jaramillo subchron. Spell and McDougall (1992) have used new ^{40}Ar–^{39}Ar ages to revise the geomagnetic polarity timescale for the mid–late Pleistocene. Figure 7.4 shows their new timescale compared with the earlier version by Mankinen and Dalrymple.

Zheng *et al.* (1992) took over 500 oriented hand samples from a 195 m loess–red clay section near Xian, China. Just below the Brunhes–Matuyama boundary they found an excursion that does not become fully normal. They called this excursion BMpc (meaning precursor to the Brunhes–Matuyama reversal). This precursor has been observed in deep-sea cores (Clement *et al.*, 1982) and also in other Chinese loess sections (Cao *et al.*, 1985; Liu *et al.*, 1988). Zheng *et al.* found five other normal polarity subchrons/excursions in the Matuyama chron – the Jaramillo, Cobb Mountain, Olduvai and Réunion 1 and 2. Subchrons were also seen in the Gauss and Gilbert chrons (see Figure 7.5). No reversed events (such as the Laschamp or Blake) were seen above the Brunhes–Matuyama boundary.

As mentioned earlier, recent ^{40}Ar–^{39}Ar dating and astronomical data indicate that the ages of polarity transitions in the Pliocene and Pleistocene are older than the values currently in use. McDougall *et al.* (1992) noticed that timescales based on the palaeomagnetic polarity of Pliocene and Pleistocene strata in the Turkana Basin, East Africa, assuming constant sediment accumulation rates, were younger with respect to those derived from K–Ar and ^{40}Ar–^{39}Ar dating of units in the section. They originally suggested two possible reasons for this discrepancy – that the K–Ar ages were measured in materials that were incorporated into the section after the time of their eruption, or that the sediments became magnetized a considerable time after their deposition. They later (1992) considered a third possibility – that the calibration of the geomagnetic polarity timescale was in error. Most of the dates on which the timescale was based are from basalts, while

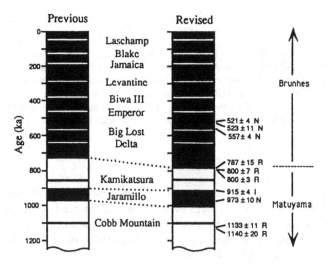

Figure 7.4 Left: geomagnetic polarity timescale for the mid- to late Pleistocene from data given in Mankinen and Dalrymple (1979) and Champion *et al.* (1988). Right: revised timescale based on previous data (Mankinen and Dalrymple, 1979; Mankinen *et al.*, 1881; Mankinen and Grommé, 1982) and new ^{40}Ar–^{39}Ar ages (Spell and McDougall, 1992). Black represents normal polarity intervals; white represents reversed polarity. After Spell and McDougall (1992).

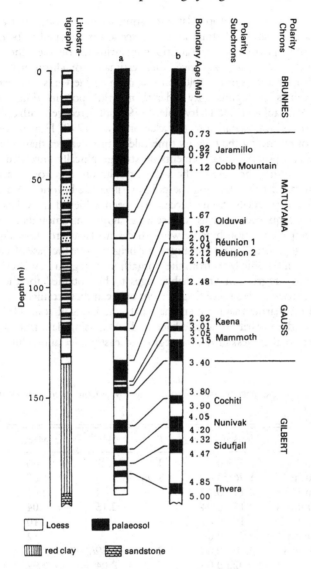

Figure 7.5 A comparison of the magnetic polarity stratigraphy derived by (a) Zheng *et al.* (1992) and (b) Mankinen and Dalrymple (1979). After Zheng *et al.* (1992).

most of the dates in the Turkana Basin have been measured on anorthoclase. McDougall *et al.* point out that dates on whole rock basalt can give ages which are too young, although this is not usually the case for dates from young volcanic alkali feldspars. This may be the result of small, but significant, amounts of radiogenic argon being lost from basalts. They thus reviewed all the data from the Turkana Basin, using isotopic dates to estimate times of polarity transitions. The rate of accumulation of strata in a local section is assumed to be constant, and the ages of magnetic polarity transitions are estimated by linear interpolation between dated levels. Table 7.2 gives their estimates for chron and subchron boundaries from the top of the Olduvai subchron to the Gauss–Gilbert boundary. Their values are significantly older than earlier polarity timescales (Mc-Dougall, 1979; Mankinen and Dalrymple, 1979), but agree well with calibrations based on astronomical calculations (Shackleton *et al.*, 1990; Hilgen, 1991a,b).

The hope of extending the polarity timescale to times earlier than 5 Ma, using K–Ar dating and polarity measurements on stratigraphically unrelated samples, is not very promising, because of the lack of resolution in isotopic dating. The same difficulties exist in extending the polarity timescale beyond 5 Ma to try to detect additional reversals. Some success has been achieved in Iceland, where there are thick sequences of lavas representing significant amounts of time and where stratigraphic relationships between successive lavas are clear. Figure 7.6 shows K–Ar ages plotted against stratigraphic height above the base of a 3500-m-thick sequence of basaltic lavas from the Borgarfjordur region in western Iceland – a least-squares regression line enables ages to be obtained for the polarity interval boundaries. There is good agreement between these results and extrapolations based on marine magnetic anomaly data. McDougall *et al.* (1976) were able to extend the polarity timescale back to 12 Ma, using data from a 2330 m lava sequence in the Neskaupstadur region of eastern Iceland. Unless similar

Table 7.2. Ages of polarity transition boundaries (Ma) from cited sources and as estimated here

	Turkana Basin	S[a]	H[b]	M&D[c]	McD[d]
Top Olduvai	1.78±0.04	1.77	1.79	1.66	1.76
Bottom Olduvai	1.96±0.03	1.95	1.95	1.88	1.91
Top Réunion II	2.11±0.04		2.14	2.01	2.07
Bottom Réunion II	2.15±0.04		2.15	2.04	2.07
Top Réunion I	2.19±0.04			2.10	2.23
Bottom Réunion I	2.27±0.04			2.12	2.23
Matuyama–Gauss	2.60±0.06	2.60	2.59/2.62	2.47	2.47
Top Kaena	3.02±0.06		3.04	2.92	2.91
Bottom Kaena	3.09±0.06		3.11	2.99	3.00
Top Mammoth	3.21±0.06		3.22	3.08	3.07
Bottom Mammoth	3.29±0.06		3.33	3.18	3.17
Gauss–Gilbert	3.57±0.05		3.58	3.41	3.41

[a]Shackleton *et al.* (1990).
[b]Hilgen (1991a,b).
[c]Mankinen and Dalrymple (1979).
[d]McDougall (1979).
After McDougall *et al.* (1992).

Table 7.3. Dated magnetic reversals older than 5 Ma

Magnetic anomaly	Material and method	Isotopic age (Ma)	References
[a]C5N(y)	K–Ar on lavas New Zealand	8.87	Evans (1970)
(Used as calibration point for DNAG scale)			
[a]C5N(y)	K–Ar on lavas Iceland	9.64	McDougall *et al.* (1984)
[a]C5N(o)	K–Ar on lavas Iceland	10.30	McDougall *et al.* (1976)
[a]C5N(o)	K–Ar biotite (tuff) Utah, S. Dakota	10.30	McDowell *et al.* (1973)
[a]C5N(o)	K–Ar on lavas Iceland	10.47	Saemundsson *et al.* (1980)
[a]C5N(o)	K–Ar on lavas Iceland	11.07	McDougall *et al.* (1984)
C9N(o)	K–Ar/Rb–Sr biotite Gubbio, Italy	28.0	Montanari *et al.* (1985)
[a]C12N(y)	K–Ar biotite (tuff) Wyoming	32.4	Prothero *et al.* (1982)
(Used as calibration point for DNAG scale)			
C12R(upper)	K–Ar biotite (tuff) Gubbio, Italy	32.0	Montanari *et al.* (1985)
[a]C13N(y)	K–Ar biotite (tuff) Wyoming	34.6	Prothero *et al.* (1982)
(Used as calibration point for DNAG scale)			
C13N(y)	K–Ar biotite (tuff) Gubbio, Italy	35.4	Montanari *et al.* (1985)
C16N(y)	K–Ar/Rb–Sr biotite Gubbio, Italy	36.0	Montanari *et al.* (1985)
C17N(upper)	K–Ar biotite (tuff) Gubbio, Italy	36.4	Montanari *et al.* (1985)
[a]C21N(y)	More than 20 K–Ar lavas and tuffs Wyoming	49.3	Flynn (1986)

[a]Dated material is from a non-marine sequence.
After Berggren *et al.* (1985a).

thick sequences of lavas of the right age can be found, it is unlikely that further extension of the polarity timescale by direct K–Ar dating and polarity measurements can be made. A review of the magnetostratigraphy and geochronology of Iceland has been given by McDougall *et al.* (1984). Table 7.3 lists older isotopic ages that may be used to calibrate the magnetic anomaly timescale.

The lack of correlation sometimes found between different polarity chronologies (lava sequences, marine sedimentary cores, ocean-floor anomalies) reflects the difficulty of accurate dating, the loss or absence of the geological record in the terrestrial volcanic sequence, assumptions about the uniformity of the rates of sea-floor spreading and the accumulation of volcanic and sedimentary rocks, and the lack of fine-scale resolution of the magnetic record of ocean-floor anomalies so that brief reversals in the marine record may be missed. Radiometric dates on rocks older than about 6.0 Ma have an uncertainty greater than the length of many magnetic reversals. Besides fluctuations in continuous deposition, unconformities exist that represent variable periods of non-deposition, erosion or both. It is impossible to estimate the amount of time represented by such unconformities. A change in magnetic polarity in two stratigraphic adjacent units gives no information about the age difference between the units. A series of reversals also tells little about the total amount of time represented by, or missing from, the record.

Attempts have been made to extend the polarity timescale to marine sedimentary sections uplifted on land. One of the first investigations was carried out by Kennett *et al.* (1971) on a Pliocene to early Quaternary section at Mangaopari Stream in the North Island of New Zealand. The principal significance of this work was its demonstration that magnetostratigraphy could be applied to land-based marine sections and, hence, help in establishing the chronology of late Cenozoic climatic history. Kennett (1980) has cautioned against the uncritical acceptance of the magnetostratigraphy from land-based marine sections, since several factors may complicate the record, particularly normal palaeomagnetic overprinting resulting from post-depositional chemical changes in the sediment. The situation greatly improved during the 1970s, with the development of more rapid and sensitive magnetometers, digital spinners and, later, cryogenic instruments making possible the measurement of large numbers of weakly magnetized samples. In 1977 a series of papers was published on the geomagnetic record preserved in marine sediments at Gubbio, Italy (Alvarez *et al.*, 1977; Arthur and Fischer, 1977; Lowrie and Alvarez, 1977; Premoli Silva, 1977; Roggenthen and Napoleone, 1977). A long, continuous sequence of magnetic polarity zones was observed in sediments of Middle Cretaceous to Palaeocene age, which were intercalibrated with planktonic foraminiferal zones. The Gubbio section is particularly valuable, since it contains an unbroken sequence across the Cretaceous–Tertiary boundary (Luterbacher and Premoli Silva, 1964). In most of the documented late Cretaceous–early Tertiary sections of the world, the boundary between the periods is represented by a hiatus. The magnetostratigraphy at Gubbio also closely matches the polarity sequence inferred from marine magnetic anomaly profiles. This is confirmed by a series of further biostratigraphic and magnetic stratigraphic investigations of late Mesozoic and Cenozoic pelagic carbonate rocks (see, e.g., Alvarez and Lowrie, 1978: Channell *et al.*, 1979; Lowrie *et al.*, 1980a,b, 1982; Vandenberg and Wonders, 1980; Napoleone *et al.*, 1983).

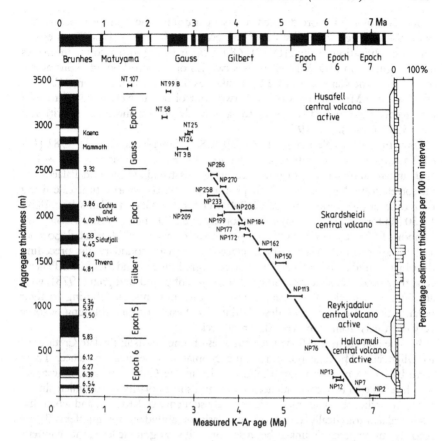

Figure 7.6 K–Ar ages, with precision limits plotted against aggregate stratigraphic thickness above the base of a 3500-m-thick sequence of basaltic lavas from western Iceland. Polarity log for the sequence shown adjacent to the thickness axis: black indicates normal polarity; white, reversed polarity. Polarity timescale, after Talwani *et al.* (1971), given at top of diagram. Regression line through data is least-squares fit with Gilbert–Gauss epoch boundary fixed at 3.32 Ma at 2520 m. Ages indicated for polarity interval boundaries on polarity log derived from regression. The proportion of sediment present for each 100 m of section is shown on right, and period of activity of central volcanoes in the region is also indicated. After McDougall *et al.* (1977).

7.5 The Cenozoic–late Cretaceous timescale (0–83 Ma)

All timescales for the Cenozoic to the late Cretaceous start with the profile of Heirtzler *et al.* (1968) (HDHPL 68), which was obtained across the relatively slow (20 mm/a) south Atlantic ridge. Surveys over fast-spreading ridges show a fine structure that cannot be recognized over slow-spreading ridges. Most of the changes that have been made to HDHPL 68 are the substitution of well-defined segments showing a more detailed reversal pattern for the equivalent part of the original sequence. The timescale of LaBrecque *et al.* (1977) (LKC77) is essentially a modification of the HDHPL 68 timescale taking into account anomalies not recognized in HDHPL 68. It uses two isotopic control points – one at the older end of the Gauss chron C2A(o), taken as 3.4 Ma, and the other at C29(o), taken as 66.7 Ma. The ages of these two control points were adjusted to take into account the revised values for the atomic abundance and decay activities of ^{40}K (Steiger and Jäger, 1977).

Later timescales (Ness *et al.*, 1980 [NLC 80]: Lowrie and Alvarez, 1981 [LA 81]) assumed that biostratigraphic ages are a more accurate measure of absolute ages and recalibrated the magnetic reversal ages to adjust the mid-Palaeogene miscorrelations. Ness *et al.* (1980) proposed a revised polarity timescale using four calibration points: an age of 3.40 Ma for the Gilbert–Gauss boundary: 10.30 Ma for the older boundary of anomaly 5: 54.90 Ma for the older boundary of anomaly 24, consistent with the placement of anomaly 24 as basal Eocene: and an age of 66.70 Ma for the Cretaceous–Tertiary boundary just preceding the older boundary of anomaly 29. For geological time boundaries in the Palaeocene they used the geological timescale of Hardenbol and Berggren (1978), with absolute ages recalculated, using revised decay constants. In referring to their polarity timescale, Ness *et al.* described it as 'at best a critical reshuffling of some very old cards from some very different decks'.

Lowrie and Alvarez (1981) fixed the ages of nine points in the late Cretaceous to Oligocene–Miocene portion of the geomagnetic reversal sequence. These were based on magnetostratigraphic studies in the Gubbio pelagic limestones which revealed a reversal sequence of long and short polarity zones which almost exactly matches the main magnetic anomaly sequence of LKC 77, and which has been palaeontologically dated on the basis of abundant foraminifera. These records can be precisely linked because, for each stratigraphic level, the magnetic and palaeontological information both come from the same small sample. The use of so many calibration tie-points increases the possibility of introducing spurious accelerations in the rate of sea-floor spreading. The geological timescale of Harland *et al.* (1982) [GTS 82] is a modification of LA81 using 11 biostratigraphic control points.

The Cenozoic and late Cretaceous timescale of Berggren *et al.* (1985a,b) [BKF 85] starts with the LKC77 polarity sequence and assumes a minimum number of changes in the rate of sea-floor spreading that will satisfy the constraints of calibration tie-points. Three linear segments were used to obtain a timescale by linear regression analysis. Their age calibration tie-points are for anomalies 2A (3.40 Ma), 5y (8.87 Ma), 12y (32.4 Ma), 13y (34.6 Ma), 21y (49.5 Ma) and 34y (84.0 Ma). Harland *et al.* (1990) in their revised geological timescale [GTS 89] criticize this piecewise linear fit to the data on the grounds that it is not clear what the basis for choosing the control points is and that only a small fraction of the available data is used to calibrate the timescale. GTS 89 is based on a slightly modified HDHPL 68 timescale. Isotopic and biostratigraphically dated magnetic anomalies were combined for calibrating the timescale. Three linear segments were used to fit the data – the first for the interval 0–3.3 Ma, the second in the interval from about 9 to 50 Ma, and the third from about 50 to 85 Ma. No data in the interval from 3.4 to about 9 Ma were used. There are no dated magnetic anomalies in the 50–85 Ma interval and a single line was fitted to the data in the Palaeocene–Campanian interval. The Tertiary–Cretaceous boundary was fixed at 65 Ma and the Campanian–Santonian boundary at 83 Ma. The first magnetostratigraphic timescale HDHPL 68 was based on a simple extrapolation from anomaly 2A and gives age estimates that are within 10% of the absolute age estimates of later timescales based on much more detailed magnetostratigraphic correlations and radiometric calibration data. This agreement shows that the assumption of a constant sea-floor spreading rate is a very good first approximation.

Cande and Kent (1992) have constructed a new geomagnetic polarity timescale (CK 92) for the late Cretaceous and Cenozoic, using marine magnetic profiles from the world's ocean basins. A composite polarity sequence was first obtained based mainly on data from the south Atlantic. Anomaly spacings in the south Atlantic were constrained by a combination of finite rotation poles and averages of stacked profiles. Fine-scale information derived from magnetic profiles on faster-spreading ridges in the Pacific and Indian Oceans was then inserted into the south Atlantic sequence. Cande and Kent assumed that spreading rates in the south Atlantic were smoothly varying, but not necessarily constant from anomaly 34 to the (zero age) ridge axis. A spline function was then used to fit nine calibration points to the composite polarity sequence. The nine calibration points (see Table 7.4) were selected from a large number of individual isotopic dates – preference being given to those data which can be tied to the magnetic anomaly sequence via marine magnetobiostratigraphic correlations and constraints from biostratigraphic correlation of sediments overlying oceanic basement (Cande *et al.*, 1989). The selected tie-points in the Neogene largely conform to the estimates of Harland *et al.* (1990). The choice of tie-points in the Palaeogene were influenced by the recent assessment of correlations and age estimates by Berggren *et al.* (1992). Age assignments for the late Cretaceous are similar to those in Berggren *et al.* (1985c) and Harland *et al.* (1990). A brief description of the calibration points selected by Cande and Kent (1992) is given below.

(1) After the zero-age ridge axis, the youngest calibration point in almost all geomagnetic polarity timescales has been the old end of anomaly 2A – the Gauss–Gilbert boundary. Cande and Kent (1992), however, chose the Matuyama–Gauss boundary for the first tie-point (2.6 Ma), as determined from astronomical calibration (Shackleton *et al.*, 1990; Hilgen, 1991a).

(2) An age of 14.8 Ma was assigned to the younger end of chron C5 Bn based on isotopic age constraints on the correlative N9/N10 foraminifera zone boundary of Berggren *et al.* (1985c) and Miller *et al.* (1985).

(3) The Miocene–Oligocene boundary is correlated to the middle part of chron C6 Cn (Berggren *et al.*, 1985a) and has an estimated age of 23.8 Ma (Harland *et al.*, 1990).

(4) The Oligocene–Eocene boundary has been correlated to a level within the

Table 7.4. Age calibrations for geomagnetic polarity timescale

Chron	South Atlantic distance[a] (km)	Age (Ma)
C2An(0.0)	41.75	2.6
C5Bn(0.0)	290.17	14.8
C6Cn.2r(0.0)	501.55	23.8
C13r(.14)	759.49	33.7
C21n(.33)	1071.62	46.8
C24r(.66)	1221.20	55.0
C29r(.3)	1364.37	66.0
C33n(.15)	1575.56	74.5
C34n(0.0)	1862.32	83.0

[a] 1.29 km subtracted to account for Central Anomaly offset.
After Cande and Kent (1992).

upper part of chronozone C 13r (Nocchi *et al.*, 1986). An age of 33.7 Ma has been assigned to this boundary based on a critical evaluation of the bio- and magnetostratigraphically well-controlled isotopic dates from the Palaeogene sequence in the Apennines (Odin *et al.*, 1991). This is 2.7 Ma younger than the age estimated by Harland *et al.* (1990).

(5) An age of 46.8 Ma was assigned to C2in. This was based on a K–Ar date on biotites from sediments in DSDP hole 516F (Bryan and Duncan, 1983) which are magnetobiostratigraphically constrained to the upper part of chron C21n (Berggren *et al.*, 1992).

(6) New ^{40}Ar–^{39}Ar dates on volcanic ash from earliest Eocene marine deposits in Denmark and the North Sea basin yield an age of 55 Ma for the nannofossil NP9/NP10 boundary (Swisher and Knox, 1991). This level approximates the position of the Palaeocene–Eocene boundary and occurs about ⅔ down in chron C24r (Berggren *et al.*, 1985a,c).

(7) An age of 66 Ma was assigned to the Cretaceous–Palaeogene boundary – ³/₁₀ down in chron C29r mainly on the basis of magnetostratigraphic sections (Preisinger *et al.*, 1986; Harland *et al.*, 1990). This estimate was also supported by laser fusion ^{40}Ar–^{39}Ar dates on single crystals of sanidine (Swisher and Dingus, written communication to Cande and Kent, 1992).

(8) The Maastrichian–Campanian boundary, in the late part of chronozone C33n (Harland *et al.*, 1990) was assigned an age of 74.5 Ma. This was based on a biostratigraphically correlative level on the western interior of North America for which Obradovich and Cobban, (1975) obtained constraining K–Ar dates on bentonites.

(9) Obradovich *et al.* (1986) confirmed an age of ~ 84 Ma which he and

Cobban (1975) had suggested earlier for the Campanian–Santonian boundary. This was based on biostratigraphically controlled ^{40}Ar–^{39}Ar dates based on a bentonite from the western interior of North America. Cande and Kent (1992) chose an age of 83 Ma for the stratigraphically younger C34n level as their last tie-point.

The timescale of Cande and Kent (CK92) is given in Figure 7.7 with geological correlations to anomalies at the stage level adjusted to the new derived ages, based on Berggren *et al.* (1985c). Table 7.5 gives the ages of the normal polarity intervals to a resolution of 1 ka.

All previous timescales show much less regular variations in spreading rate in the south Atlantic. In the Neogene there are two main differences between the CK92 timescale and previous versions. Since Cande and Kent (1992) used the astronomical calibration of 2.60 Ma for the Matuyama–Gauss boundary, which is about 5% older than the estimate of 2.48 Ma based on K–Ar isotopic dates (Mankinen and Dalrymple, 1979), their CK92 timescale gives proportionally older ages for other geomagnetic boundaries in the younger part of the timescale. The other main difference in the Neogene is the age of anomaly 5 (chron C5n), which is ∼0.5 Ma older than in the chronologies of Berggren *et al.* (1985b) and Harland *et al.* (1990). The new age range for chron C5n is in good agreement with the most recent estimate based on isotopic dating of Icelandic lavas (McDougall *et al.*, 1984). Also, chrons C9 through C24 are 2–3 Ma younger than in the timescales of Berggren *et al.* (1985a) and Harland *et al.* (1990), reflecting the revised ages of calibration points in the Palaeocene–Oligocene. Differences in the late Cretaceous (chrons C30n to C34n) are within 1 Ma.

Cande and Kent (1992) also identified a number of 'tiny wiggles' in the records from the Cenozoic. Most of the smaller-amplitude anomalies have not been confirmed in magnetostratigraphic sections. They may be very short polarity intervals (Blakely, 1974) or longer-period (50–200 ka) intensity variations of the dipole field (Cande and LeBrecque, 1974).

McIntosh *et al.* (1992) have determined a terrestrial volcanic calibration of the late Eocene–Oligocene timescale, using ^{40}Ar–^{39}Ar dated ignibrite sequences in New Mexico, Colorado and Texas. They were able to put age constraints for several polarity reversals that occurred during three periods of intense volcanism: 36.8–33.5 Ma, 32.7–31.4 Ma and 29.1–26.9 Ma. They were based on relative lengths of polarity intervals. Their preferred correlation yields calibration ages for chron 10R (28.0–29.0 Ma) and chron 13R (34.4–33.1 Ma) which indicate an Eocene–Oligocene boundary near 33.4 Ma. This is in good agreement with the value assigned (33.7 ± 0.4 Ma) by Cande and Kent for one of their tie-points in their CK92 polarity timescale.

Table 7.5. Normal polarity intervals (after Cande and Kent, 1992)

Normal polarity interval (Ma)	Polarity chron	Normal polarity interval (Ma)	Polarity chron
0.000–0.780	C1n	22.599–22.760	C6Bn.1n
0.984–1.049	C1r.1n	22.814–23.076	C6Bn.2n
1.757–1.983	C2n	23.357–23.537	C6Cn.1n
2.197–2.229	C2r.1n	23.678–23.800	C6Cn.2n
2.600–3.054	C2An.1n	23.997–24.115	C6Cn.3n
3.127–3.221	C2An.2n	24.722–24.772	C7n.1n
3.325–3.553	C2An.3n	24.826–25.171	C7n.2n
4.033–4.134	C3n.1n	25.482–25.633	C7An
4.265–4.432	C3n.2n	25.807–25.934	C8n.1n
4.611–4.694	C3n.3n	25.974–26.533	C8n.2n
4.812–5.046	C3n.4n	27.004–27.946	C9n
5.705–5.946	C3An.1n	28.255–28.484	C10n.1n
6.078–6.376	C3An.2n	28.550–28.716	C10n.2n
6.744–6.901	C3Bn	29.373–29.633	C11n.1n
6.946–6.981	C3Br.1n	29.737–30.071	C11n.2n
7.153–7.187	C3Br.2n	30.452–30.915	C12n
7.245–7.376	C4n.1n	33.050–33.543	C13n
7.464–7.892	C4n.2n	34.669–34.959	C15n
8.047–8.079	C4r.1n	35.368–35.554	C16n.1n
8.529–8.861	C4An	35.716–36.383	C16n.2n
9.069–9.149	C4Ar.1n	36.665–37.534	C17n.1n
9.428–9.491	C4Ar.2n	37.667–37.915	C17n.2n
9.592–9.735	C5n.1n	37.988–38.183	C17n.3n
9.777–10.834	C5n.2n	38.500–39.639	C18n.1n
10.940–10.989	C5r.1n	39.718–40.221	C18n.2n
11.378–11.434	C5r.2n	41.353–41.617	C19n
11.852–12.000	C5An.1n	42.629–43.868	C20n
12.108–12.333	C5An.2n	46.284–47.861	C21n
12.618–12.649	C5Ar.1n	48.947–49.603	C22n
12.718–12.764	C5Ar.2n	50.646–50.812	C23n.1n
12.941–13.094	C5AAn	50.913–51.609	C23n.2n
13.263–13.476	C5ABn	52.238–52.544	C24n.1n
13.674–14.059	C5ACn	52.641–52.685	C24n.2n
14.164–14.608	C5ADn	52.791–53.250	C24n.3n
14.800–14.890	C5Bn.1n	55.981–56.515	C25n
15.038–15.162	C5Bn.2n	57.800–58.197	C26n
16.035–16.318	C5Cn.1n	61.555–61.951	C27n
16.352–16.515	C5Cn.2n	63.303–64.542	C28n
16.583–16.755	C5Cn.3n	64.911–65.732	C29n
17.310–17.650	C5Dn	66.601–68.625	C30n
18.317–18.817	C5En	68.745–69.683	C31n
19.083–20.162	C6n	71.722–71.943	C32n.1n
20.546–20.752	C6An.1n	72.147–73.288	C32n.2n
21.021–21.343	C6An.2n	73.517–73.584	C32r.1n
21.787–21.877	C6AAn	73.781–78.781	C33n
22.166–22.263	C6AAr.1n	83.000–(118.0)	C34n
22.471–22.505	C6AAr.2n		

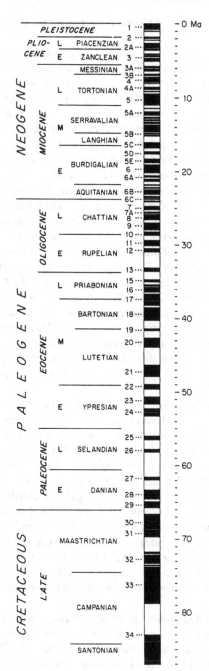

Figure 7.7 New geomagnetic polarity timescale for the late Cretaceous and Cenozoic. Correlations of geological stages with polarity reversal pattern taken from Berggren *et al.* (1985c). Polarity events shorter than 30 ka duration have been omitted from the reversal pattern. After Cande and Kent (1992).

7.6 The late Jurassic and early Cretaceous timescale (83~160 Ma)

The standard magnetic reversal model for the M sequence has been derived from the Hawaiian lineations assumed to have formed at a constant rate of sea-floor spreading (Larson and Hilde, 1975). The sequence has been extended beyond M25 in the Pacific by Cande *et al.* (1978) and in the Atlantic by Bryan *et al.* (1980). The chronological control used by Larson and Hilde was based on biostratigraphic assessment of basal sediments of five DSDP holes drilled over identified M anomalies. More exact magnetobiostratigraphic correlations have since become available (see Kent and Gradstein, 1985, for a detailed discussion). However, numerical age estimates for the late Jurassic and early Cretaceous are still poor, because of the lack of reliable radiometric dates. The M sequence oceanic crust has not been dated – there is no direct evidence of the M marine anomalies. The reliability of Larson and Hilde's (1975) Hawaiian lineations has been reviewed by Channell *et al.* (1987). Thick exposures of pelagic carbonate rocks of the Umbrian sequence have been extensively investigated in the Contessa and Bottaccione gorges near Gubbio in the northern Apennines of Italy. The magnetic stratigraphy of these sections matches well with the marine magnetic anomaly sequence in the Palaeogene – and in the late Cretaceous (Lowrie *et al.*, 1982; Napoleone *et al.*, 1983) – see Figure 7.8. Kent and Gradstein (1985) in their timescale [KG 86] used the age estimates of GTS 82 for the Barremian–Aptian boundary (119 Ma) and the Oxfordian–Kimmeridgian boundary (156 Ma) as tie-points to calibrate the M sequence. Harland *et al.* based their revised timescale [GTS 89] on the lengths of polarity intervals given by Kent and Gradstein and on more recent biostratigraphic data – their results are shown in Table 7.1.

Ogg and Steiner (1984), Bralower (1987) and Channell *et al.* (1987) have reviewed the positions of the M sequence anomalies and their relation to the biostratigraphically dated Early Cretaceous stage boundaries in sections in the Umbrian Apennines and at a site in the southern Alps. In particular, Bralower correlated nannofossil datum planes with polarity chrons CMO–CM11 in four Italian sections with published magnetic stratigraphies. Although the correlation of nannofossil events to polarity events were, in general, consistent, the polarity patterns (the relative thicknesses of polarity zones) did not resemble the Larson and Hilde Hawaiian model, casting doubt on the assumption of a constant spreading rate for the Hawaiian lineations and on the correlation of nannofossil events with polarity chrons. For the CM13 to CM22 interval, Bralower *et al.* (1989) correlated nannofossil events to polarity chrons at six European land sections and two DSDP sites for which the magnetostratigraphies had already been published.

Channell and Erba (1992) later identified polarity chrons CMO–CM11 in pelagic limestone sections near Bresica in northern Italy. The polarity chrons were correlated to nannofossil and foraminiferal events. The Polaveno section records polarity chrons CM3 to CM11, in excellent agreement with the Larson and Hilde polarity model (see Figure 7.9). This is the first complete magnetostratigraphic record in this interval and indicates that at least this part of the Hawaiian polarity model represents an approximately constant spreading rate.

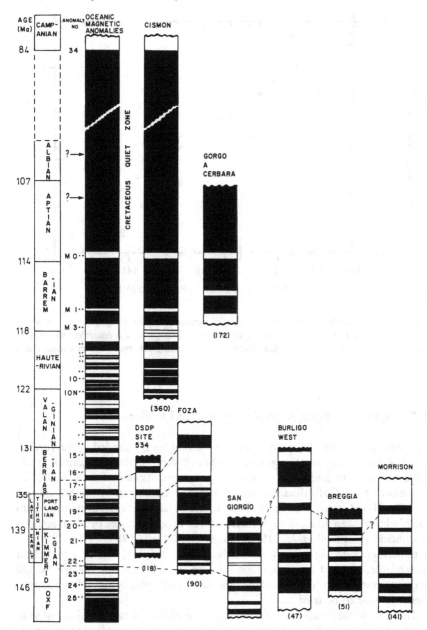

Figure 7.8 Early Cretaceous and late Jurassic magnetostratigraphic sections and their correlation with the geomagnetic polarity interpreted from M-sequence oceanic magnetic anomalies (Larson and Hilde, 1975). The absolute ages associated with stage boundaries are adopted uncorrected from Van Hinte (1976a,b). The six polarity stratigraphies at the base of the figure have been expanded for clarity. Dashed tie-lines are primarily biostratigraphic and secondarily magnetostratigraphic, and correlate the magnetozones to the oceanic sequences. From Channell *et al.* (1982).

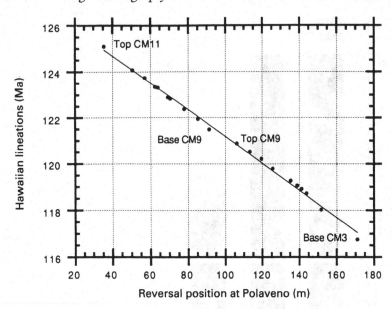

Figure 7.9 Distance between reversals in the Polaveno section plotted against age of reversals in the Larson and Hilde (1975) Hawaiian block model. The age scale is equivalent to sea-floor distance between reversals in the Hawaiian constant-spreading-rate model. After Channell and Erba (1992).

7.7 Magnetostratigraphy in the Jurassic and earlier

The older end of the M sequence of marine anomalies is bounded by the Jurassic Quiet Zone – magnetostratigraphic studies on land sections (Channell *et al.*, 1982) indicate an interval of predominantly normal polarity for the Collovian and most of the Oxfordian. However, from the Bathonian to the Sinemurian (~ 170–200 Ma), frequent reversals are found in land sections (principally from the Mediterranean area), though correlation of magneto zones between sections is difficult (Channell *et al.*, 1982; Steiner and Ogg, 1984). Thus, for pre-Kimmeridgian geochronology (older than about 160 Ma), the general lack of correlatable, lineated magnetic anomalies means that it is not possible to use magnetostratigraphy to calibrate the timescale – biostratigraphic and radiometric methods have to be used.

There are some detailed magnetostratigraphic studies of earlier individual sedimentary sequences with continuous or nearly continuous records of the magnetic field. Irving and Parry (1963) discovered a reversed polarity superchron in New South Wales, Australia, lasting some 50 Ma between the upper Carboniferous and the upper Permian to Lower Triassic. This long-lasting reversed polarity period has been named the Kiaman and has been confirmed by a number of later studies. The Kiaman in Australia is followed by a normal polarity zone and

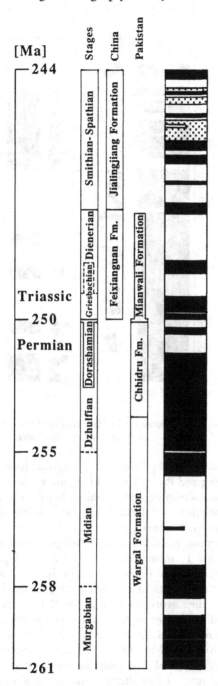

Figure 7.10 Composite magnetostratigraphic polarity profile for the Upper Permian (Nammal Gorge) to Lower Triassic, South China. Timescale not linear. After Haag and Heller (1991).

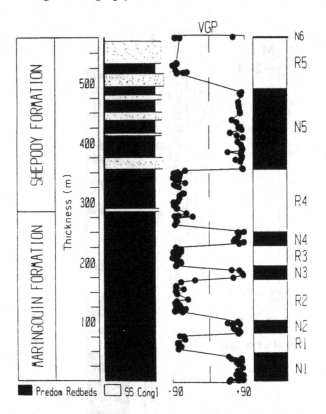

Figure 7.11 Magnetostratigraphic column for the Maringouin and Shepody formations. VGPs are normalized relative to formation means. After Divenere and Opdyke (1990).

subsequently by a zone of mixed polarity, which together have been termed the Illawara reversals (Irving and Parry, 1963). Haag and Heller (1991) have combined lower Triassic magnetostratigraphic sections based on results from marine limestone sections in South China (Heller *et al.*, 1988; Steiner *et al.*, 1989) with new Permian data from the Nammal gorge, Salt Range, north-west Pakistan, to obtain an upper Permian to lower Triassic magnetostratigraphic reference polarity profile across the Permian–Triassic boundary (see Figure 7.10). Haag and Heller found several normal polarity zones in the lower upper Permian – current assumption is that rocks of this age belong to the Kiaman reversed superchron. However, their work shows that the Kiaman must end and the Illawara zone of mixed polarity must begin in or prior to the lowermost upper Permian. The upper Permian at Nammal together with the lower Triassic sections in southern China cover ~ 20 Ma. Nearly 30 polarity changes are observed, resulting in an average reversal frequency similar to that observed during the early Tertiary. The reversal rate after the end of the Kiaman increases in a manner similar to that after the end of the Cretaceous long normal superchron (~ 1/Ma). In the lower Triassic the frequency seems to double.

Divenere and Opdyke (1990) took 235 oriented cores in a stratigraphic sequence of 565 m of red sediments from the Namurian Maringouin and Shepody formations in New Brunswick, Canada, in order to investigate the reversal patterns of the Earth's magnetic field prior to the onset of the Permo-Carboniferous reversed superchron. They were able to obtain the first coherent magnetic stratigraphy from the late Palaeozoic of North America, with five discrete reversed polarity and five discrete normal polarity zones (Figure 7.11). Later Divenere and Opdyke (1991) analysed 105 oriented cores from 340 m of the upper Mauch Churk Formation at Pottsville, Pennsylvania, in an attempt to correlate the section with their previous study in New Brunswick. They found three normal and three reversed polarity zones and tentatively correlated the bottom part of the Maringouin section with the top of Mauch Churk. If substantiated, this would place the Mississippian–Pennsylvanian boundary within the Maringouin Formation.

8
The Earth's magnetic field and other geophysical phenomena

8.1 Introduction

In this chapter possible relationships between the Earth's magnetic field and other geophysical phenomena will be discussed. This will include climate, which is a rather grey area, and, in spite of the increasing interest and vast literature on the subject, little convincing evidence for correlations has been obtained. Another grey area is possible relationships between magnetic reversals, impacts and faunal extinctions, which will be discussed in Section 8.4. Finally, the possibility of a connection between events at the Earth's surface and those in the core, where the Earth's magnetic field is generated, will be reviewed. In the past it has always been considered unlikely that the solid, rocky, mantle and the fluid, mainly iron, outer core could influence one another.

8.2 The orbital climatic theory of Milankovitch

The first suggestion that orbital variations might affect climate was made in 1830 by John Herschel. The idea was revived many times, more recently by Milankovitch in 1941, with whose name it is now associated. The theory holds that regular, predictable changes in the orientation of the Earth's axis of rotation and the shape of its orbit affect the distribution of sunlight over the Earth's surface. The tilt of the Earth's axis away from the plane of its orbit (its obliquity ε) varies between 22.1° and 24.5° (it is now 23.5°). Since it is the obliquity which causes seasons, a cyclic variation in the obliquity will in turn produce a cycle in the strength of the contrast between seasons. The Earth's axis of rotation also precesses, describing a small circle among the stars. This is measured as the angular distance ω of perihelion from the autumnal equinox. Precession also affects the contrast of the seasons, since it determines at what point in the Earth's orbit winter and summer occur. Winters occurring near the Earth's closest approach to the Sun would be warmer on average than those occurring at its farthest point. Finally, the eccentricity *e* of the Earth's orbit, which is a measure of the Sun's departure from the geometric centre of the Earth's orbit, is not constant, varying between 0 and 0.06. Integrated over all latitudes and over an entire year, the energy influx depends only on *e*. However, the geographical and seasonal pattern of irradiation essentially depends only on ε and on *e*sinω.

Each of the orbital elements is a quasi-periodic function of time (see Figure 8.1). Although the curves have a large number of sinusoidal components, calculated spectra are dominated by a small number of peaks. The most important term in the series expansion for eccentricity has a period calculated by Berger (1977b) to be 413 ka. Eight of the next twelve terms range from 95 ka to 136 ka. In low-resolution spectra these terms contribute to a peak that is often loosely referred to as the 100 ka eccentricity cycle. In contrast, the spectrum of the obliquity is relatively simple, being dominated by components with periods near 41 ka. The main components of precession have periods near 23 and 19 ka. In low-resolution spectra these are seen as a single peak near 22 ka.

Berger *et al.* (1992) have calculated the effects that changes in orbital characteristics over geological time, such as the shortening of the Earth–moon distance and of the length of the day back in time, have had on these periodicities. They showed that they induced a shortening of the fundamental periods for the obliquity and precision from 54 to 35, 41 to 29, 23 to 19 and 19 to 16 ka over the last 500 Ma (see Table 8.1). Berger *et al.* also found that changes in the periods of precession, obliquity and eccentricity due to chaotic motion in the solar system are negligible and can be neglected. The geological record shows that cyclicity is a pronounced characteristic of nearly all Cretaceous and Cenozoic pelagic carbonate rock sequences. Orbital cycles have also been observed in pre-

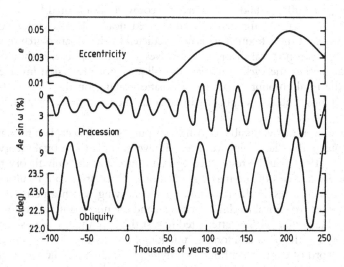

Figure 8.1 Variations in orbital geometry as a function of time. After Imbrie and Imbrie (1980).

Table 8.1. Estimated past values of the main astronomical periods, derived taking into account the slow variations of the Earth–moon and planetary systems

Age (Ma)	Periods (ka)			
0	19	23	41	54
50	18	22	39	52
100	18	22	38	50
150	18	21	37	48
200	18	21	36	46
250	17	21	35	45
300	17	20	34	42
350	17	20	32	40
400	16	19	31	38
450	16	19	30	36
500	16	18	29	35

After Berger *et al.* (1992).

Cretaceous sedimentary records. Most of the periods found in pre-Quaternary geological data are approximately 100, 41 and 21 ka. This apparent difference between the calculated values of Berger *et al.* and those deduced from proxy data is most probably related to the difficulty of obtaining a precise, absolute timescale for the Cretaceous, Jurassic and Triassic.

Over the past 600 ka, almost all climatic records are dominated by variance components in a narrow frequency band centred near a 100 ka cycle. Yet a climatic response at these frequencies is not predicted by the Milankovitch theory – or any other astronomical theory that involves a linear response. It should be stressed that neither the cyclic variation in obliquity nor precession involves a change in the total amount of sunlight insolation falling on the Earth – they merely affect how much sunlight a particular latitude receives at a particular season.

There have been many possible explanations put forward to explain the cause of fluctuations in the Pleistocene ice sheets. However, only the orbital theory of Milankovitch (1941) can be tested geologically, since it is the only theory that predicts the frequencies of the major fluctuations. There are two difficulties in carrying out such a test – the uncertainty in identifying which aspects of the radiation budget are critical to climatic change, and the uncertainty of geological chronology. Hays *et al.* (1976) attempted to test the hypothesis by considering secular changes in the orbit as a forcing function of a system whose output is the geological record of climate, without identifying, or evaluating, the mechanism through which climate is modified by changes in the global pattern of incoming radiation. Most of their climatic analysis is based on the assumption that the climatic system responds linearly to orbital forcing. They used two cores from the southern Indian Ocean and for geological data for the last 450 ka chose the $\delta^{18}O$ composition of planktonic foraminifera, an estimate of summer sea-surface temperature at the core site derived from a statistical analysis of radiolarian assemblages and the percentage of *Cycladophora davisiana*, the relative abundance of a radiolarian species not used in the estimation of sea-surface temperatures. They carried out both frequency-domain and time-domain tests. Over the frequency range 10^{-4}-10^{-5} cycles/a they found that the climatic variance of these three measures of global climate is concentrated in three discrete spectral peaks at periods of 23 ka, 42 ka and 100 ka, which correspond to the dominant periods of the Earth's solar orbit (precession, obliquity and eccentricity). Moreover, the ratio between the two frequencies (obliquity and precession) detected in the cores does not differ significantly from the predicted ratio (~ 1.8). In the time-domain analysis they used two filters, one centred at a frequency of 0.025 cycles/ka (the 40 K filter) and one at 0.043 cycles/ka (the 23 K filter). They found that over the past 300 ka each of the 40 K components of the geological record showed a constant phase relationship with obliquity and the 23 K components a constant phase relationship with precession. The dominant 100 ka climatic component has an average close to, and in phase with, orbital eccentricity. Hays *et al.* suggested that such a correlation probably requires a non-linear response of the climatic system to orbital forcing. They concluded that changes in the Earth's orbital geometry are the fundamental cause of the succession of Quaternary ice ages.

Wigley (1976) has re-examined the results of Hays *et al.* (1976). He pointed out that their orbital input spectra show a split peak near the precession period of 21 ka (Table 4 of Hays *et al.* gives peaks at periods of 23.1 ka and 18.8 ka). Wigley showed that non-linear interaction between these two frequencies could be expected to give an output signal with a period 101 ka. In addition to this effect, non-linear response to amplitude modulation of the precession signal will produce a spectral peak at the frequency of amplitude modulation: i.e. at the eccentricity period of approximately 100 ka. These two mechanisms both lead to strong output power near a period of 100 ka with little or no input power at this period. The data of Hays *et al.* also show significant power at periods corresponding to precession and obliquity, so that the climate response function cannot be wholly non-linear. Wigley (1976) suggested that the response function has both linear and non-linear parts of approximately equal importance.

Berger (1977a) pointed out that the periods found by Hays *et al.* (1976) are very close to those predicted by himself (1976, 1977b) in later, more accurate calculations of the variations of the various 'Milankovitch' parameters. The roughly 24 ka and 19.5 ka periods from the core samples are not significantly different from the periods associated with the largest amplitude terms in the series expansion of the precessional parameter $e\sin\omega$; the periods around 42 ka are essentially identical with the most important term in the expansion of the obliquity ε; and the peaks around 106 ka, containing most of the variance, might be regarded either as a contribution from the eccentricity, where a weighted mean of the main amplitude terms has a period of 110.753 ka or as a beat effect of precessional periods, as suggested by Wigley (1976). Berger (1976) also pointed out that the most important term in the series expansion of the eccentricity has a periodicity of 412.085 ka – too long to be seen in the analysis of Hays *et al.* – and the convergence of the series is very slow. If longer cores can be obtained, Berger (1977a) forecast that the 100 ka peak would be split into two peaks at 95 ka and 123 ka, with the main period of 413 ka being clearly evident.

Kominz *et al.* (1979) confirmed the conclusion of Hays *et al.* (1976) that changes in the Pleistocene climate are a result of forcing by periodic fluctuations of the Earth's obliquity and precession. They based their conclusion on a spectral and cross-spectral analysis of the $\delta^{18}O$ records of core V28–238 from the equatorial Pacific and a detailed composite Indian Ocean record.

As already mentioned, most published climatic records that are more than 600 ka old do not exhibit a strong 100 ka cycle – it is virtually absent from isotopic records of planetary ice volume from earlier parts of the Pleistocene (Shackleton and Opdyke, 1976). The problem of explaining the 100 ka cycle with a simple non-linear mechanism of the type suggested by Hays *et al.* (1976) or Wigley (1976) poses a second problem. It is difficult to introduce substantial 100 ka power into the response without also introducing power reflecting the 413 ka eccentricity cycle in amounts that are much greater than have been detected in most climatic records.

Imbrie and Imbrie (1980) argued for a fundamental change in research strategy: instead of using numerical models of climate to test the astronomical theory, one should use the geological record as a criterion against which to judge the performance of physically motivated models of climate. They further pointed out that even an excellent correlation between climate and a particular insolation curve $Q(t)$ is no assurance that physical mechanisms operating at the latitude and season represented by $Q(t)$ actually dominate the climatic response. This ambiguity exists because a curve nearly identical with $Q(t)$ may be expressed as a linear combination of curves for other latitudes and seasons.

Imbrie and Imbrie (1980) fitted simple non-linear mathematical models to $\delta^{18}O$ curves. They found that reasonable fits were obtained if the timescale for ice sheet growth is about 27 ka and for decay about 7 ka. Oerlemans (1980) considered the problem of the 100 ka cycle in a similar way. Experiments with a northern hemisphere ice sheet model showed that the 100 ka cycle and its sawtooth shape may be explained by ice sheet/bedrock dynamics alone, as first suggested by Weertman (1961). This means that the occurrence of a 100 ka peak in the power spectrum of the $\delta^{18}O$ record is unrelated to variations in eccentricity. Work by Kominz and Pisias (1979) strongly supports this view: they found that coherence spectra of an oxygen isotope record and the orbital parameters showed no significant peak around a period of 100 ka. Oerlemans' work may, however, support the Milankovitch theory. The strength of the 100 ka periodicity is not constant and almost vanishes about 1 Ma ago. This would be difficult to understand if the cause were purely internal, but less so if climatic changes were the result of orbital forcing. The lack of a 100 ka periodicity in the early climate record may be related to the fact that perennial sea-ice cover in the Arctic only began around 700 ka ago (Margolis and Herman, 1980). Further work has shown that the 100 ka climate cycle might reflect either an oscillation driven by non-linear interactions occurring within the system itself or an interaction occurring between the system and the astronomically forced responses (Imbrie and Imbrie, 1980; Maasch and Saltzman, 1990; Gallee *et al.*, 1992). In this regard Liu (1992) showed that variations in the frequency of the obliquity cycle can give rise to strong 100 ka forcing of climate.

Kukla *et al.* (1981) noted that a specific orbital configuration (high obliquity combined with the June perihelion) marked the beginning of the past three interglacials, from which they inferred that the primary cause of the glacial cycle may be astronomical. They thus introduced a purely empirical astronomical climate index (ACLIN) combining the three orbital variables to predict the major climate changes in the late and middle Pleistocene and the near future. ACLIN closely correlates with the major climatic events revealed by independently dated proxy climate indicators of the past 130 ka – it also successfully differentiates the interglacials and shows a 100 ka periodicity.

A record of orbital variations preserved in the sediments at the Deep Sea Drilling Program's (DSDP) site 158 in the equatorial Pacific has been found by Moore *et al.* (1982). The orbital cycles appear as variations in the amount of calcium carbonate in sediments deposited between 5 and 8.5 Ma ago. Moore *et al.* also studied the carbonate cycles in sedimentary cores deposited over the past 2 Ma. They identified three different cycles of changing carbonate content, one having a period of about 400 ka. Using cross-spectral analysis, they demonstrated a correlation between the eccentricity cycle and the carbonate content of the cores, showing that the two were in phase over the last 8 Ma. Two other orbital cycles, which had already been identified in sediments of the past half million years, also showed up in the 2 Ma-long Pleistocene carbonate record. One cycle had a period of 100 ka and was in phase with a 100 ka cycle in eccentricity. A minor cycle of 41 ka matched the variations in the tilt of Earth's axis.

Although the 100 ka cycle dominated climatic variability during the Pleistocene ice ages, accounting for more than half of the variability due to the three orbital cycles and 29% of the total variability, it had a minor effect 5–8 Ma ago (see Figure 8.2). Even after the 400 ka cycle was used to adjust or 'tune'* the dating of the core, the 100 ka cycle accounted for six times less variability than it did during the Pleistocene. The variability of the 400 ka cycle remained unchanged, which means that the variability of the carbonate record, and presumably of the climate, was twice as great during the past 2 Ma as it was 7 Ma ago.

The oceanic $\delta^{18}O$ chronology for the late Quaternary was further revised by Johnson (1982) and Imbrie *et al.* (1984), who generated a stacked $\delta^{18}O$ record (SPECMAP) for the past 800 ka and used astronomical data to fine-tune the geological timescale. As discussed in Section 7.4, Shackleton *et al.* (1990) and Hilgen (1991a) calibrated the geomagnetic timescale back to 2.5 Ma, using

---

* By tuning a model one means adjusting the model parameters over a range of physically reasonable values until the optimum model variant is identified.

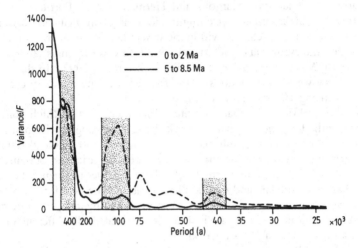

Figure 8.2 Variance spectra showing how much stronger the effects of the 100 ka cycle were over the past 2 Ma (broken curve) than 5–8 Ma ago (full curve). The larger the peak, the more variability in the carbonate content of the sediment (and presumably in climate) is attributable to cycles having that period. The variability attributable to the 400 ka cycle remained the same, but that in the 100 ka cycle was about six times greater during the more recent period. Variability having periods between 40 ka and 100 ka also increased dramatically. After Moore *et al.* (1982).

orbital theory. The agreement of their astronomically calibrated (polarity) timescale with revised radiometric dates by Baksi *et al.* (1992) and Tauxe *et al.* (1992) gives strong support for the Milankovitch theory. In their analysis Shackleton *et al.* used oxygen isotope data from both benthonic and planktonic foraminifera from ODP site 677 and assumed that there has been a consistent phase relationship between astronomical forcing and climatic response over the past 2.6 Ma. Hilgen (1991a) used sapropel-bearing sequences exposed in southern Italy and Crete, and assumed inferred phase relationships between orbital precession and eccentricity cycles and sapropel cycles. In a later paper (1991b) he extended the astronomically calibrated polarity timescale to the Miocene–Pliocene boundary. He correlated the detailed record of $CaCO_3$ cycles in the Rossello composite section in Sicily to the astronomical record using inferred phase relations between these $CaCO_3$ cycles and the corresponding orbital cycles together with calibration points provided by his previously established astronomical calibration of sapropelitic layers which occur in the topmost part of the $CaCO_3$ record.

Stothers (1987) carried out a time series spectral analysis on the times of geomagnetic reversals for the last 20 Ma and found some evidence of high spectral peaks for two long periodicities of 0.4 Ma and 1.3 Ma. He speculated that they may be associated with variations of the Earth's orbital eccentricity, as predicted by Berger (1977a). He suggested that variations of latitudinal and annual mean insolation would affect the Earth's globally averaged temperature and, hence, the surface distribution of ice and water. This shift of latitudinal mass loading could alter the Earth's moment of inertia and, hence, its rate of rotation, which could disturb the geodynamo and lead to reversals. This scenario is similar to that of Muller and Morris (1986) – see Section 8.4. The geomagnetic response would be non-linear with an unknown time lag, so that its strength would be unlikely to be directly proportional to the orbital forcing and thus difficult to predict. For this and other reasons, including statistical significance tests, Stothers does not believe that slow orbital eccentricity periodicities appear in the record of reversals of the Earth's magnetic field.

Some doubt has been cast recently on the Milankovitch theory of ice ages. Winograd *et al.* (1992) obtained a long, continuous high-resolution climate record from a calcite vein in Devil's Hole, a water-filled fissure in the desert floor of Nevada. Water that fell as rain or snow over the surrounding mountains seeped into the ground and into Devil's Hole. As the water flowed through, it deposited a layered crust of calcite on its walls. Oxygen isotope studies in the calcite give a continuous climate history from 600 to 60 ka.

Winograd *et al.* found that warm interglacial periods tended to last 20 ka, rather than 10 ka, an increasing duration of the quasi-100 ka cycle as we come forward in time and the occurrence of a well-developed glacial cycle at 3.5–4.5 ka, when the Milankovitch theory indicates that none should occur. In particular, the onset of the last interglacial began ~ 128 ka ago according to the Milankovitch theory, whereas the Devil's Hole record indicates that it began much earlier, about 140 ka ago. This discrepancy is not resolved. Shackleton (1993) and Edwards and Gallup (1993) suggested mechanisms that could invalidate the [230]Th dates that Ludwig *et al.* (1992) obtained for the Devil's Hole calcite vein. Ludwig *et al.* (1993a,b) answer these criticisms and maintain that the Devil's Hole record presents a challenge to the Milankovitch hypothesis. They believe that the Pleistocene glacial cycles originate from internal non-linear feedbacks within the atmosphere–ice sheet–ocean system. Broecker (1992) comments 'I remain confused. The geochemist in me says that the Devil's Hole chronology is the best we have. And the palaeoclimatologist in me says that the correlation between the accepted marine chronology and Milankovitch cycles is just too convincing to be put aside.' In a later paper (1993), Imbrie *et al.* show that applying the Devil's Hole chronology to ocean cores requires physically implausible changes in sedimentation rate. Moreover, spectral analysis of the Devil's Hole record shows clear evidence of orbital influence. They therefore conclude that it is inappropriate to transfer the Devil's Hole chronology to the marine record and that the evidence for the Milankovitch theory is still strong.

Herbert and D'Hondt (1990) observed Milankovitch cyclicity in well-dated late Cretaceous–early Tertiary marine sediments, and Herbert (1992) later extended the palaeomagnetic calibration from 65–80 Ma to 125 Ma. The lower Cretaceous Maiolica Formation of northern and central Italy shows rhythmic bedding in sections whose magnetic reversal sequence is clearly correlated to the Upper M-series marine anomalies. The Maiolica Formation shows a regular modulation of bedding thickness and carbonate content (termed 'bundling'). Each bundle is built up of a number of pairs (couplets) of carbonate-rich and carbonate-poor beds. Herbert showed that carbonate couplets have an estimated mean period of 23.5 ka and the modulations in bedding thickness a mean period of 117 ka.

8.3 The Earth's magnetic field, reversals and climate

From an analysis of deep-sea sediment cores, Wollin *et al.* (1977, 1978) claimed that, over the past 2 Ma, high eccentricity of orbit has corresponded with low magnetic field intensity and warm climate, and they concluded tentatively that the orbital eccentricity partially modulates both the magnetic field and climate. Chave and Denham (1979) have argued that no connection between climate and magnetic intensity has been established, the correlation found by Wollin *et al.* being due to their failure to remove the effect of the varying input of magnetic material which is largely climatically controlled. It is easy then to see why the ancient magnetic field might correlate with climate. Ruddiman and McIntyre (1976) showed that the flows of biogenic and terrigenous matter (i.e. material which contains magnetic constituents) into oceanic sediments vary by factors of up to at least 3, depending upon climatic conditions. During glacial periods the relative contribution of terrigenous material increases. Generally, then, the NRM of sediments produced during glacial times should be higher than that of inter-glacial sediments even if the strength of the geomagnetic field were to remain constant, and whatever the variations in the Earth's eccentricity. Chave and Denham supported their contention by estimating the variation of palaeointensity along a core from the North Atlantic deposited between 127 ka and 60 ka ago. They used, not raw NRM values, but values of NRM normalized against an imposed anhysteretic remanent magnetization, ARM. Whereas both NRM and ARM vary with the quantity of magnetic material present, the ratio between the two should not. The palaeointensity they obtained showed no correlation at all with climate as represented by faunal variations throughout the core.

Harrison (1974) had earlier maintained that those cores which show correlations between direction and/or intensity of magnetization and climatic indicators are not accurately recording the relevant parameters of the Earth's magnetic field – the correlation being caused by climatic effects which have a direct influence on the magnetization of the sediments. For example, Bonatti and Gartner (1973) showed, in a core from the Caribbean, that there are several geochemical changes produced directly by climatic effects which could easily affect the authogenic production of magnetic material in such a way as to cause the correlation between intensity and climate claimed by Wollin *et al.* (1971a).

Kent (1982) has also cautioned against accepting the apparent correlation of palaeomagnetic intensity with climatic records in deep-sea sediments. He showed a pronounced dependence of the NRM intensity on sediment composition. The observed correlation between NRM intensity and climate (e.g. the $\delta^{18}O$ record) can be largely accounted for as the result of an intermediary lithological effect. Decreased carbonate content during glacials results in increased concentrations of magnetic material, which in turn contribute to higher NRM intensities. Kent also found cases where high NRM intensities characterized sediments deposited during glacial intervals in one region, but were associated with sediments deposited during interglacials in another.

Doake (1977) has suggested that there is a connection between ice ages and reversals of the Earth's magnetic field. The moment of inertia of the Earth will change by variations in the size of polar ice sheets and the resulting redistribution of water masses. To conserve angular momentum, the Earth's rate of rotation must also change. Doake therefore suggested that the generation mechanism of the Earth's magnetic field may be affected by changing conditions at the CMB. He carried out simple order-of-magnitude calculations which indicated that a continuous change in the speed of rotation over a sufficiently long time could alter conditions at the CMB sufficiently to perturb the magnetic field and perhaps cause it to reverse its polarity. It is doubtful, however, whether the effect is strong enough. Moreover, although such a result seems plausible, conclusive evidence for a correlation between glacial periods and changes in the magnetic field is hard to obtain, because of the difficulty in identifying and accurately dating variations in the two parameters.

In a later paper Doake (1978) analysed statistically the probability of a correlation between climatic changes and field reversals. He compared dates of climatic episodes as recorded in deep-sea cores with dates of palaeomagnetic polarity transitions during the upper Pliocene (1.5–4.3 Ma ago). Dating of the climatic record was obtained by determining the ages of three biostratigraphic and two palaeoclimatic horizons by reference to palaeomagnetically dated sequences in New Zealand – the core was taken from a site on the west side of the North Island. Any such correlation will be affected by dating errors which can arise from misidentification of stratigraphic horizons when relating magnetic and climatic data and from non-uniform sedimentation rates when determining relative timescales between horizons. While there is uncertainty in the dating, Doake's results appear to show a correlation between climatic change and reversals of the Earth's magnetic field – the chance that the number of observed coincident dates will be random can be as low as 3×10^{-4}. However, because of possible dating errors, it is not clear in the association of magnetic events and climatic change which is cause and which is effect.

The Blake event, about 110 ka ago, first detected in deep-sea cores by Smith and Foster (1969), has since been reported in a long sediment core from Lake Biwa, Japan. Similar events were observed lower in the core by Kawai *et al.* (1972) and estimated to be about 180 ka old (the Biwa I event) and about 295 ka (the Biwa II event). Yaskawa (1974) constructed an age against depth curve for the Lake Biwa core, using ^{14}C age determinations in the upper part and fission track dating of volcanic ash layers between depths of 40 and 100 m. Excursions that are believed to correlate with the Lake Biwa events have also been detected in deep-sea cores (Wollin *et al.*, 1971b; Yaskawa, 1974) – see Figure 8.3. The magnetic events coincide with lows in organic carbon in the Lake Biwa core which are believed to be related to periods of cool climate and low productivity of the lake (Kawai *et al.*, 1975). The Blake event apparently occurred about the same time as the sudden build-up of ice and climatic cooling that immediately followed and ended the last interglacial stage. Studies of the oxygen isotope ratio

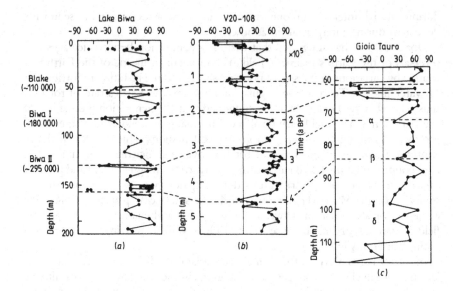

Figure 8.3 (a) Magnetic inclination plotted against depth in the Lake Biwa core: results of preliminary measurements of samples taken at 5 m intervals (Yaskawa, 1974). Estimated dates of excursions based on fission-track dating of volcanic ash layers are given at left. (b) Magnetic inclination plotted against depth in deep-sea core V20–108 (Wollin *et al.*, 1971b). Timescale is based on the assumption of constant sedimentation rates between the top of core (zero age) and the Brunhes–Matuyama boundary. (c) Magnetic inclination plotted against depth for the Gioia Tauro sequence. Major unconformity occurs at about 55 m. Correlation of double event between 60 and 65 m designated as the Blake event by Creer *et al.* (1980), with the Blake event and Biwa I event, assumes low sedimentation rates or a hiatus in the sequence between the two events. The other correlations are after Creer *et al.* (1980). After Rampino (1981).

in deep-sea cores by Shackleton (1976, 1977) and Johnson (1978) and of changes in sea level recorded in tropical coral reef terraces by Matthews (1972) suggest that, about 110 to 115 ka ago, sea level fell 60–70 m in less than 10 ka. The rate of growth of ice sheets would have produced a volume of ice of about 28 × 10⁶ km³ (equivalent to the present Antarctic and Greenland ice sheets) in less than 10 ka. Climatic changes on timescales of 10^3–10^5 a may be strongly modulated by changes in the Earth's orbital geometry and their effects on the seasonal distribution of insolation. A climatic cooling at the time of the Blake event would be predicted by the Milankovitch insolation mechanism, although the ice build-up was extremely rapid.

Creer *et al.* (1980) carried out palaeomagnetic and palaeontological studies on a 250 m core from Gioia Tauro, Italy. They found a number of possible excursions, including the Blake event, which they suggested was a double event lasting about 50 ka (between about 105 and 155 ka ago). They also attempted to correlate these excursions with similar events reported from Lake Biwa (Yaskawa, 1974) and in the deep-sea core V20–108 (Ninkovich *et al*, 1966). There are, however, problems with their interpretation of the Gioia Tauro record, the dating of which was based on magnetostratigraphy back to the Matuyama–Gauss transition and assumptions regarding the rate of sedimentation. Although a short double Blake event was reported by Denham (1976) in deep-sea cores, its duration is less than 13 ka in the Lake Biwa core and in the original finding in a deep-sea core by Smith and Foster (1969). A further problem is that the Biwa I event was not seen in the Gioia Tauro core. Rampino (1981) has offered an alternative interpretation of the Gioia Tauro section which avoids these difficulties. He suggested that the long, double Blake event in the Gioia Tauro core is the result of changes in the sedimentation rate and perhaps an undetected hiatus. A change in the nature or rate of sedimentation is suggested by the inferred difference in magnetic mineral content between the interval recording the 'Blake event' and the overlying and underlying sediments. Rampino obtained revised dates for the excursions in the Gioia Tauro core based on those seen in the V20–108 and Lake Biwa cores (see Figure 8.3). The revised estimates of the ages of the excursions show an approximate cycle of about 100 ka which is similar to that of the variation in the eccentricity of the Earth's orbit (see Figure 8.4). The excursions (if real) seem to occur at times of maximum eccentricity, giving some support to a possible connection between orbital parameters of the Earth and magnetic fluctuations.

The Lake Mungo excursion about 30 ka ago (see Section 4.3) is also associated with a time of apparent rapid cooling immediately following a major climatic warming and retreat of ice. This brief interval of warming (less than 5 ka) is not well recorded in most deep-sea cores and its extent is somewhat controversial. The Biwa I magnetic excursion also appears to have coincided with a time of rapid growth of ice sheets, as indicated by oxygen isotope curves of deep-sea sediment cores (see Figure 8.5). Furthermore, it took place when the Milankovitch mechanism would suggest growth of global ice sheets. However, the expansion of the ice sheets was again very rapid, as seen in the oxygen isotope curves. (It must be pointed out that these magnetic events have not been detected in the same deep-sea cores that were used for isotopic studies, so that any correlation of climate with magnetic events is indirect.) On the other hand, the time of the Biwa II event was apparently not accompanied by rapid changes in ice volume,

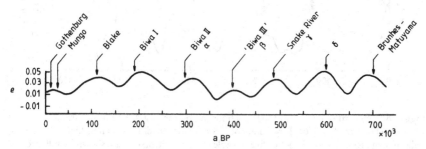

Figure 8.4 Eccentricity (*e*) of the Earth's orbit over the past 725 ka (after Kominz *et al.*, 1979). Arrows indicate inferred ages of magnetic excursions. Age of Snake River event is (480 ± 50) ka. Age of the δ event is uncertain, as it has only been reported in the Gioia Tauro section. After Rampino (1981).

as indicated in oxygen isotope studies of most deep-sea cores (see Figure 8.5). However, the high deposition rate of the Lake Biwa core (~ 50 cm/1 ka), indicates a marked decrease in productivity, interpreted as a regional cooling at the time of the Biwa II event, but it must be confessed that the details of climatic changes that occurred about that time are not well known. An as yet unnamed magnetic excursion that has been detected in deep-sea cores and dated at about 400 ka also apparently correlates with a time of rapid ice growth and immediately follows a time of major melting of northern hemisphere ice sheets (Figure 8.5). Indications of such an event are also found in the Lake Biwa core (Figure 8.3). Although not well documented at present, this adds further support to the possible association of climate and geomagnetism.

It must be stressed that coincidence of events, such as changes in global ice volume, magnetic excursions, and short-term rapid glaciations and climatic coolings, do not in themselves prove cause-and-effect relationships. Also, other magnetic excursions that have been reported apparently did not occur at the same times as major ice volume changes and rapid cooling, as recorded in oxygen isotope curves of deep-sea cores (Verosub and Banerjee, 1977). Again many of the reported excursions are poorly dated, and the reality of several magnetic events has been questioned.

It is worth noting that the last four (and possibly five) times of maximum eccentricity of the Earth's orbit appear to be closely followed by magnetic excursions (see Figure 8.4). Eccentricity maxima also appear to coincide with brief periods (around 10 ka) of minimum ice cover (the interglacial periods) that followed rapid melting of northern hemisphere ice. Some excursions (for example, the Lake Mungo excursion) are not associated with a maximum of orbital eccentricity, but they do seem to have followed times of rapid ice melting.

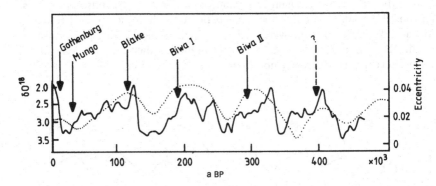

Figure 8.5 Record of oxygen isotope variations during the past 475 ka in sub-Antarctic deep-sea cores (Hays *et al.*, 1976). Low values of $\delta^{18}O$ indicate small ice volumes; high $\delta^{18}O$ values indicate large ice volumes. Times of five possible magnetic excursions are shown by full arrows. (These excursions were not detected in deep-sea cores used to construct record of $\delta^{18}O$.) Time of a sixth magnetic excursion, detected in some deep-sea cores and unnamed at present, is shown by broken arrow. Dotted line is plot of the eccentricity of the Earth's orbit. After Rampino (1979).

It appears that if the excursion occurs at an unfavourable time in the Milankovitch insolation cycle (as with the Gothenburg excursion about 13.5 ka ago), then only a brief glacial readvance is observed, but if the geomagnetic trigger occurs at a 'sensitive' phase of the Milankovitch cycle, then a full glacial stage may be initiated.

8.4 Reversals, impacts and mass extinctions

There has been much discussion in recent years on impacts of extraterrestrial bodies with the Earth and their possible connection with reversals of the Earth's magnetic field and mass extinctions. Glass and Heezen (1967) pointed out that the great field of tektites covering Australia, Indonesia and a large part of the Indian Ocean fell about 700 ka ago at about the time of the last magnetic reversal. They suggested that the fall of the body from which the tektites were formed killed the now extinct radiolaria and gave a jolt to the Earth, disturbing motions in the core and causing the dynamo to reverse. But, as Bullard (1968) pointed out, it is 'difficult to believe that the fall of a large meteorite could selectively kill certain species of radiolaria all over the world and yet spare the kangaroos near the point of fall'. De Menocal *et al.* (1990) have examined the relative stratigraphic positions of interglacial stage 19.1 (determined by oxygen isotope studies), the Brunhes–Matuyama reversal and the widespread tektite layer in eight deep-sea sediment cores. They found that the Brunhes–Matuyama reversal occurred 6 ± 2 ka after the stage 19.1 datum, which in turn occurred 9 ± 3 ka after the tektite-strewn field. They concluded that there is no connection between geomagnetic field reversals, climate change and impact events.

Durrani and Khan (1971) and Glass and Zwart (1979) suggested that there could be an association between the Ivory Coast microtektites and the onset of the Jaramillo normal polarity subchron about 0.97 Ma ago. This prompted Schneider and Kent (1990) to re-evaluate the palaeomagnetic stratigraphy of two critical deep-sea cores containing Ivory Coast microtektites. They concluded that the event which produced them (the Bosumtwi impact crater in eastern Ghana? – see Shaw and Wasserburg, 1982) most likely occurred during the Jaramillo subchron, but ~ 30 ka after its onset and ~ 40 ka before its termination, thus giving no support for any causal relationship between the two events. Furthermore, there appears to be no record of tektites or microtektites associated with the time of other field reversals (there have been more than 70 in the last 20 Ma). It has also been difficult to explain how an impact on the Earth's surface could physically disturb motions in the core and, hence, affect the Earth's magnetic field. In this regard the K/T boundary which, it has been suggested (Alvarez *et al.*, 1980), may mark a large impact event, does not coincide with a polarity reversal, excursion or any change in reversal rate. Muller and Morris (1986) suggested that the impact of a large extraterrestrial body with the Earth would raise enough dust from the crater and soot from fires to lower the temperature of continental land areas and start a little ice age. The moment of inertia of the

Earth is then changed by the redistribution of tropical ocean water to ice at high latitudes and this in turn alters the rotation rate of the Earth, which they speculate would lead to a change in shear at the CMB, causing a reversal or excursion. This scenario is difficult to test, since it depends on several interacting non-linear processes with possible time lags of different magnitudes at different times. Gubbins (1983) investigated the possibility that changes in pressure resulting from a large impact might change the rate of freezing at the ICB and, hence, cause reversals, and concluded that the effect is probably too small.

Again, as discussed in Section 3.4, the mean rate of occurrence of reversals has increased since the Cretaceous up to the present. If all reversals had been caused by impacts, then the Earth would have to have been hit at an ever-increasing rate by extraterrestrial objects since the Cretaceous, which is not borne out by the limited amount of data (Raup and Sepkoski, 1984; Shoemaker, 1984). Geological evidence also argues against reversals being triggered by changes in the moment of inertia resulting from changes in ice cover, as in the Muller–Morris model. The present ice age occurs at a time of rapid reversals, in agreement with their model, but the Gondwana ice ages occurred during a time of almost no reversals (the Kiaman).

Schneider *et al.* (1992) later re-examined the record of two high-sedimentation rate deep-sea sediment cores from marginal seas of the Indonesian archipelago, which record the Australasian impact with well-defined microtektite layers, the Brunhes–Matuyama polarity reversal and global climate measured with oxygen isotope variations in planktonic foraminifera. Both cores show the impact to have preceded the reversal of the magnetic field by about 12 ka. Moreover, both records indicate that the intensity of the field was increasing near the time of impact and that it continued to do so for about 4 ka afterwards. Again the oxygen isotope record shows no indication of climate cooling following the impact – in fact, the microtektite layer occurred in the latter part of glacial Stage 20 and was followed by a smooth *warming* trend to interglacial Stage 19 (see Figure 8.6). This more detailed study of Schneider *et al.* confirms the earlier conclusions of de Menocal *et al.* (1990) and contradicts the suggestion by Muller and Morris (1986) that an impact can trigger a geomagnetic polarity reversal by means of rapid climate cooling.

There have been many attempts in the past to look for periodicities in the reversal record. Crain *et al.* (1969) computed a Fourier power spectrum of the proportion of normal polarity measurements, using a compilation by Simpson (1966) with an effective sampling interval of about 15 Ma. They observed periodicities of 300 and 80 Ma. Later Crain and Crain (1970) carried out a spectral analysis of Simpson's mixed measurements and obtained periods of about 150 and 40 Ma. Fourier spectral analysis is, however, of limited usefulness in the treatment of short time series. Maximum-entropy spectral analysis is a more powerful method, and Ulrych (1972) applied it to the analysis of polarity ratios. He took the data compiled by McElhinny (1971), fitted them to a ninth-order polynomial and interpolated with an effective sampling interval of 25 Ma. He obtained periodicities of 700 ± 100 Ma and 250 ± 50 Ma. Irving and Pullaiah (1976) also carried out a maximum-entropy spectral analysis, using overlapping

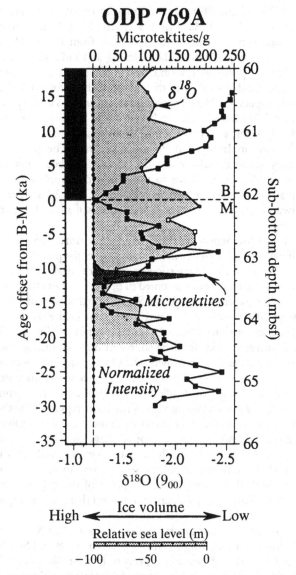

Figure 8.6 Comparison of global ice volume record with palaeomagnetic and microtektite stratigraphies. The relative age scale is derived assuming a constant sedimentation rate of 11 cm/ka over the interval. The relative sea-level scale is constructed, using a 0.11‰ shift in $\delta^{18}O$ for each 10 m sea-level change. Microtektite abundance data (●) are shown as number per gram of wet sediment. Normalized palaeointensity (■) is from ARM normalization of discrete sample data, scaled to a range of 0–250 (to match the microtektite abundance scale). To facilitate comparison of these records, the Brunhes–Matuyama (B–M) boundary (the level of the abrupt reversal of palaeomagnetic directions) and the ARM-normalized palaeointensity results have been shifted upward by 16 cm to compensate for the depression of the magnetic lock-in zone. After Schneider *et al.* (1992)

averages of 5, 10, 25, 50 and 100 Ma, and found three major components with mean periodicities of 297 ± 34 Ma, 113 ± 5 Ma and 57 ± 1 Ma. The latter two periodicities are not very evident in the spectrum from the 50 Ma averages, although there is some indication for them. Irving and Pullaiah were unable to reproduce the 700 Ma periodicity found by Ulrych (1972), and are sceptical about estimates of periods longer than the length of the record (600 Ma). Negi and Tiwari (1983) applied the Walsh power spectrum analysis to the geomagnetic record for the past 570 Ma and claim to have found periodicities in the reversal frequency of 285, 114, 64, 47 and 32 Ma. The maximum power is in the 285 Ma term, as was the case in Irving and Pullaiah's analysis. The period of about 300 Ma is common to all the above analyses and, as had been pointed out by Crain and Crain (1970), corresponds to the period of complete revolution of the solar system around the Milky Way galactic centre. The 32 Ma period found by Negi and Tiwari corresponds to that of the sun's oscillation above and below the galactic plane.

The claim by Raup and Sepkoski (1984) of an approximate 26 Ma periodicity in the pattern of mass extinctions of marine animals over the past 250 Ma resulted in a spate of papers in *Nature* suggesting an extraterrestrial cause. In the past mass extinctions have generally been assumed to be the result of changing geography and climate. The paper by Raup and Sepkoski led to speculation that the claimed periodicity of reversals of the Earth's magnetic field is associated with a periodicity in mass extinctions caused by extraterrestrial bodies that impacted the Earth with a similar periodicity. The source of material hitting the Earth is purported to be the distant ring of comets (the Oort cloud). Davis *et al.* (1984) and Whitmire and Jackson (1984) proposed the existence of an unseen companion star to the sun (which they called Nemesis) in a highly eccentric orbit. When near perihelion, it passes through the dense inner region of the Oort cloud and by perturbing the cometary orbits initiates an intense comet shower leading to a series of terrestrial impacts. An alternative suggestion for the initiation of comet showers was put forward by Schwartz and James (1984) and Rampino and Stothers (1984) – perturbations of the solar system by the sun's oscillation (half-period ∼ 33 Ma) above and below the plane of the galaxy. Alvarez and Muller (1984), in the same group of papers, reported that ancient impact craters on the Earth occur in a 28.4 Ma cycle.

In less than 6 months another group of papers appeared in *Nature* criticizing the above mechanisms for initiating a comet shower. Torbett and Smoluchowski (1984) found that the orbit of Nemesis is likely to be unstable, as did Clube and Napier (1984), who, although arguing strongly against the Nemesis proposal, favoured the model invoking oscillations of the sun about the galactic plane. There appear to be just as many problems if the disturbing force on the Oort cloud is a tenth planet X instead of a close, unseen companion of the sun, as suggested by Whitmire and Matese (1985). Thaddeus and Chanan (1985) showed that molecular clouds are not sufficiently tightly bunched around the galactic plane to make much difference to the chance of the Earth's encountering a cloud near the plane than at the extremities of its oscillations where the clouds have begun to thin out.

In assessing the value of all these speculative cyclicities, it is necessary to reassess the evidence put forward by Raup and Sepkoski (1984) in support of their claim of a 26 Ma periodicity in the pattern of mass extinctions. As stressed by Hoffman (1985), the evidence for their claim is strongly dependent on arbitrary decisions concerning the absolute dating of stratigraphical boundaries, the culling of the database and the definition of what is mass extinction as opposed to background extinction. Hoffman maintains that the evidence for regular cyclicity is inconclusive under other plausible geological timescales and other acceptable definitions of mass extinctions.

There has been much argument over the evidence for periodicity in mass extinctions. The arguments revolve around the data used (see, e.g., *Nature*, 321, 533–6, 1986) and the statistical methods used for testing for periodicity (see, e.g., *Science*, 241, 94–9, 1988). The question will not be discussed further here, since this book is concerned with reversals of the Earth's magnetic field. (The possibility of a 30 Ma periodicity superimposed on the predominantly stochastic signal in the record of reversals of the Earth's magnetic field has been discussed in Section 3.4.) It must be pointed out, however, that in a later paper Sepkoski (1989) has presented new data in support of periodicity in mass extinction events. In the interval from the Permian to Recent, these new data comprise some 13 000 generic extinctions providing a more sensitive indicator of species-level extinctions than previously used familial data. Time series analysis of the generic data show nine strong peaks that are nearly uniformly spaced at 26 Ma intervals over the last 270 Ma (see Figure 8.7). Most of the extinction peaks correspond to events recognized in detailed biostratigraphic studies conducted at the species level. Sepkoski (1989) also answers many of the criticisms that have been made of the taxonic database, sampling intervals, chronometric timescales and statistical methods that have been used in previous analyses.

Lutz (1987), in a discussion of the statistical analysis of geological time series, argued that periodic functions should also be compared to distributions reflecting constrained episodicity in which events tend to have some minimum spacing resulting from an embedded recovery time. In an analysis of Raup and Sepkoski's (1984) data, he concluded that the record of extinction events was not random, but could reflect either a periodic forcing with a severe wobble or an episodic forcing with a long recovery time. Sepkoski (1989) also points out that, although his new data strengthen the case for periodicity, they offer little new insight into the driving mechanism – in fact, the data suggest that many of the periodic events may not have been catastrophic, occurring instead over several stratigraphic stages or substages.

Grieve *et al.* (1985/86) have reviewed the evidence for periodic cometary impacts on Earth. They pointed out that the record used is incomplete and may not be representative. (Alvarez and Muller, 1984, in their report on cyclicity in the crater record used craters having ages between 5 and 250 Ma old, dating uncertainties of 20 Ma or less and diameters greater than 10 km. Only 13 craters met these criteria, so that their conclusions must be viewed with caution.) Estimates of crater ages are of variable accuracy, and, in some cases, different dating methods give different results. Grieve *et al.* showed that it is possible to obtain

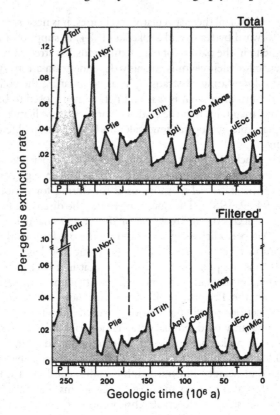

Figure 8.7 Per-genus extinction rate (in units of extinctions per genus per Ma) for 49 sampling intervals from the mid-Permian (Leonardian) to Recent. A 26 Ma periodicity, represented by the vertical lines, is superimposed in best-fit position to illustrate the conformity of the data to the hypothesis of periodic extinction. The upper time series, labelled Total, is for the entire data set of 17 500 genera, whereas the lower, 'filtered', time series is for a subset of 11 000 from which genera confined to single stratigraphic intervals have been excluded. Systems and stages (but not sampling intervals) are indicated along the abscissa of each graph. The timescale is from Harland *et al.* (1982), with the durations of Jurassic stages rescaled, using Westermann (1984), in order to eliminate the unrealistic uniformity of estimated durations. After Sepkoski (1989).

a number of periodicities of different magnitude (29 Ma, 21 Ma, 18.5 Ma, 13.5 Ma) and phase, depending on which craters are considered the most representative sample of the record.

If further studies indicate that there are no periodicities in the cratering and mass extinction records, there still could be a causal relationship between mass extinctions and impacts – perhaps as a result of comet showers. In 1980 Alvarez *et al.* suggested that the mass extinctions at the Cretaceous–Tertiary (K–T) boundary were the result of a major impact about 65 Ma ago. Their evidence was anomalous amounts of iridium and other siderophile elements seen first at Gubbio, Italy, and subsequently in other locations around the world. Their suggestion generated much controversy, which still persists – no attempt will be made to summarize the various arguments for and against their suggestion. Probably the majority of Earth scientists accept their hypothesis. It has been given final credence by the discovery of a 180-km-wide impact crater (Chicxulub) in Yucatan, Mexico, whose outline is based on magnetic and gravity anomalies. ^{40}Ar–^{39}Ar dating of drill-core samples of a glassy melt rock from beneath a massive impact breccia gives an age of 64.98 ± 0.05 Ma. The age is indistinguishable from those obtained on tektite glass from Beloc, Haiti, and Arroyo el Mimbral, north-eastern Mexico, almost 2000 km away (Swisher *et al.*, 1992). These tektite glass pellets are believed to be remnants of the debris that was flung 1000 km away when the asteroid or comet hit the Earth. Sharpton *et al.* (1992) analysed Chicxulub melt rocks and found that they contained anomalously high levels of iridium consistent with the Ir-enriched K–T boundary layer. Shocked breccia clasts in the core are similar to shocked lithic fragments found worldwide in the K–T boundary enjecta layer. They also showed that the melt rocks acquired a remanent magnetization as they cooled at a time when the Earth's magnetic field was reversed. The only time consistent with the ^{40}Ar–^{39}Ar dates is chron 29R, which includes the K–T boundary.

Another contender for the site of the impact is the smaller (diameter 35 km) Manson crater in Iowa. The latest dating of the crater gives an age almost the same as that for the Chicxulub impact. It is thus possible that a large comet broke up with two or more pieces hitting the Earth, or that at least two comets struck the Earth from a comet shower. As mentioned earlier, not everybody agrees that the mass extinctions that occurred at the K–T boundary were the result of an impact, but were caused by a long period of intense volcanic activity. Their case is strengthened by the massive outpouring of the Deccan flood basalts, which cover about one-sixth of India, and has been dated at around 65 Ma. Negi *et al.* (1993) suggested that this volcanism could have been triggered by a bolide which hit the Earth near Bombay. Their suspected impact site was based on an oval-shaped unusual positive gravity anomaly, and nearby unusual geothermal and structural features. Pandey and Negi had earlier (1987) argued that pro-longed volcanic activity, instead of impact cratering, may have been the primary cause of climatic and other environmental changes sufficiently severe to lead to mass extinctions on a global scale. In support of their suggestion, they compiled a list of major global magmatic episodes over the last 250 Ma. Analysis showed intervals of significant enhancement of volcanic activity with a periodicity of 33 Ma.

Claeys *et al.* (1992) have found glass spherules, similar to microtektites at Senzeilles, Belgium, near the Frasnian–Famennian boundary (~ 367 Ma ago) when 70% of all invertebrate species in the oceans died out. They suggest that the mass extinctions were caused by an impact. There is some indication that the extinctions took place over millions of years around the boundary, which, if substantiated, could indicate multiple impacts.

Rampino and Caldeira (1993) compiled a list of major geological events of the past 250 Ma (mass extinctions, anoxic events, evaporite deposits, flood basalts, sea-floor spreading, sequence boundaries, orogenic events). Analysis shows evidence for a statistically significant component with a periodicitiy of 26.6 Ma and a recent maximum close to the present time. The cycle may not be strictly periodic, but a periodicity of ~ 30 Ma is robust to probable errors in dating of the geological events. The intervals of geological change appear to involve jumps in sea-floor spreading associated with episodic continental rifting, volcanism, enhanced orogeny, global sea-level changes and climate fluctuations. Rampino and Caldeira reanalysed the data while eliminating each type of geological event one at a time. The dominant period remained at 26.6 ± 0.4 Ma, showing that this period is not dependent upon any one particular type of geological event.

Finally, it must be pointed out that time series studies by themselves are insufficient evidence for a periodic causative force producing terrestrial evidence. Additional data in the impact record alone would not resolve the more general question of any relationship between large-scale impacts and mass extinctions or other geological phenomena. Statistical arguments in themselves are insufficient – as Kitchell and Pena (1984) have pointed out, they may reflect stochastic processes which exhibit a pseudo-periodic behaviour.

8.5 Reversals, plumes and tectonics

There have been many suggestions in the past of possible relationships between events at the Earth's surface and those in the core. Heirtzler (1970) speculated that increased earthquake activity and related increased upper-mantle activity may be linked with magnetic changes. He suggested that a sufficiently large earthquake may cause a wobble of the Earth's spin axis and create a magnetic reversal. Kennett and Watkins (1970) extended this idea by suggesting that an increase in upper-mantle activity should be manifested by increased volcanic activity, and thus evidence of such activity should be found at times of magnetic reversals. They presented evidence from late Cenozoic deep-sea sediment sequences from the Southern Ocean indicating increased amounts of volcanic ash close to the times of reversals. The theory was expanded by suggesting that if increased episodes of explosive volcanicity have occurred at specific times in the past, these possibly also affected the global climate and, in turn, faunal changes. Chappel (1975) has critically examined many of these suggestions and, on theoretical grounds, dismissed most of the proposed relationships.

The characteristic time associated with individual reversals (20 ka) differs by several orders of magnitude from that associated with changes in reversal frequency (50 Ma). It seems probable, therefore, that the two phenomena have different causes. Individual reversals are more likely to have their origin in the fluid motions and electric currents in the Earth's core – their characteristic times are those intrinsic to the geodynamo. Changes in reversal frequency, on the other hand, are probably due to changes in the rate at which energy becomes available to generate turbulence in the core or to a change in conditions at the CMB. Changes in the rate of supply of energy could result from processes such as the slow decay of radioactive heat sources in the core or changes in the amount of latent heat of fusion supplied by growth of the solid inner core. The characteristic times of such processes are difficult to estimate but probably would be comparable with the age of the Earth. A time period of the order of 50 Ma, on the other hand, is of the same order as that associated with geological events – the formation of new ocean basins and mountain ranges. It is also a reasonable time to associate with convection in the mantle. The basic question is whether convection takes place through the mantle and, if so, whether it could affect the characteristics of the geodynamo.

It is interesting to speculate on a possible connection of the suggested 300 Ma periodicity in the reversal record with the timescale of plate tectonics. The change from reversed to normal bias of the geomagnetic field coincident with the Palaeozoic–Mesozoic boundary is also one of the greatest breaks in the geological record. Since only small variations in core motions are required to reverse the polarity of the dipole field, these could be caused by changes in the boundary conditions at the CMB. Such boundary conditions could be affected by motions in the lower mantle and it is therefore not implausible, as Hide (1967) has suggested, that reversals are correlated to some extent with other phenomena that may be affected by motions in the mantle. Irving (1966) has also suggested that the magnetic field would reverse frequently during times of active convection and tectonism (e.g. in the late Cenozoic). It is difficult to believe that every reversal is related to some event at the Earth's surface – the timescale of tectonic events is much longer than that for reversals. On the other hand, it is not unreasonable to associate a marked change in the frequency of reversals (approximate timescale 50 Ma) with other phenomena.

Hide (1967) suggested that changes in the radial velocity of the fluid motions in the Earth's core might in some cases be impressed from outside. Horizontal temperature variations of only a few degrees and topological features, 'bumps' only a few kilometres high at the CMB, might affect core motions, perhaps causing reversals. Gradual changes in the radius of the core and in the strength of the mechanism that drives core motions might also produce occasional reversals. Hide therefore suggested that there are two types of reversals: 'forced' reversals due to changes impressed from outside the core, and 'free' reversals that would arise even in the absence of impressed changes. Each type of reversal would be characterized by its own timescale. Major geological events are associated with large-scale motions in the mantle. If these motions penetrate to a sufficient depth to produce horizontal variations in the physical conditions that prevail at the CMB, then 'forced' reversals should be strongly correlated with other worldwide geological phenomena. 'Free' reversals, however, should show no such correlations, being determined by random processes in the fluid core.

The CMB separates two dynamic systems with very different compositions and material properties. There has been much argument about the structure of the bottom 200–300 km of the mantle (called by Bullen the D″ layer). Constraints on the properties of this transition zone come from a number of geophysical disciplines – seismology, mineral physics, geomagnetism and geodynamics. Seismology has provided the longest record of observations, but there is still no agreement on their interpretation. Some seismologists maintain that the velocity of both P and S waves decreases as the CMB is approached, others that the velocities remain almost constant and yet others that the velocities continue to increase slightly. The main unresolved problem is whether this transition zone is a thermal or a chemical boundary layer. Decreased velocity gradients have been used to infer a thermal boundary layer with a temperature drop of 800 K. Long-wavelength topography on the CMB has been inferred from an analysis of seismic waves – Morelli and Dziewonski (1987) found up to 10 km of relief with scale lengths of 3000–6000 km. The depressed regions tend to correlate with zones of higher seismic velocities in D″, suggesting that the topography is dynamically supported by the downwelling of cool mantle material. Bloxham and Gubbins (1987) suggested that the mantle controls flow in the OC by thermal coupling. Regions of upwelling material in the core are associated with seismically slow (presumably hot) regions in D″, and core downwellings are associated with seismically fast (cold) regions in the mantle. Gubbins and Richards (1986) have also looked at the mechanical and thermal effects that topography on the CMB would have on flow in the core and on the geodynamo. If there is topographical relief on the CMB, the isotherms will no longer coincide with gravitational potentials, so that there will be both a lateral temperature gradient as well as mechanical interaction with the flow. As discussed later, reversals of the Earth's magnetic field may be the result of slowly changing temperature variations in the mantle, modulating the core flow regime. Experiments at high pressures and temperatures by Knittle and Jeanloz (1989, 1991) indicate that chemical reactions at the CMB over geological time have created chemical heterogeneities of silicate-rich and iron-alloy-rich zones in the D″ layer. The iron-enriched zones would be accompanied by a large increase in the electrical conductivity.

The D″ layer may well have a more complex structure – a thin (<100 km) basal, low-velocity, high-attenuation zone that can be explained as a thermal boundary layer (with a temperature drop of 800 K) and, superimposed on this, another ~200 km layer of subducted slab material defined by increased S wave velocity (Lay and Helmberger, 1983). Since the viscous relaxation time for unsupported topography at the CMB is <1 Ma a dynamic mechanism is needed to sustain it and account for its lateral scale and regional variability. Loper and Eltayeb (1986) have studied the stability of a D″ thermal boundary layer and shown that, if the viscosity drop across the layer is $> 10^3$, convection can take place confined to the low-viscosity sublayer. Olson *et al.* (1987) have carried out numerical experiments which confirm this and have shown that small-scale convection cells develop within the low-viscosity sublayer on a timescale of tens of Ma and with horizontal scales of tens of kilometres. Groups of these convection cells merge to form plumes, also on a timescale of tens of Ma. If the D″ layer contains significant compositional heterogeneity, it would inhibit plume escape and increase this time. Boundary-layer convection can also support small-scale roughness on the CMB with half-widths as small as 20 km and amplitudes of ~2 km. Perhaps the best solution is that D″ is a heterogeneous chemical boundary layer embedded in a thermal boundary layer. Thermal instabilities in the form of plumes as well as large-scale mantle convection may reorganize and entrain the compositional heterogeneity. Both flow regimes could affect the topography on the CMB with a wide spectrum of scale lengths.

Because of the low viscosity of the OC, the roughly 100 Ma variation in reversal rate may reasonably reflect the timescale for changes in the boundary conditions at the CMB. Variations in the temperature gradient across the OC can critically alter core convection and therefore the magnetic field. It is also possible that lateral variations in the temperature gradient at the base of the mantle change in a 100 Ma timescale in such a way as to produce a significant alteration in core convection. The timescale of 100 Ma is also approximately the characteristic timescale for mantle convection. Thus, heat transfer in the mantle may also have a long-term effect on reversal rate. Because tectonic processes at the Earth's surface are directly related to heat transfer in the mantle, one can speculate on the possibility of correlations (with appropriate lags) between changes in reversal rate and other geological phenomena.

A layer at the base of the mantle with a temperature gradient of some 800 K would be highly unstable unless its composition makes it more dense than the overlying mantle – chemical reactions between the molten iron alloy of the core and solid silicate minerals from the mantle may create a layer of different composition and density from those of the overlying mantle. Ahrens and Hager (1987) and Sleep (1988) showed that a stable dense layer could undergo internal convection, or may be compositionally stratified. Kellogg and King (1993) showed that development of a layer at the base of the mantle depends both on the composition of the material forming at the CMB and on the rate at which material diffuses into the mantle. If the material is less than 3–6% denser than the overlying mantle, the reactant material will be swept away by upwelling plumes. On the other hand, more dense material forms an internally convecting layer that entrains material from above.

Wilson (1963, 1965) introduced the idea of stationary mantle 'hot spots' across which lithospheric plates drift, in order to explain the regular progression in age along the chain of volcanic Hawaiian islands. Later Morgan (1971, 1972) suggested that such hot spots were the surface expression of narrow plumes originating deep in the mantle. Stacey (1975) and Jones (1977) placed the source in the D'' layer at the base of the mantle, and Stacey and Loper (1983) interpreted the D'' layer as a thermal boundary layer with a temperature increment across it of ~840 K. In a companion paper Loper and Stacey (1983) showed that narrow, long-lived plumes are a necessary consequence of lower boundary heating of a medium with strongly temperature-dependent viscosity. Jones (1977) was the first to suggest that long-term variations in reversal frequency may be the result of fluctuating temperatures at the CMB caused by intermittent breakdown of a static D'' layer. His model assumes that the geodynamo is driven by thermal convection in the OC and that the D'' layer is a thermal boundary layer. McFadden and Merrill (1984) disagree with the details of Jones's model, although conceding that his general ideas may be correct. In Jones's model, the thermal boundary layer becomes unstable and breaks down by the formation of blobs or plumes. The destruction of such a boundary layer takes up only a small fraction of the time in each thermal cycle, the build-up of a super adiabatic temperature gradient by thermal conduction requiring the longer time. Thus, there would be a relatively rapid decrease in mean polarity length following the long Cretaceous normal interval with a subsequently longer period of time in which the mean polarity lengths *in*crease. This is not what is observed. A *de*crease in the length of polarity intervals since the Cretaceous appears to have continued up to ~10 Ma ago.

Loper and McCartney (1986) have developed these ideas further. They first point out that the D″ layer is not static and, as was shown by Yuen and Peltier (1980), must actively convect. They assume that the rate of reversals is related to the rate of supply of energy to the dynamo, which, in their model, is directly related to the rate of cooling of the core, which in turn is controlled by the D″ layer. When the D″ layer is thick, the temperature gradient across it is small, and the energy supply to the dynamo is low, so that it is in a quiet state with few reversals. On the other hand, when the D″ layer is thin, the temperature gradient across it is large and the energy supply to the dynamo is greater, so that it is in a more disturbed state with frequent reversals. It is interesting that Sheridan (1983), who has developed a model based on similar ideas, has suggested a correlation just the opposite to that of Loper and McCartney. In his model, reversal frequency is low during periods of plume eruption and high when plumes are absent.

Loper and McCartney (1986) and Loper *et al.* (1988) carried out analogue experiments in the laboratory in which a layer of dyed water representing the heated material in the D″ layer is placed below viscous corn syrup representing the cold mantle. This initial configuration is dynamically unstable – a complete overturn is prevented by placing a silk membrane between the two fluids. Figures 8.8 and 8.9 illustrate some of the flows that were observed, which suggest that in the D″-plume system they may be unsteady and quasi-periodic. They estimated a thermal relaxation time of ~22 Ma, which is reasonably close to the postulated 30 Ma periodicity in reversals. However, the thermal conductivity of the D″ layer is not well known and their estimate of 22 Ma is very uncertain.

Any surface effects would depend on the amount of material and the history of its ascent. The laboratory experiments described above indicate that hot, low-viscosity material rises much faster within a pre-existing plume. This should cause an increase in non-explosive volcanism, because it can vent easily to the surface. If a diapir of low-mantle material rises outside an established plume, it rises at a much slower rate, allowing it to grow much larger as it is fed by material from below resulting in a very explosive event. An example of such an event at the K – T boundary could be the Deccan flood basalts, which at that time were situated over the Réunion hotspot. It must be emphasized that much of the plume model, like others, is highly speculative – we really know very little about what happens when a diapir of lower-mantle material reaches the lithosphere undersurface. The existence of plumes rising from the CMB seems to be on a firm basis, but it is difficult to assess their effect on the surface of the Earth.

It has been implicitly assumed that the core senses a change of thermal boundary conditions coincident with the change in plume activity and that the change

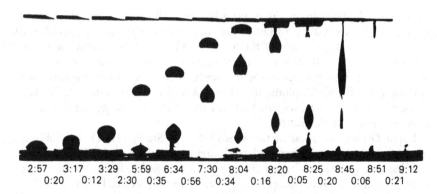

2:57   3:17   3:29   5:59   6:34   7:30   8:04   8:20   8:25   8:45   8:51   9:12
0:20   0:12   2:30   0:35   0:56   0:34   0:16   0:05   0:20   0:06   0:21

Figure 8.8 This and Figure 8.9 each show a sequence of 12 photographs of one run of an experiment described in the text to study the evolution of mantle plumes. The elapsed time is shown immediately beneath each photograph in minutes and seconds, while the second line of numbers gives the time difference between adjacent photographs. After Loper and McCartney (1986).

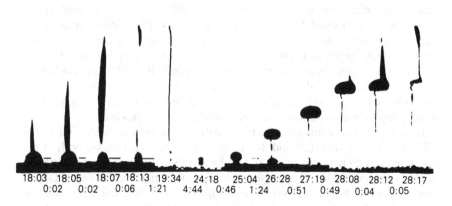

18:03   18:05   18:07   18:13   19:34   24:18   25:04   26:28   27:19   28:08   28:12   28:17
0:02   0:02   0:06   1:21   4:44   0:46   1:24   0:51   0:49   0:04   0:05

Figure 8.9 During the time interval 12.16–17.00, a second plume was established and the first became inactive – this sequence is not shown. The first five photographs show a sharply peaked pulse of fluid which rapidly rises through the syrup. Between the fourth and sixth photographs, a steady plume prevailed. The last seven photographs show the growth of a third dome of fluid which rises and grows until it intersects an inactive plume conduit. This provides a pathway for the buoyant water to rise rapidly to the surface. After Loper and McCartney (1986).

in behaviour of the dynamo precedes the surface events by the length of time it takes the plume to rise to the surface. Loper (1992) has shown that this assumption is incorrect. The core responds to a thermal signal which can change only on a long timescale and this timescale must be subtracted from the plume rise time to obtain the correct phase lag between core and surface events. It is very likely that this timescale exceeds the plume rise time, so that the surface event would precede the change of dynamo behaviour, calling into question the entire concept of the correlation of plume flux and reversal frequency. Loper showed that the natural length scale of the lower mantle is ~ 150 km and the natural thermal timescale ~ 1000 Ma – this timescale is far longer than that associated with the long time changes in the frequency of reversals of the magnetic field. An impulsive change in plume flux does not lead to an immediate appreciable change in heat flux – rather, the heat flux from the core changes on a timescale close to the age of the Earth.

Loper favours the ideas of Gubbins (1987) or Stacey (1991). In Gubbins's model, reversals are triggered by lateral variations in the heat flux from the core. The increase in reversal frequency during the past 86 Ma is due to the growth or migration of the patch of opposite polarity currently located beneath Southern Africa during that same time interval. Loper believes that the polarity patch is caused by a persistent thermal anomaly in the lowermost mantle beneath Southern Africa and that this hot spot was initiated by the rise of a starting plume that created the Deccan traps some 65 Ma ago. If the rise time of the plume was several tens of millions of years, it might have initiated a thermal anomaly at the base of the mantle at the end of the Cretaceous normal interval. In Stacey's model, dense mantle material differentiates into what he calls crypto-continents which accumulate at the base of the D″ layer, where they are rafted about by the horizontal flow that feeds the plumes. This rafting creates areas clear of the dense material – the crypto-continents act as insulating blankets, causing lateral variations in the heat flux from the core. Stacey estimates that the timescale for redistribution of the crypto-continents is ~100 Ma, which is in accord with the observed long-term variation in the frequency of reversals. Furthermore, the dense material may have an appreciable electrical conductivity which may have an additional influence on the dynamo process.

Larson (1991a) estimated that the formation of oceanic crust showed a 50–75% increase between 120 and 80 Ma ago. He found this increase both in the spreading rate from ocean ridges and in the age distribution of oceanic plateaux. Larson then noted that this increase, which was concentrated in the Pacific Ocean, coincides with the Cretaceous normal superchron (CNS) during which the Earth's field showed no reversals. He interpreted this as a result of a superplume that originated about 125 Ma ago near the CMB which rose by convection through the mantle and erupted beneath the Pacific basin.

Gaffin (1987) had reported earlier a negative correlation between long-term eustatic sea-level change and geomagnetic reversal rate for the past 150 Ma, and suggested that sea-floor creation rate and/or subduction rate are closely associated with the frequency of reversals. (Larson argued that subduction did not vary significantly during the past 150 Ma.) It should be noted that Larson calculated the rate of mid-ocean crust production as a function of time, using Kominz's (1984) tables of ridge crest lengths versus spreading rates, which she used to calculate sea-level variations.

Jacobs (1981) found a correlation between oceanic heat flow and the frequency of reversals of the Earth's magnetic field, the onset of the CNS coinciding with the beginning of increased heat flow. The maximum in the heat flow occurred at the end of the CNS, returning to its previous value about 40 Ma ago. Thus, the beginning and end of the CNS were marked by changes in the oceanic heat flow, although there is a 40 Ma time lag between the resumption of reversal activity and the return of heat flow to its earlier value.

Courtillot and Besse (1987) used the velocity of true polar wander (TPW) (the migration of the hot spot reference frame in the mantle relative to the rotation axis of the earth) as a measure of the strength of mantle convection. They found a period of very low TPW from about 170 to 100 Ma ago, coincident with a significant burst of continental break-up. This occurred well before the beginning of the CNS, although during a time of continuing reduction in the frequency of reversals. Shortly after the beginning of the CNS, TPW recovered rather abruptly to approximately its present high value. There is thus quite a long phase lag between Courtillot and Besse's measure of mantle convection and the CNS. Their period of very slow TPW coincides with the period of increased heat flow found by Jacobs with a phase lag of ~ 60 Ma.

It does not seem reasonable that every reversal is associated with an event at the CMB. On the other hand, it is very plausible that a major change in the pattern of the frequency of reversals, such as the CNS, is connected with changing conditions in the mantle. One would dearly like to have more observational evidence for any other changes in reversal frequency. Marine magnetic anomalies provide the greatest single source of information about magnetic reversals but only for the past 160 Ma, the age range of ocean floor that preserves a record of reversals. However, land measurements have shown that for a period of 60 Ma from about 290 to 235 Ma ago, the Earth's field appears to have been reversed almost all of the time. Panella (1972) has estimated the length of the synodic month back to about 500 Ma from tidally controlled periodical growth patterns on molluscs and stromatolites. He found that the deceleration rate of the Earth's rotation has not been constant. There were two breaks, one beginning in the Pennsylvanian and ending in the Permian, after which it remained essentially constant until about 75 Ma ago in the upper Cretaceous, when it returned to its earlier value of 2 ms/century. The first break coincides almost exactly with the long 60 Ma period when the field was reversed almost all of the time. The second break occurred just after the end of the CNS.

In a later paper Larson (1991b) developed his ideas further. Figure 8.10 contains a revised and expanded version of his earlier (1991a) ocean-crust production curve and shows its correlation with long-term eustatic sea level, high-latitude sea-surface palaeotemperatures, magnetic reversal stratigraphy and times of black shale deposition and world oil resources. He further argued that the mid-Cretaceous super-plume is only the most recent of a longer history of aperiodic releases of heat from the CMB and that it should be possible to recognize other superplume episodes by identifying other intervals of constant magnetic polarity. In this connection, he suggested that an earlier superplume was responsible for the long period of constant reversed polarity during Pennsylvanian–Permian time (323–248 Ma), which was also a time of increased coal generation and gas accumulation accompanied by an intracratonic transgression of epicontinental seas.

Cox (1991) has pointed out the difficulty in proving these correlations. The record of geomagnetic reversal frequency is fairly well established, but the estimation of plume activity through summing the production of oceanic plateaux, sea-mount chains and flood basalts is much more uncertain. A further problem is the relative timing of the superplume and superchron, as pointed out by Loper (1992). Larson's papers (1991a,b) provoked comments and reply – particularly on his calculation of the rates of sea-floor generation for the past 150 Ma (see *Geology*, **20**, 475, 1992).

Marzocchi *et al.* (1992) also point out that the time series of mean sea level and of geomagnetic reversals are both characterized by non-stationary behaviour, which means that there is a trend in the rates of occurrence. Hence, a quantitative study of their correlation based only on the analysis of cross-correlation as carried out by Gaffin (1987) can be misleading and meaningless. Marzocchi *et al.* carried out a statistical analysis of a possible correlation between reversals and mean sea level over the last 150 Ma. They found a highly significant negative correlation between long-term sea-level variations and the rate of reversals. An increase in the reversal rate is followed, after 6–9 Ma, by a decrease in the long-term sea level. They found no correlation between the frequency of short-term sea-level fluctuations and variations in the geomagnetic reversal rate.

Larson and Olson (1991) have also shown that there is an inverse relationship between geomagnetic reversal frequency and the production of oceanic plateaux, sea-mount chains and continental flood basalts. This inverse correlation is especially striking during the long Cretaceous normal superchron, when mantle-plume activity as expressed by crustal production was a maximum. They explain this correlation by the following sequence of events (see Figure 8.11). Mantle plumes rise from the D″ layer, thinning the layer. This thinning increases the temperature gradient across D″, increasing the rate of heat conducted across the CMB. The core then convects more vigorously to restore the abnormal heat loss. Larson and Olson show that it is the increase in the vigour of core convection that causes a *decrease* in magnetic reversal frequency, in accord with predictions for both α^2 and $\alpha\omega$ dynamos. In their model reversals are oscillations intrinsic to the dynamo process and are not the result of instabilities in the OC. This is just the opposite view to that of McFadden and Merrill (1986), Loper and McCartney (1986) and Courtillot and Besse (1987), in which each reversal is associated with a separate instability. A superchron then represents a low-energy state for the dynamo corresponding to an inactive period in the D″ layer with little heat transfer.

Fuller and Weeks (1992) suggest possible ways in which to resolve the contradicting predictions of the effect changed convective vigour would have on reversal rate. If the temperature at the CMB changes, the flow in the outermost core presumably must change (Zhang and Gubbins, 1992). At present this flow is intimately linked to the secular variation, so that if the flow were fundamentally different during the superchron, the secular variation should also be different and this could perhaps be investigated. Again, Olson's and Hagee's (1990) dynamo model predicts an inverse correlation between field intensity and reversal rate, and this should be able to be tested. It is interesting in this regard that Pal and Roberts (1988) found that the dipole moment is appreciably greater during fixed-polarity superchrons. However, their data for the Cretaceous normal

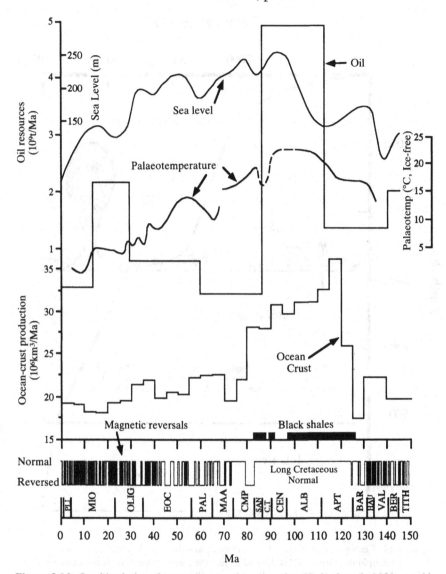

Figure 8.10 Combined plot of magnetic reversal stratigraphy (Harland *et al.*, 1990), world ocean-crust production (modified from Larson, 1991a), high-latitude sea-surface palaeotempera- tures (Savin, 1977; Arthur *et al.*, 1985), long-term eustatic sea level (Haq *et al.*, 1988), times of black shale deposition (Jenkyns, 1980) and world oil resources (Irving *et al.*, 1974; Tissot, 1979) plotted on geological timescale calibration of Harland *et al.* (1990). After Larson (1991b).

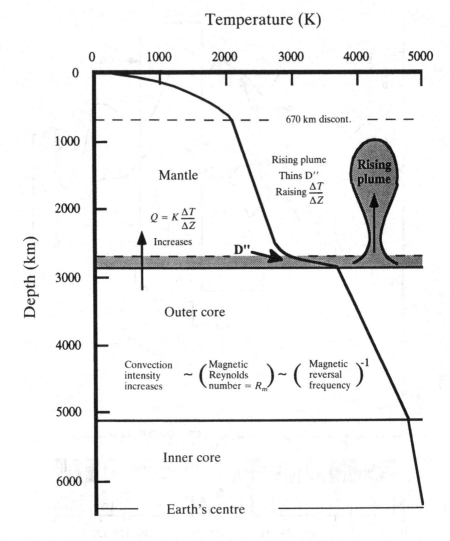

Figure 8.11 Response of core–mantle boundary conductive heat flow and outer core convection to a superplume episode. D″ is thinned to fuel the plumes, raising its vertical temperature gradient. This raises conduction across the core–mantle boundary, which in turn raises convection intensity in the outer core. Convection intensity, characterized by a magnetic Reynolds number, is inversely proportional to reversal frequency. After Larson and Olson (1991).

superchron are limited and those for the Permian reversed superchron not very convincing. Fuller and Weeks stress that the data and interpretations are themselves controversial, and that we do not even know whether the main geodynamo is deep within the core and merely perturbed by the surface flow, or whether the surface flow is a more fundamental aspect of the dynamo.

Weinstein (1993) has shown that phase changes in the mantle (such as those at ~ 410 and 670 km) may affect the amount of flow between the upper and lower mantle. He carried out two-dimensional calculations for a fluid layer heated from below and internally with two isochemical phase changes, and showed that the transition region of the Earth's mantle may act as a capacitor for subducting slabs. Slab material accumulates in the transition region until a threshold level of thermal buoyancy is reached and is then rapidly discharged into the lower mantle. In Weinstein's model this occurs as a catastrophic burst of convection lasting ~ 10 Ma with elevated heat transfer rates lasting ~ 100 Ma. In Larson's and Olson's (1991) model the magnetic superchron begins before flood basalt eruption commences. In Weinstein's model heat flow at the CMB increases, owing to the accumulation of cold material which falls on it. However, hot material reaches the upper surface before cold material reaches the bottom, i.e. flood basalt generation begins before the onset of the superchron. Weinstein suggested that perhaps a superchron represents a period when portions of the upper-mantle become gravitationally unstable, leading to a transition from two-layer convection to vigorous whole mantle convection.

Honda *et al.* (1993) carried out three-dimensional numerical simulations of mantle convection, including both the major phase transitions at ~ 410 and 670 km. They showed that the instability produced bears no resemblance to local boundary-layer instabilities that are observed in models of convection without phase changes. Their three-dimensional simulations confirm their earlier work in two dimensions (Steinbach and Yuen, 1992) and are similar to those of Weinstein (1993). Cold sheets of mantle material collide, merge and form a strong down-flow that is stopped temporarily by the transition zone. The accumulated cold material gives rise to a strong gravitational instability that causes the cold mass to sink rapidly into the lower mantle. This leads to a massive exchange between the upper and lower mantle, with global instability in the plume system.

8.6 Magnetic reversals and faunal extinctions

Uffen (1963) proposed that if the Earth's magnetic field were reduced to a very low value during a polarity change, the solar wind and a large proportion of cosmic rays would be able to reach the surface of the Earth. He suggested that the biological effects of the increase in radiation might affect the course of evolution and bring about the extinction and transformation of species at the time of a polarity change. Uffen argued on palaeontological grounds that rates of evolution were exceptionally high at times when the Earth's magnetic field was undergoing many changes in polarity. The estimated time interval of about 10^4 a when the Earth's dipole field is greatly reduced during a reversal is almost instantaneous on the geological timescale but is very many generations for most living organisms. There are, of course, other causes of mutation, but the exposure to increased radiation would be sudden on the geological timescale and could contribute to both heritable variation and natural selection. The most striking unexplained fact from palaeontology is that biological evolution has progressed in bursts. Quite adequate explanations have been given for the extinction of whole species by climatic or other environmental changes, but the sudden appearance of new forms of life on which the selection pressures of the time would act has not been adequately explained.

A number of investigations (see, e.g., Harrison and Funnell, 1964; Hays and Opdyke, 1967; Watkins and Goodell, 1967) have shown that the extinction of certain species of marine organisms occurred at about the time of the most recent reversal of the Earth's magnetic field. However, the reality of any correlation is, in principle, statistical, and there are just not sufficient data. Moreover, Hays et al. (1969) found that most polarity changes during the last 4 Ma that have extinctions associated with them in Antarctic cores are not the same as those that have extinctions associated with them in equatorial cores from the Pacific. Finally, since only a few organisms become extinct near a polarity change, and some have survived such changes before becoming extinct, it is clear that a polarity change alone is probably insufficient to cause an extinction.

Simpson (1966) attempted to correlate accelerations in the rate of evolution during the Phanerozoic with reversals of the Earth's magnetic field. He compared the change in the percentage of new species as a function of existing species with the occurrence of reversely magnetized rocks. However, as McElhinny (1973) has pointed out, it is the *frequency* of reversals that is the important parameter. McElhinny compared his estimate of the variation in reversal frequency with Newell's (1963) estimate of the rate of organic evolution. He found that accelerations in evolution seemed to occur at times when the reversal rate was either a minimum or a maximum, indicating no significant correlation.

A fundamental difference in the polarity history occurred during the latest Cretaceous through the entire Cenozoic, which is represented by a rapidly oscillating reversal sequence. Helsley and Steiner (1974) suggested that major geological eras, defined on the basis of major changes in the fossil record, may end with a low frequency of reversals and begin with a high reversal frequency. They speculated that the major changes in life forms that take place at these times may have resulted from evolutionary adaptation during the long periods of constant polarity in which the magnetic field is used in the regulation of some necessary or vital function; and then this changes to the disadvantage of some forms. Regardless of the mechanism involved, it is clear that major crises in the Earth's biota at the Palaeozoic–Mesozoic boundary and at the Mesozoic–Cenozoic boundary are associated with major changes in the tempo of polarity changes in the Earth's magnetic field. The faunal extinctions near the end of the Cretaceous and Permo–Carboniferous periods coincide with the close of long intervals of dominantly one magnetic polarity. Kent (1977) has examined the magnetostratigraphy in relation to planktonic foraminiferal changes over an apparently continuous Cretaceous–Tertiary boundary section in the pelagic limestones at Gubbio, northern Italy. He found that this planktonic foraminiferal change and, hence, the Cretaceous–Tertiary boundary was not coincident with any particular polarity change in the Gubbio section, thus ruling out any simple explanation based on a relationship between faunal change and possible effects on organisms of such a reversal.

Quantitative estimates of the effects on the Earth and atmosphere at a polarity change have been given by Black (1967), Waddington (1967) and Harrison (1968). The surface of the Earth is shielded from cosmic rays by the geomagnetic field and the atmosphere. At the poles all particles can reach the atmosphere – at the equator, only particles with energies greater than 15 GeV. If the present geomagnetic field were reduced to zero during a polarity transition, the increased radiation dose would amount to 10% at the equator and zero at the poles. If the decrease took place over 1000 years, this amounts to an increase in dose of 0.01% per year. The rate of change of dose with sunspot cycle is 0.4% per year, so the possible effects are negligible. Even complete dumping of the energetic particles in the Van Allen radiation belts would not give rise to the necessary increased dosages. A further possibility is that the solar wind might produce ozone in the high atmosphere which would absorb radiation and produce large changes in climate. However, Black (1967) has shown that any ozone produced in this way is but a small fraction of that produced by ultraviolet light. On the other hand, Reid *et al.* (1976) have suggested that the reduced magnetic field during a reversal could permit solar proton events seriously to deplete stratospheric ozone. Crutzen *et al.* (1975) had shown that a few such intense events a year could materially increase the amount of NO in the stratosphere. Ozone is catalytically destroyed by NO, thus reducing the efficiency of the ozone shield which protects the surface of the Earth from harmful solar ultraviolet radiation.

The only possibility of a causal connection between magnetic reversals and the extinction of certain species seems to be that there is a climatic change at the time of a reversal (see Section 8.3). There is some evidence (Harrison, 1968) that the magnetic field of the Earth has some control over the temperature of the upper atmosphere; also, removal of the Earth's magnetic field would cause large increases in ionization at certain levels in the upper atmosphere. What climatic changes, if any, would be brought about by such effects is not known. However, a causal connection between magnetic reversals and climatic fluctuations seems unlikely, since the timescale of climatic variations is a good deal shorter than the interval between reversals.

In 1975 Blakemore found that certain aquatic bacteria are magnetotactic and tend to swim along the field lines of the Earth's magnetic field. This is in contrast to most other bacteria, which are chemotactic or phototactic. Even when magnetotactic bacteria are killed, they line up along the direction of the Earth's magnetic field. When viewed in a transmission electron microscope, a chain of particles (opaque to the electron beam) is seen more or less parallel to the long axis of the cell. These particles, called magnetosomes, consist of magnetite (Fe_3O_4), an oxide of iron. It has been shown that single-domain particles of magnetotactic bacteria synthesize magnetosomes and carry them with them. Matsuda *et al.* (1983) have studied the morphology and structure of magnetosomes, using high-resolution electron microscopy. They found them to be single crystals with a hexagonal prism shape truncated by {111} planes. They are believed to consist of magnetite, since the spacings and orientations of all the observed fringes agree with those of magnetite. Their polarity can be changed by exposure to a brief but strong magnetic field in the opposite direction to that of the ambient field.

Their original polarity can be restored by a second exposure to a magnetic field antiparallel to the first. Presumably, during a reversal of the Earth's magnetic field the polarity of the magnetotactic bacteria in both hemispheres becomes reversed as well.

Laboratory experiments have shown that it is the vertical component of the geomagnetic field that is the relevant factor in determining the polarity of bacterial populations in natural environments. In the northern hemisphere most magnetotactic bacteria are north-seeking, i.e. they tend to move downwards, keeping to the sediments and away from surface water. Since magnetotactic bacteria are anaerobic or microaerophilic, a tendency to migrate downwards would help to avoid the toxic effects of the greater concentration of oxygen in surface waters.

Blakemore *et al.* later (1980) found several types of magnetotactic bacteria in sediments of the southern hemisphere. They were consistently south-seeking, i.e. they moved downwards like those in the Northern hemisphere. Electron microscopy showed that they contained internal chains of electron-opaque particles similar to those observed in magnetotactic bacteria from the northern hemisphere. Like the northern hemisphere bacteria, their magnetic polarity can be permanently reversed in the laboratory and they cannot be demagnetized. Frankel *et al.* (1981) found approximately equal numbers of north-seeking and south-seeking bacteria in freshwater and marine sediments at Fortaliza, Brazil, close to the geomagnetic equator. This confirms the hypothesis that the vertical component of the geomagnetic field determines the predominant polarity of magnetotactic bacteria. Because of the horizontal orientation of the magnetic field at the geomagnetic equator, the motion of magnetotactic bacteria there will be directed horizontally. This would be advantageous to bacteria of either polarity in reducing detrimental upward migration as compared with random motion.

If the geomagnetic field is reversed, magnetotactic bacteria would be directed away from their natural environment into one in which they might be unable to survive. This seems to give a wholly convincing explanation of the extinction phenomenon. It is all the more convincing because it is an effect against which a species cannot protect itself on the basis of past experience. Presumably, the mechanism operates within some range of *intensity* of the magnetic field, but if it takes its *direction* from the field, it is incapable of knowing when that direction has changed relative to anything else.

All this seems reasonable if there is a sufficiently sudden change in the direction of the field. Whether a species has any capability of adjusting to a gradual change is not known. It does not follow, of course, that the mode of extinction of the smallest living organisms has any connection with that of larger creatures. However, since 1975 magnetite has been found in many other organisms besides bacteria – chitons (a class of marine molluscs), honeybees, butterflies, homing pigeons, dolphins, tuna fish and sea turtles.

It is noteworthy that the two great extinctions about 225 and 65 Ma ago occurred in the mixed magnetic intervals 230–204 Ma and 70–50 Ma ago. During each of these there were many polarity reversals, any one of which could have been catastrophic for any species for which some crucial instinctive behaviour was linked to the geomagnetic field. The former of these two mixed intervals followed the Permian reversed superchron: the latter followed the Cretaceous normal superchron. In both cases the biosphere had had time to evolve as though polarity never changed before a sequence of 'rapid' changes began.

Funaki *et al.* (1989) investigated the possibility of identifying the fine magnetic structure of iron–nickel grains in the St Séverin chondrite by north-seeking magnetotactic bacteria (NSB) which migrate to the south pole only. They found that the bacteria are sensitive magnetic sensors and that the directions of the magnetic lines of force radiating from the grains could be observed. The magnetic coercive forces could also be measured by the bacteria. They later (1992) used NSB to identify the south pole on a polished surface of magnetic-rich pyroxenite. The NSB formed clusters at various places on the magnetite grains in the natural state, but did not form any clusters after AF demagnetization to 60 mT. When the sample acquired SIRM to IT, very dense clusters were formed on grains at the opposite side of the applied field direction. The effect of magnetic anistropy was estimated from the formation of the clusters, and that of VRM by the variation of density and size of the clusters. These methods can in principle be applied to terrestrial rocks having relatively strong NRM. Thus, magnetotactic bacteria may give useful information for rock magnetic and palaeomagnetic studies as a biomagnetometer.

References

Chapter 1

Barraclough, D. R. (1978). Spherical harmonic models of the geomagnetic field. *Geomag. Bull.*, **8**, Inst. Geol. Soc. London.

Bartels, J. (1936). The eccentric dipole approximating the Earth's magnetic field. *Terr. Mag.*, **41**, 225.

Bingham, D. K. and Stone, D. B. (1972). Secular variation in the Pacific Ocean region. *Geophys. J. Roy. Astr. Soc.* **28**, 337.

Brunhes, B. (1906). Reserches sur la direction d'aimentation des roches volcaniques. *J. Phys.*, **5**, 705.

Busse, F. H. (1972). Comments on paper by G. Higgins and G. C. Kennedy. 'The adiabatic gradient and the melting point gradient in the core of the Earth'. *J. Geophys. Res.*, **77**, 1589.

Busse, F. H. (1978). Magnetohydrodynamics of the Earth's dynamo. *Ann. Rev. Fluid Mech.*, **10**, 435.

Cain, J. C., Hendricks, S. J., Langel, R. and Hudson, W. V. (1967). A proposed model for the International Geomagnetic Reference Field – 1965. *J. Geomag. Geoelect.*, **19**, 335.

Cande, S. C. and Kent, D. V. (1992). A new geomagnetic timescale for the late Cretaceous and Cenozoic. *J. Geophys. Res.*, **97**, 13, 917.

Chauvin, A., Gillot, P.-Y. and Bonhommet, N. (1991). Palaeointensity of the Earth's magnetic field recorded by two late Quaternary volcanic sequences at the Island of La Réunion (Indian Ocean). *J. Geophys. Res.*, **96**, 1981.

Coe, R. S., Grommé, C. S. and Mankinen, E. A. (1978). Geomagnetic palaeointensities from radiocarbon dated lava flows in Hawaii and the question of the Pacific non-dipole low. *J. Geophys. Res.*, **83**, 1740.

Coupland, D. H. and Van der Voo, R. B. (1980). Long term non-dipole components in the geomagnetic field during the last 130 Myr. *J. Geophys. Res.*, **85**, 3529.

Cox, A. (1975). The frequency of geomagnetic reversals and the symmetry of the non-dipole field. *Rev. Geophys. Space Phys.*, **13**, 35.

Creer, K. M., Georgi, D. T. and Lowrie, W. (1973). On the representation of the Quaternary and late Tertiary geomagnetic fields in terms of dipoles and quadrupoles. *Geophys. J. Roy. Astr. Soc.*, **33**, 323.

Creer, K. M. and Tucholka, P. (1982). Construction of type curves of geomagnetic secular variation for dating lake sediments from east central North America. *Can. J. Earth Sci.*, **19**, 1106.

Creer, K. M. and Tucholka, P. (1983). On the current state of lake sediment palaeomagnetic research. *Geophys. J. Roy. Astr. Soc.*, **74**, 223.

Dagley, P., Wilson, R. L., Ade-Hall, J. M., Walker, G. P. L., Haggerty, S. E., Sigurg-

eirsson, T., Watkins, N. D., Smith, P. J., Edwards, J. and Grasty, R. L. (1967). Geomagnetic polarity zones for Icelandic lavas. *Nature*, 216, 25.

David, P. (1904). Sur la stabilité de la direction d'aimantation dans quelques roches volcaniques. *C. R. Acad. Sci.*, 138, 41.

Doell, R. R. (1972). Palaeosecular variation of the Honolulu volcanic series, Oahu, Hawaii. *J. Geophys. Res.*, 77, 2129.

Doell, R. R. and Cox, A. (1971). Pacific geomagnetic secular variation. *Science*, 171, 248.

Doell, R. R. and Cox, A. (1972). The Pacific geomagnetic secular variation anomaly and the question of lateral uniformity in the lower mantle. In *The Nature of the Solid Earth*, ed. E. C. Robertson, pp. 245–84. New York: McGraw-Hill.

Duncan, R. A. (1975). Palaeosecular variation at the Society Islands, French Polynesia. *Geophys. J. Roy. Astr. Soc.*, 41, 245.

Dyson, F. W. and Furner, H. (1923). The Earth's magnetic potential. *Mon. Not. R. Astron. Soc.*, 1 (Geophysics Supplement), 76.

Elsasser, W. M. (1954). Dimensional values in magnetohydrodynamics. *Phys. Rev.*, 95, 1.

Finch, H. F. and Leaton, B. R. (1957). The Earth's magnetic field – epoch 1955.0. *Mon. Not. R. Astron. Soc.*, 7 (Geophysics Supplement), 314.

Fritsche, H. (1899) *Die elemente des erdmagnetismus für die epochen 1600, 1650, 1700, 1780, 1842 und 1885 und ihre säkularen änderungen, berechnet mit hilfe der aus allen brauchbaren beobachtungen abgeleiteten koeffizienten der Gaussischen allgemeinen theorie des erdmagnetismus* (The elements of terrestrial magnetism for the epochs 1600, 1650, 1700, 1780, 1842 and 1885 and their secular variation, calculated with the aid of the coefficient of the Gaussian general theory of terrestrial magnetism derived from all useful observations). St Petersburg.

Gauss, C. F. (1839) Allgemeine Theorie des Erdmagnetismus, Leipzig.

Gubbins, D. (1977). Energetics of the Earth's core. *J. Geophys.*, 43, 453.

Gubbins, D. and Masters, T. G. (1979). Driving mechanisms for the Earth's dynamo. *Adv. Geophys.*, 21, 1.

Gubbins, D., Masters, T. G. and Jacobs, J. A. (1979). Thermal evolution of the Earth's core. *Geophys. J. Roy. Astr. Soc.*, 59, 57.

Gubbins, D. and Roberts, P. H. (1987). Magnetohydrodynamics of the Earth's core. In *Geomagnetism*, Vol. 2, ed. J. A. Jacobs, pp. 1–183. New York: Academic Press.

Hanna, R. and Verosub, K. L. (1989). A review of lacustrine palaeomagnetic records from western North America: 0–40,000 years BP. *Phys. Earth Planet. Int.*, 56, 76.

Holcomb, R., Champion, D. E. and McWilliams, M. (1986). Dating recent Hawaiian flows using palaeomagnetic secular variation. *Geol. Soc. Am. Bull.*, 97, 829.

Hurwitz, L., Fabiano, E. B. and Peddie, N. W. (1974). A model of the geomagnetic field for 1970. *J. Geophys. Res.*, 79, 1716.

Isaacson, L. B. and Heinrichs, D. F. (1976). Palaeomagnetism and secular variation of Easter Island basalts. *J. Geophys. Res.*, 81, 1476.

James, R. W. and Winch, D. E. (1967). The eccentric dipole. *Pure Appl. Geophys.*, 66, 77.

Kalinin, Yu. D. and Rozanova, T. S. (1983). Asymmetry of the westward drift of the geomagnetic field relative to the Earth's equatorial plane. *Geomag. Aeron.*, 23, 823.

Kovacheva, M. (1980). Summarized results of the archaeomagnetic investigation of the geomagnetic field variation for the last 8,000 yr in southeastern Europe. *Geophys. J. Roy. Astr. Soc.*, 61, 57.

Krause, F. and Rädler, K. H. (1980). *Mean Field Magnetohydrodynamics and Dynamo Theory*. Oxford: Pergamon.

Langel, R. A. (1987). The mainfield. In *Geomagnetism*, Vol. 1, ed. J. A. Jacobs, pp. 249–512. New York: Academic Press.

Langel, R. A., Estes, R. H., Mead, G. D., Fabiano, E. B., and Lancaster, E. R. (1980). Initial geomagnetic field model from Magsat Vector data. *Geophys. Res. Lett.*, 7, 793.

Leaton, B. R., Malin, S. R. C. and Evans, M. J. (1965). An analytical representation of the estimated geomagnetic field and its secular change for the epoch 1965.0. *J. Geomag. Geoelect.*, 17, 187.

Levi, S. and Karlin, R. (1989). A sixty thousand year palaeomagnetic record from Gulf of California sediments: secular variation, late Quaternary excursions and geomagnetic implications. *Earth Planet. Sci. Lett.*, 92, 219.

Livermore, R. A., Vine, F. J. and Smith, A. G. (1983). Plate motions and the geomagnetic field. I. Quaternary and late Tertiary. *Geophys. J. Roy. Astr. Soc.*, 73, 153.

Livermore, R. A., Vine, F. J. and Smith, A. G. (1984). Plate motions and the geomagnetic field. II. Jurassic to Tertiary. *Geophys. J. Roy. Astr. Soc.*, 79, 939.

Loper, D. E. (1978). The gravitationally powered dynamo. *Geophys. J. Roy. Astr. Soc.*, 54, 389.

Lund, S. P. and Banerjee, S. K. (1975). Late Quaternary palaeomagnetic field secular variation from two Minnesota Lakes. *J. Geophys. Res.*, 90, 803.

Lund, S. P., Liddicoat, J. C., Lajoie, K. R., Henyey, T. L. and Robinson, S. W. (1988). Palaeomagnetic evidence for long term (10^4 year) memory and periodic behaviour on the Earth's core dynamo process. *Geophys. Res. Lett.*, 15, 1101.

McElhinny, M. W. and Merrill, R. T. (1975). Geomagnetic secular variation over the past 5 m.y. *Rev. Geophys. Space Phys.*, 13, 687.

McWilliams, M. O., Holcomb, R. T. and Champion, D. E. (1982). Geomagnetic secular variation from ^{14}C dated flows in Hawaii and the question of the Pacific non-dipole low. *Phil. Trans. Roy. Soc. A*, 306, 211.

Malkus, W. V. R. (1973). Convection at the melting point, a thermal history of the Earth's core. *Geophys. Fluid Dynam.*, 4, 267.

Merrill, R. T. and McElhinny, M. W. (1977). Anomalies in the time-averaged palaeomagnetic field and their implications for the lower mantle. *Rev. Geophys. Space Phys.*, 15, 309.

Merrill, R. T. and McElhinny, M. W. (1983). *The Earth's Magnetic Field*. New York: Academic Press.

Meynadier, L., Valet, J.-P., Weeks, R., Shackleton, N. J. and Hagee, V. L. (1992). Relative geomagnetic intensity of the field during the last 140 ka. *Earth Planet Sci. Lett.*, 114, 39.

Moffatt, H. K. (1970). Turbulent dynamo action at low magnetic Reynolds number. *J. Fluid Mech.*, 41, 435.

Moffatt, H. K. (1978). *Magnetic Field Generation in Electrically Conducting Fluids*. Cambridge: Cambridge University Press.

Nelson, J. H., Hurwitz, L. and Knapp, D. G. (1962). Magnetism of the Earth. *US Department of Commerce Coast Geod. Surv.* Publ. 40–1.

Newitt, L. R. and Dawson, E. (1984). Secular variation in North America during historical times. *Geophys. J. Roy. Astr. Soc.*, 78, 277.

Olson, P. (1983). Geomagnetic polarity reversals in a turbulent core. *Phys. Earth Planet. Int.*, 33, 260.

Olson, P. and Hagee, V. L. (1990). Geomagnetic polarity reversals, transition field structure and convection in the outer core. *J. Geophys. Res.*, 95, 4609.

Opdyke, N. D., Glass, B., Hays, J. D. and Foster, J. (1966). Palaeomagnetic study of Antarctic deepsea cores. *Science*, 154, 349.

Parker, E. N. (1955). Hydromagnetic dynamo models. *Astrophys. J.*, 122, 293.

Parker, E. N. (1970). The generation of magnetic fields in astrophysical bodies. I. The dynamo equations. *Astrophys. J.*, 162, 665.

Parker, E. N. (1971). The generation of magnetic fields in astrophysical bodies. IV. The solar and terrestrial dynamos. *Astrophys. J.*, 164, 491.

Parker, E. N. (1979). *Cosmical Magnetic Fields, Their Origin and Their Activity.* Oxford: Clarendon Press.

Parkinson, W. D. (1983). *Introduction to Geomagnetism.* Edinburgh: Scottish Academic Press.

Peddie, N. W. (1982). International Geomagnetic Reference Field: the third generation. *J. Geomag. Geoelect.*, **34**, 309.

Peddie, N. W. and Fabiano, E. B. (1976). A model of the geomagnetic field for 1975. *J. Geophys. Res.*, **81**, 2539.

Peng Lei and King, J. W. (1992). A late Quaternary geomagnetic secular variation record from Lake Waiau, Hawaii, and the question of the Pacific non-dipole low. *J. Geophys. Res.*, **97**, 4407.

Roberts, P. H. (1987). Origin of the mainfield: dynamics. In *Geomagnetism*, Vol. 2, ed. J. A. Jacobs, pp. 251–306. New York: Academic Press.

Roberts, P. H. and Gubbins, D. (1987). Origin of the mainfield: kinematics. In *Geomagnetism*, Vol. 2, ed. J. A. Jacobs, pp. 185–299. New York: Academic Press.

Roberts, P. H. and Soward, A. M. (1978). *Rotating Fluids in Geophysics.* New York: Academic Press.

Roberts, P. H. and Stix, M. (1972). α-Effect dynamos by the Bullard–Gellman formalism. *Astron. Astrophys.*, **18**, 453.

Roperch, P., Bonhammet, N. and Levi, S. (1988). Palaeointensity of the Earth's magnetic field during the Laschamp excursion and its geomagnetic implications. *Earth Planet. Soc. Lett.*, **88**, 209.

Schloessin, H. H. and Jacobs, J. A. (1980). Dynamics of a fluid core with inward growing boundaries. *Can. J. Earth Sci.*, **17**, 72.

Schmidt, A. (1895). Mitteilungen über eine neue berechnung des erdmagnetischen potentials. (Notes on a new calculation of the terrestrial magnetic potential.) *Abh. K. Bayerischen Akad. Wiss.* **19**, Div. 1, 1.

Schneider, D. A. and Kent, D. V. (1988). Inclination anomalies from Indian Ocean sediments and the possibility of a standing non-dipole field. *J. Geophys. Res.*, **93**, 11, 621.

Schneider, D. A. and Kent, D. V. (1990a). The time-averaged palaeomagnetic field. *Rev. Geophys.*, **28**, 71.

Schneider, D. A. and Kent, D. V. (1990b). Testing models of the Tertiary palaeomagnetic field. *Earth Planet. Sci. Lett.*, **101**, 260.

Senanayake, W. E., McElhinny, M. W. and McFadden, P. L. (1982). Comparisons between the Thelliers' and Shaw's palaeointensity methods using basalts less than 5 million years old. *J. Geomag. Geoelect.*, **34**, 141.

Skiles, D. D. (1970). A method of inferring the directions of drift of the geomagnetic field from palaeomagnetic data. *J. Geomag. Geoelect.*, **22**, 441.

Sproul, D. R. and Banerjee, S. K. (1989). The Holocene palaeosecular variation record from Elk Lake, Minnesota. *J. Geophys. Res.*, **94**, 9369.

Steenbeck, M. and Krause, F. (1966). The generation of stellar and planetary magnetic fields by turbulent dynamo action. *Z. Naturf. a*, **21**, 1285.

Tric, E., Valet, J. P., Tucholka, P., Paterne, M., Labeyrie, L., Guichard, F., Tauxe, L. and Fontugne, M. (1992). Palaeointensity of the geomagnetic field during the last 80,000 years. *J. Geophys. Res.*, **97**, 9337.

Vestine, E. H. and Kahle, A. B. (1966). The small amplitude of magnetic secular variation in the Pacific area. *J. Geophys. Res.*, **71**, 527.

Vestine, E. H., Laporte, L., Cooper, C., Lange, I. and Hendrix, W. (1947a). Description of the Earth's main magnetic field and its secular change, 1905–1945. *Carnegie Inst. Wash. Publ.* No. 578.

Vestine, E. H., Lange, I., Laporte, L. and Scott, W. E. (1947b) The geomagnetic field, its description and analysis. *Carnegie Inst. Washington Publ. No. 580.*

Wells, J. M. (1973). Non-linear spherical harmonic analysis of palaeomagnetic data. *Meth. Computat. Phys.*, **13**, 239.

Wilson, A. L. (1970). Permanent aspects of the Earth's non-dipole magnetic field over Upper Tertiary times. *Geophys. J. Roy. Astr. Soc.*, **19**, 417.

Wilson, A. L. (1971). Dipole offset – the time averaged palaeomagnetic field over the past 25 million years. *Geophys. J. Roy. Astr. Soc.*, **22**, 491.

Wilson, A. L. (1972). Palaeomagnetic differences between normal and reversed field sources, and the problem of far-sided and right-handed pole positions. *Geophys. J. Roy. Astr. Soc.*, **28**, 295.

Wilson, R. L. and Ade-Hall, J. M. (1970). Palaeomagnetic indications of a permanent aspect of the non-dipole field. In *Palaeogeophysics*, ed. S. K. Runcorn, p. 307. New York: Academic Press.

Wilson, R. L. and McElhinny, M. W. (1974). Investigations of the large scale palaeomagnetic field over the past 25 Myr: eastward drift of the Icelandic spreading ridge. *Geophys. J. Roy. Astr. Soc.*, **39**, 571.

Yukutake, T. (1962). The westward drift of the magnetic field of the Earth. *Bull. Earthquake Res. Inst.*, **41**, 1.

Yukutake, T. and Tachinaka, H. (1969). Separation of the Earth's magnetic field into drifting and standing parts. *Bull. Earthquake Res. Inst.*, **47**, 65.

Chapter 2

Ade-Hall, J. and Watkins, N. D. (1970). Absence of correlation between opaque petrology and natural remanence polarity in Canary Island lavas. *Geophys. J. Roy. Astr. Soc.*, **19**, 351.

Balsey, J. R. and Buddington, A. F. (1954). Correlation of reverse remanent magnetism and negative anomalies with certain minerals. *J. Geomag. Geoelect.*, **6**, 176.

Balsey, J. R. and Buddington, A. F. (1958). Iron-titanium oxide minerals, rocks and aeromagnetic anomalies of the Adirondack area, New York. *Econ. Geol.*, **53**, 777.

Bhimasankaram, V. L. S. (1964). Partial self-reversal in pyrrhotite. *Nature*, **202**, 478.

Burek, P. J. (1970). Magnetic reversals; their application to stratigraphical problems. *Bull. Am. Assoc. Petrol. Geol.*, **54**, 1120.

Burns, C. A. (1989). Timing between a large impact and a geomagnetic reversal and the depth of NRM acquisition in deep-sea sediments. In *Geomagnetism and Palaeomagnetism*, ed. F. J. Lowes *et al.*, p. 253. Amsterdam: Elsevier.

Butler, R. F. (1992). *Palaeomagnetism: Magnetic Domains to Geologic Terranes.* Oxford: Blackwell Scientific.

Carmichael, C. M. (1959). Remanent magnetism of the Allard Lake ilmenites. *Nature*, **183**, 1239.

Carmichael, C. M. (1961). The magnetic properties of ilmenite-haematite crystals. *Proc. Roy. Soc. A*, **263**, 508.

Channell, J. E. T. (1978). Dual magnetic polarity measured in a single bed of Cretaceous pelagic limestones from Sicily. *Z. Geophys.*, **44**, 613.

Creer, K. M., Hedley, I. G. and O'Reilly, W. (1975). Magnetic oxides in geomagnetism. In *Magnetic Oxides*, ed. D. J. Crack. New York: Wiley.

Creer, K. M., Mitchell, J. G. and Valencio, D. J. (1971). Evidence for a normal geomagnetic field polarity event at 263 ± 5 Myr BP within the late Palaeozoic reversal interval. *Nature Phys. Sci.*, **233**, 87.

Creer, K. M. and Petersen, N. (1969). Thermochemical magnetisation in basalts. *J. Geophys.*, **35**, 501.

de Menocal, P. B., Ruddiman, W. F. and Kent, D. V. (1990). Depth of post-depositional remanence acquisition in deep-sea sediments: a case study of the Brunhes–Matuyama reversal and oxygen isotopic stage 19.1. *Earth Planet. Sci. Lett.*, **99**, 1.

Denham, C. R. (1981). Viscous demagnetisation and the longevity of palaeomagnetic polarity messages. *Geophys. Res. Lett.*, **8**, 137.

Domen, H. (1969). An experimental study on the unstable natural remanent magnetisation of rocks as a palaeomagnetic fossil. *Bull. Fac. Educ. Yamaguchi Univ.*, **18**(2), 1.

Elston, D. P. and Purucker, M. (1979). Detrital magnetisation in red beds of the Moenkopi Formation. *J. Geophys. Res.*, **84**, 1653.

Ellwood, B. B. (1984). Bioturbations: some effect on remanent magnetization acquisition. *Geophys. Res. Lett.*, **11**, 653.

Everitt, C. W. F. (1962). Self-reversal in a shale containing pyrrhotite. *Phil. Mag.*, **7**, 831.

Fujiwara, Y. and Ohtake, T. (1975). Palaeomagnetism of late Cretaceous alkaline rocks in the Nemuro Peninsula, Hokkaido, NE Japan. *J. Geomag. Geoelect.*, **26**, 549.

Fuller, M. D. (1970). Geophysical aspects of palaeomagnetism. *Crit. Rev. Solid State Phys.*, **1**, 137.

Gorter, E. W. and Schulkes, J. A. (1953). Reversal of spontaneous magnetisation as a function of temperature in Li Fe Cr spinels. *Phys. Rev.*, **90**, 487.

Graham, J. W. (1949). The stability and significance of magnetism in sedimentary rocks. *J. Geophys. Res.*, **54**, 131.

Haag, M., Heller, F., Allenspach, R. and Roche, K. (1990). Self-reversal of natural remanent magnetization in andesitic pumice. *Phys. Earth Planet. Int.*, **65**, 104.

Haag, M., Heller, F., Lutz, M. and Reusser, E. (1993). Domain observations of the magnetic phases in volcanics with self-reversed magnetization. *Geophys. Res. Lett.*, **20**, 675.

Hamano, Y. (1980). An experiment on the post-deposition remanent magnetisation in artificial and natural sediments. *Earth Planet Sci. Lett.*, **51**, 221.

Havard, A. D. and Lewis, M. (1965). Reversed partial thermo-magnetic remanence in natural and synthetic titano-magnetites. *Geophys. J. Roy. Astr. Soc.*, **10**, 59.

Hedley, I. G. (1968). Chemical remanent magnetization of the FeO OH, Fe_2O_3 system. *Phys. Earth Planet. Int.*, **1**, 103.

Heller, F. (1980). Self-reversal of natural remanent magnetisation in the Olby–Laschamp lavas. *Nature*, **284**, 334.

Heller, F., Carracedo, J. C. and Soler, V. (1986). Reversed magnetisation in pyroclastics from the 1985 eruption of Nevado de Ruiz, Columbia. *Nature*, **324**, 241.

Heller, F. and Egloff, R. (1974). Multiple reversals of natural remanent magnetisation in a granite-aplite dyke of the Bergell Massif (Switzerland). *J. Geomag. Geoelect.*, **26**, 499.

Heller, F., Markert, H. and Schmidbauer, E. (1979). Partial self-reversal of natural remanent magnetisation of an historical lava flow of Mt Etna (Sicily). *J. Geophys.*, **45**, 235.

Heller, F. and Petersen, N. (1982a). The Laschamp excursion. *Phil. Trans. Roy. Soc. A*, **306**, 169.

Heller, F. and Petersen, N. (1982b). Self-reversal explanation for the Laschamp–Olby geomagnetic field excursion. *Phys. Earth Planet. Int.*, **30**, 358.

Helsley, C. E. and Steiner, M. B. (1974). Palaeomagnetism of the lower Triassic Moenkopi Formation. *Geol. Soc. Am. Bull.*, **85**, 457.

Helsley, C. E. and Herrero-Bervera, E. (1985). Reply. *J. Geophys. Res.*, **90**, 2063.

Herrero-Bervera, E. and Helsley, C. E. (1983). Palaeomagnetism of a polarity transition in the Lower(?) Triassic, Chugwater Formation, Wyoming. *J. Geophys. Res.*, **88**, 3506.

Hildebrand, J. A. and Staudigel, H. (1986). Seamount magnetic polarity and Cretaceous volcanism of the Pacific Basin. *Geology*, **14**, 456.

Hoffman, K. A. (1975). Cation diffusion processes and self-reversal of thermoremanent magnetisation in the ilmenite–hematite solid solution series. *Geophys. J. Roy. Astr. Soc.*, **41**, 65.

Hoffman, K. A. (1982). Partial self-reversal in basalts containing mildly low temperature oxidized titanomagnetite. *Phys. Earth Planet. Int.*, **30**, 357.

Hoffman, K. A. (1984). Late acquisition of 'primary' remanence in some fresh basalts: a cause of spurious palaeomagnetic results. *Geophys. Res. Lett.*, **11**, 681.

Hoffman, K. A. (1992). Self-reversal of thermoremanent magnetisation in the Ilmenite–Hematite system: order–disorder, symmetry, and spin alignment. *J. Geophys. Res.*, **97**, 10, 883.

Hoffman, K. A. and Slade, S. B. (1986). Polarity transition records and the acquisition of remanence: a cautionary note. *Geophys. Res. Lett.*, **13**, 483.

Irving, E. (1964). *Palaeomagnetism and its Application to Geological and Geophysical Problems.* New York: Wiley.

Ishikawa, Y. and Syono, Y. (1963). Order–disorder transformation and reverse thermo-remanent magnetism in the $FeTiO_3–Fe_2O_3$ system. *J. Phys. Chem. Solids*, **24**, 517.

Kennedy, L. P. (1981). Self-reversed thermoremanent magnetisation in a late Brunhes dacite pumice. *J. Geomag. Geoelect.*, **33**, 429.

Kristjansson, J. and McDougall, I. (1982). Some aspects of the late Tertiary geomagnetic field in Iceland. *Geophys. J. Roy. Astr. Soc.*, **68**, 273.

Kropaček, V. (1968). Self-reversal of spontaneous magnetisation of natural cassiterite. *Stud. Geophys. Geod. Ceskoslov. Accad. Ved.*, **12**, 108.

Larson, E. E. and Strangway, D. W. (1966). Magnetic polarity and igneous petrology. *Nature*, **212**, 756.

Larson, E. E. and Strangway, D. W. (1968). Discussion: 'Correlation of petrology and natural magnetic polarity in Columbia Plateau basalts by R. L. Wilson and N. D. Watkins'. *Geophys. J. Roy. Astr. Soc.*, **15**, 437.

Larson, E. E. and Walker, T. R. (1982). A rock magnetic study of the Lower Massive Sandstone, Moenkopi Formation (Triassic) Gray Mountain area, Arizona. *J. Geophys. Res.*, **87**, 4819.

Larson, E. E. and Walker, T. R. (1985). Comment on 'Paleomagnetism of a polarity transition in the Lower(?) Triassic Chugwater Formation, Wyoming by Emilio Herrero-Bervera and C. E. Helsley'. *J. Geophys. Res.*, **90**, 2060.

Larson, E. E., Walker, T. R., Patterson, P. E., Hoblitt, R. P. and Rosenbaum, I. G. (1982). Palaeomagnetism of the Moenkopi Formation, Colorado Plateau; basis for long term model of acquisition of chemical remanent magnetism in red beds. *J. Geophys. Res.*, **87**, 1081.

Lawson, C. A., Nord, G. L., Jr., Dowty, E. and Hargraves, R. B. (1981). Antiphase domains and reverse thermoremanent magnetism in ilmenite-hematite materials. *Science*, **213**, 1372.

Liebes, E. and Shive, P. N. (1982). Magnetisation acquisition in two Mesozoic red sandstones. *Phys. Earth Planet. Int.*, **30**, 396.

Løvlie, R. (1974). Post-depositional remanent magnetisation in a re-deposited deep-sea sediment. *Earth Planet. Sci. Lett.*, **21**, 315.

Lund, S. P. and Karlin, R. (1990). Introduction to the Special Section on Physical and Biogeochemical processes responsible for the magnetisation of sediments. *J. Geophys. Res.*, **95**, 4353.

McClelland, E. (1987). Self-reversal of chemical remanent magnetization: a palaeomagnetic example. *Geophys. J. Roy. Astr. Soc.*, **90**, 615.

McClelland, E. and Goss, C. (1993). Self-reversal of chemical remanent magnetization on the transformation of maghemite to haematite. *Geophys. J. Int.*, **112**, 517.

McElhinny, M. W. (1969). The palaeomagnetism of the Permian of south-east Australia and its significance regarding the problem of intercontinental correlations. *Spec. Publ. Geol. Soc. Aust.*, **2**, 61.

Maillot, J. M. and Evans, M. E. (1992). Magnetic intensity variations in red beds of the Lodeve Basin (southern France) and their bearing on the magnetization acquisition process. *Geophys. J. Int.*, 111, 281.

Merrill, R. T. (1985). Correlating magnetic field polarity changes with geologic phenomena. *Geology*, 13, 487.

Merrill, R. and Grommé, C. S. (1969). Non-reproducible self-reversal of magnetization in diorite. *J. Geophys. Res.*, 74, 2014.

Moberly, R. and Campbell, J. F. (1984). Hawaiian hot-spot volcanism mainly during geomagnetic normal intervals. *Geology*, 12, 459.

Nagata, T., Uyeda, S. and Akimoto, S. (1952). Self-reversal of thermoremanent magnetism of igneous rocks. *J. Geomag. Geoelect.*, 42, 22.

Nagata, T., Akimoto, S. and Uyeda, S. (1953). Self-reversal of thermoremanent magnetism of igneous rocks (III). *J. Geomag. Geoelect.*, 5, 168.

Nagata, T. and Uyeda, S. (1959). Exchange interaction as a cause of reverse thermoremanent magnetism. *Nature*, 184, 890.

Néel, L. (1949). Théorie du trainage magnétique des ferromagnétiques au grains fins avec applications aux terres cuites. *Ann. Geophys.*, 5, 99.

Néel, L. (1951). L'inversion de l'aimantation permanente des roches. *Ann. Geophys.*, 7, 90.

Néel, L. (1955). Some theoretical aspects of rock magnetism. *Phil. Mag. Suppl. Adv. Phys.*, 4, 191.

Niitsuma, N. (1977). Zone magnetisation model and depth lag of NRM in deep-sea sediments. *Rock Magn. Paleogeophys.*, 4, 65.

Okada, M. and Niitsuma, N. (1989). Detailed paleomagnetic records during the Brunhes–Matuyama geomagnetic reversal, and a direct determination of depth lag for magnetization in marine sediments. *Phys. Earth Planet. Int.*, 56, 133.

O'Reilly, W. and Banerjee, S. K. (1967). The mechanism of oxidation in titano magnetites: a magnetic study. *Min. Mag.*, 36, 29.

Ozima, M. and Ozima, M. (1967). Self-reversal of remanent magnetisation in some dredged submarine basalts. *Earth Planet. Sci. Lett.*, 3, 213.

Ozima, M., Funaki, M., Hamada, N., Aramaki, S. and Fujii, T. (1992). Self-reversal of thermoremanent magnetization in pyroclastics from the 1991 eruption of Mt. Pinatubo, Phillippines. *J. Geomag. Geoelect.*, 44, 979.

Petersen, N. and Bleil, U. (1973). Self-reversal of remanent magnetism in synthetic titanomagnetites. *Z. Geophys.*, 39, 965.

Petherbridge, J. (1977). A magnetic coupling occurring in partial self-reversal of magnetism and its association with increased magnetic viscosity in basalts. *Geophys. J. Roy. Astr. Soc.*, 50, 395.

Purucker, M. E., Shoemaker, E. M. and Elston, D. P. (1980). Early acquisition of characteristic magnetisation in red beds of the Moenkopi Formation (Triassic), Gray Mountain, Arizona. *J. Geophys. Res.*, 85, 997.

Quidelleur, X., Valet, J.-P. and Thouveny, N. (1992). Multicomponent magnetisation in paleomagnetic records of reversals from continental sediments in Bolivia. *Earth Planet. Sci. Lett.*, 111, 23.

Robertson, W. A. (1963). Palaeomagnetism of some Mesozoic intrusives and tuffs from eastern Australia. *J. Geophys. Res.*, 68, 2299.

Roperch, P., Bonhommet, N. and Levi, S. (1988). Palaeointensity of the Earth's magnetic field during the Laschamp excursion and its geomagnetic implications. *Earth Planet. Sci. Lett.*, 88, 209.

Ryall, P. J. C. and Ade-Hall, J. M. (1975). Laboratory induced self-reversal of thermoremanent magnetisation in pillow basalts. *Nature*, 257, 117.

Sasajima, S. and Nishida, J. (1974). On the self-reversal of TRM in a highly oxidized submarine basalt. *Rock Magn. Palaeogeophys.*, 2, 5.

Schult, A. (1968). Self-reversal of magnetisation and chemical composition of titanomagnetities in basalts. *Earth Planet. Sci. Lett.*, 4, 57.

Schult, A. (1976). Self-reversal above room temperature due to N-type magnetisation in basalt. *J. Geophys.*, 42, 81.

Shive, P. N., Steiner, M. B. and Huycke, D. T. (1984). Magnetostratigraphy, paleomagnetism and remanence acquisition in the Triassic Chugwater Formation of Wyoming. *J. Geophys. Res..*, 89, 1801.

Stacey, F. D. (1963). The physical theory of rock magnetism. *Adv. Phys.*, 12, 46.

Stacey, F. D. and Banerjee, S. K. (1974). *The Physical Principles of Rock Magnetism.* Amsterdam: Elsevier.

Tauxe, L., Kent, D. V. and Opdyke, N. D. (1980). Magnetic components contributing to the NRM of Middle Siwalik red beds. *Earth Planet. Sci. Lett.*, 47, 279.

Tauxe, L. and Kent, D. V. (1984). Properties of detrital remanence carried by haematite from study of modern river deposits and laboratory redeposition experiments. *Geophys. J. Roy. Astr. Sco.*, 76, 543.

Tauxe, I. and Badgley, C. (1984). Transition stratigraphy and the problem of remanence lock-in times in the Siwalik red beds. *Geophys. Res. Lett.*, 11, 611.

Tucker, P. (1980). A grain mobility model of post-depositional realignment. *Geophys. J. Roy. Astr. Soc.*, 63, 149.

Tucker, P. and O'Reilly, W. (1980). Reversed thermoremanent magnetisation in synthetic titanomagnetites as a consequence of high temperature oxidation. *J. Geomag. Geoelect.*, 32, 341.

Uyeda, S. (1955). Magnetic interactions between ferromagnetic minerals contained in rocks. *J. Geomag. Geoelect.*, 7, 9.

Uyeda, S. (1958). Thermoremanent magnetism as a medium of palaeomagnetism, with special reference to reverse thermoremanent magnetism. *Japan J. Geophys.*, 2, 1.

van Hoof, A. A. M. and Langereis, C. G. (1991). Reversal records in marine marls and delayed acquisition of remanent magnetisation. *Nature*, 351, 223.

van Velzen, A. J. and Zijderveld, J. D. A. (1990). Rock magnetism of the early Pliocene Trubi formation at Eraclea Minoa, Sicily. *Geophys. Res. Lett.*, 17, 791.

Verhoef, J., Collette, B. J. and Williams, C. A. (1985). Comment on 'Hawaiian hotspot volcanism mainly during geomagnetic normal intervals'. *Geology*, 13, 314.

Verhoogen, J. (1956). Ionic ordering and self-reversal of magnetisation in impure magnetites. *J. Geophys. Res.*, 61, 201.

Verhoogen, J. (1962). Oxidation of iron titanium oxides in igneous rocks. *J. Geol.*, 70, 168.

Walker, T. R., Larson, E. E. and Hoblitt, R. P. (1981). Nature and origin of hematite in the Moenkopi Formation (Triassic), Colorado Plateau: a contribution to the origin of magnetism in red beds. *J. Geophys. Res.*, 86, 317.

Watkins, N. D. and Haggerty, S. E. (1968). Oxidation and polarity variations in Icelandic lavas and dikes. *Geophys. J. Roy. Astr. Soc.*, 15, 305.

Westcott-Lewis, M. F. and Parry, L. G. (1971). Thermoremanence in synthetic rhombohedral iron-titanium oxides. *Aust. J. Phys.*, 24, 735.

Wilson, R. L. (1962). The palaeomagnetism of baked contact rocks and reversals of the Earth's magnetic field. *Geophys. J. Roy. Astr. Soc.*, 7, 194.

Wilson, R. L. (1966a). Further correlations between the petrology and the natural magnetic polarity of basalts. *Geophys. J. Roy. Astr. Soc.*, 10, 413.

Wilson, R. L. (1966b). Palaeomagnetism and rock magnetism. *Earth Sci. Rev.*, 1, 175.

Wilson, R. L. and Watkins, N. D. (1967). Correlation of petrology and natural magnetic polarity in Columbia Plateau basalts. *Geophys. J. Roy. Astr. Soc.*, 12, 405.

Zapletal, K. (1992). Self-reversal of isothermal remanent magnetisation in a pyrrhotite (Fe_7S_8) crystal. *Phys. Earth Planet. Int.*, 70, 302.

Chapter 3

Barbetti, M. F. and McElhinny, M. W. (1976). The Lake Mungo geomagnetic excursion. *Phil. Trans. Roy. Soc. A*, **281**, 515.

Berggren, A. B., Kent, D. V., Flynn, J. J. and van Couvering, J. A. (1985). Cenozoic geochronology. *Geol. Soc. Am. Bull.*, **96**, 1407

Bingham, D. K. and Evans, M. E. (1975). Precambrian geomagnetic field reversal. *Nature*, **253**, 332.

Blakely, R. J. (1974). Geomagnetic reversals and crustal spreading rates during the Miocene. *J. Geophys. Res.*, **79**, 2979.

Blakely, J. R. and Cox, A. (1972). Evidence for short geomagnetic polarity intervals in the early Cenozoic. *J. Geophys. Res.*, **77**, 7065.

Bogue, S. W. and Coe, R. S. (1984). Transitional paleointensities from Kauai, Honolulu and geomagnetic reversal models. *J. Geophys. Res.*, **89**, 10341

Bol'shakov, A. S. and Solodovnikov, G. M. (1980). Palaeomagnetic data on the intensity of the magnetic field of the Earth. *Izv. Earth Phys.*, **16**, 602.

Burakov, K. S., Gurary, G. Z., Khramov, A. N., Petrova, G. N., Rassanova, G. V. and Rodionov, V. P. (1976). Some peculiarities of the virtual pole positions during reversals. *J. Geomag. Geoelect.*, **28**, 295.

Chauvin, A., Roperch, P. and Duncan, R. A. (1990). Records of geomagnetic reversals from volcanic islands of French Polynesia. 2. Palaeomagnetic study of a flow sequence (1.2–0.6 Ma) from the Island of Tahiti and discussion of reversal models. *J. Geophys. Res.*, **95**, 2727.

Clement, B. M. (1991). Geographical distribution of transitional VGPs: evidence for non-zonal equatorial symmetry during the Matuyama–Brunhes geomagnetic reversal. *Earth Planet. Sci. Lett.*, **104**, 48

Clement, B. M. and Kent, D. V. (1984). Latitudinal dependency of geomagnetic polarity transition. *Nature*, **310**, 488.

Clement, B. M. and Kent, D. V. (1986/87). Short polarity intervals within the Matuyama: transitional field records from hydraulic piston cored sediments from the North Atlantic. *Earth Planet. Sci. Lett.*, **81**, 253.

Clement, B. M., Kent, D. V. and Opdyke, N. D. (1982). Brunhes–Matuyama polarity transition in three deep-sea sediment cores. *Phil. Trans. Roy. Soc. A*, **306**, 113.

Coe, R. S., Grommé, C. S. and Mankinen, E. A. (1984). Geomagnetic paleointensities from excursion sequences in lavas on Oahu, Hawaii. *J. Geophys. Res.*, **89**, 1059.

Cox, A. (1969). Geomagnetic reversals. *Science*, **163**, 237.

Cox, A. (1975). The frequency of geomagnetic reversals and the symmetry of the non-dipole field. *Rev. Geophys. Space Phys.*, **13**, 35.

Cox, A. (1981). A stochastic approach towards understanding the frequency and polarity bias of geomagnetic reversals. *Phys. Earth Planet. Int.*, **24**, 178.

Cox, A, Doell, R. R. and Dalrymple, G. R. (1964). Reversals of the Earth's magnetic field. *Science*, **144**, 1537.

Creer, K. M. (1972). The behaviour of the palaeogeomagnetic field during reversals. *Trans. Am. Geophys. Union*, **53**, 614.

Creer, K. M., Gross, D. L. and Lineback, J. A. (1976). Origin of regional geomagnetic variations recorded by Wisconsinan and Holocene sediments from Lake Michigan, USA and Lake Windermere, England. *Geol. Soc. Am. Bull.*, **87**, 531.

Creer, K. M. and Ispir, Y. (1970). An interpretation of the behaviour of the geomagnetic field during polarity transitions. *Phys. Earth Planet. Int.*, **2**, 283.

Dagley, P. and Lawley, E. (1974). Palaeomagnetic evidence for the transitional behaviour of the geomagnetic field. *Geophys. J. Roy. Astr. Soc.*, **36**, 577.

Dodson, R., Dunn, J. R., Fuller, M., Williams, I., Ito H., Schmidt, V. A. and Wu

Yu, M. (1978). Palaeomagnetic record of a late Tertiary field reversal. *Geophys. J. Roy. Astr. Soc.*, 53, 373.

Dodson, R., Fuller, M. and Kean, W. E. (1977). Palaeomagnetic records of secular variation from Lake Michigan sediment cores. *Earth Planet. Sci. Lett.*, 34, 387.

Doell, R. R. and Cox, A. (1971). Pacific geomagnetic secular variation. *Science*, 171, 248.

Doell, R. R. and Cox, A. (1972). The Pacific geomagnetic secular variation anomaly and the question of lateral uniformity in the lower mantle. In *The Nature of the Solid Earth*, ed. E. C. Robertson, pp. 245–84 New York: McGraw-Hill.

Doell, R. and Dalrymple, G. B. (1966). Geomagnetic polarity epochs; a new polarity event and the age of the Brunhes/Matuyama boundary. *Science*, 152, 1060.

Dunn, J. R., Fuller, M. Ito, H. and Schmidt, V. A. (1971). Palaeomagnetic study of a reversal of the Earth's magnetic field. *Science*, 172, 840.

Freed, W. K. (1977). The vertical geomagnetic polepath during the Brunhes/Matuyama polarity change when viewed from equatorial latitudes. *Trans. Am. Geophys. Union*, 58, 380.

Gallet, Y., Besse, J., Marcoux, J., Theveniaut, H. and Krystyn, L. (1992). Magneto-stratigraphy of the late Triassic Bolucektasi Tepe section (southwestern Turkey): implications for changes in magnetic reversal frequency. *Phys. Earth Planet Int.*, 73, 85.

Gurarii, G. Z. (1981). The Matuyama–Jaramillo geomagnetic inversion in Western Turkmenia. *Izv. Earth Phys.*, 17, 212.

Hammond, S. R., Seyb, S. M. and Theyer, F. (1979). Geomagnetic polarity transitions in two oriented sediment cores from the northwest Pacific. *Earth Planet. Sci. Lett.*, 44, 167.

Harland, W. B., Cox, A. V., Llewellyn, P. G., Picton, C. A. G., Smith, A. G. and Walters, R. (1982). *A Geologic Timescale*. Cambridge: Cambridge University Press

Harland, W. B., Armstrong, R. L., Cox, A. V., Craig, L. E., Smith, A. G. and Smith, D. G. (1990). *A Geologic Timescale 1989*. Cambridge: Cambridge University Press.

Harrison, C. G. A., McDougall, I. and Watkins, N. D. (1979). A geomagnetic field reversal timescale back to 130 million years before present. *Earth Planet. Sci. Lett.*, 42, 143.

Heirtzler, J. R., Dickson, G. O., Herron, E. M., Pitman, W. C. and LePichon, X. (1968). Marine magnetic anomalies, geomagnetic field reversals and motions of the ocean floor and continents. *J. Geophys. Res.*, 73, 2119.

Herrero-Bervera, E. and Helsley, C. E. (1983). Palaeomagnetism of a polarity transition in the Lower (?) Triassic Chugwater Formation, Wyoming. *J. Geophys. Res.*, 88, 3605.

Hillhouse, J. and Cox, A. (1976). Brunhes–Matuyama polarity transition. *Earth Planet. Sci. Lett.*, 29, 51.

Hillhouse, J. W., Cox, A., Denham, C. R., Blakely, R. J. and Butler, R. F. (1972). Geomagnetic polarity transitions. *Trans. Am. Geophys. Union*, 53, 971.

Hoffman, K. A. (1977). Polarity transition records and the geomagnetic dynamo. *Science*, 196, 1329.

Hoffman, K. A. (1984). A method for the display and analysis of transitional paleomagnetic data. *J. Geophys. Res.*, 89, 6285.

Hoffman, K. A. (1991). Long lived transitional states of the geomagnetic field and the two dynamo families. *Nature*, 354, 273.

Irving, E. and Pullaiah, G. (1976). Reversals of the geomagnetic field, magnetostratigraphy, and relative magnitude of palaeosecular variation in the Phanerozoic. *Earth Sci. Rev.*, 12, 35.

Ito, H. (1970). Polarity transitions of the geomagnetic field deduced from the natural remanent magnetisation of Tertiary and Quaternary rocks in Southwest Japan. *J. Geomag. Geoelect.*, 22, 273.

Kaporovich, I. G., Makarova, Z. V., Petrova, G. N. and Rybak, R. S. (1966). The transitional stage of the geomagnetic field in the Pliocene on the territory of Turkmenia and Azerbaidzhan. *Izv. Earth Phys.*, **1**, 59.

Kawai, N. and Nakajima, T. (1975). Vanished geomagnetism. *Proc. Japan Acad.*, **51**, 640.

Kawai, N., Otofuji, Y. and Kobayashi, K. (1976). Palaeomagnetic study of deep-sea sediments using thin sections. *J. Geomag. Geoelect.*, **28**, 395.

Kawai, N., Sato, T., Sueishi, T. and Kobayashi, K. (1977). Palaeomagnetic study of deep-sea sediments from the Melanesia Basin. *J. Geomag. Geolect.*, **29**, 211.

Kristjansson, L. (1985). Some statistical properties of paleomagnetic directions in Icelandic lava flows. *Geophys. J. Roy. Astr. Soc.*, **80**, 57.

Kristjansson, L. and McDougall, I. (1982). Some aspects of the late Tertiary geomagnetic field in Iceland. *Geophys. J.*, **68**, 273.

LaBrecque, J. L., Kent, D. V. and Cande, S. C. (1977). Revised magnetic polarity timescale for late Cretaceous and Cenozoic time. *Geology*, **5**, 330.

Laj, C., Mazaud, A., Weeks, R., Fuller, M. and Herrero-Bervera, E. (1991). Geomagnetic reversal paths. *Nature*, **351**, 447.

Liddicoat, J. C. (1981). Gauss/Matuyama polarity transition. *Trans. Am. Geophys. Union*, **62**, 263.

Liddicoat, J. C. (1982). Gauss/Matuyama polarity transition. *Phil. Trans. Roy. Soc. A*, **306**, 121.

Lowrie, W. (1982). Revised magnetic polarity timescale for the Cretaceous and Cainozoic. *Phil. Trans. Roy. Soc. A*, **306**, 129.

Lowrie, W. and Alvarez, W. (1981). One hundred million years of geomagnetic polarity history. *Geology*, **9**, 392.

Lowrie, W. and Kent, D. V. (1983). Geomagnetic reversal frequency since the Late Cretaceous. *Earth Planet. Sci. Lett.*, **62**, 305.

Lund, S. P., Liddicoat, J. C., Lajoie, K. R., Henyey, T. L. and Robinson, S. W. (1988). Paleomagnetic evidence for long term (10^4 year) memory and periodic behavior in the Earth's core dynamo process. *Geophys. Res. Lett.*, **15**, 1101.

Lutz, T. M. (1985). The magnetic reversal record is not periodic. *Nature*, **317**, 404.

Lutz, T. M. and Watson, G. S. (1988). Effects of long term variation in the frequency spectrum of the geomagnetic reversal record. *Nature*, **334**, 240.

McDougall, I., Saemundsson, K., Johannesson, H., Watkins, N. D. and Kristjansson, L. (1977). Extension of the geomagnetic polarity timescale to 65 Myr: K–Ar dating, geological and palaeomagnetic study of a 3500 m lava succession in Western Iceland. *Geol. Soc. Am. Bull.*, **88**, 1.

McElhinny, M. W. (1971). Geomagnetic reversals during the Phanerozoic. *Science*, **172**, 157.

McFadden, P. L. (1984a). 15-Myr periodicity in the frequency of geomagnetic reversals since 100 Myr. *Nature*, **311**, 396.

McFadden, P. L. (1984b). Statistical tools for the analysis of geomagnetic reversal sequences. *J. Geophys. Res.*, **89**, 3363.

McFadden, P. L. (1987). 'A periodicity of magnetic reversals?' Comment by P. L. McFadden. *Nature*, **330**, 27.

McFadden, P. L. and Merrill, R. T. (1984). Lower mantle convection and geomagnetism. *J. Geophys. Res.*, **89**, 3354.

McWilliams, M. O., Holcomb, R. T. and Champion, D. E. (1982). Geomagnetic secular variation from ^{14}C dated flows in Hawaii and the question of the Pacific non-dipole low. *Phil. Trans. Roy. Soc. A*, **306**, 211.

Mankinen, E. A., Prévot, M., Grommé, C. S. and Coe, R. S. (1985). The Steens Mountain (Oregon) geomagnetic polarity transition. 1. Directional history, duration of episodes, and rock magnetism. *J. Geophys. Res.*, **90**, 10393.

Marzocchi, W. and Mulargia, F. (1990). Statistical analysis of the geomagnetic reversal sequences. *Phys. Earth Planet Int.*, **61**, 149.

Marzocchi, W. and Mulargia, F. (1992). On the periodicity of geomagnetic reversals. *Phys. Earth Planet. Int.*, **73**, 222.

Massey, N. W. D. (1979). Keweenawan palaeomagnetic reversals at Mamainse Point, Ontario: fault repetition or three reversals? *Can. J. Earth Sci.*, **16**, 373.

Mazaud, A., Laj, C., de Seze, L. and Verosub, K. L. (1983). Evidence for periodicity in the reversal frequency during the last 100 Ma. *Nature*, **304**, 328.

Mazaud, A. and Laj, C. (1991). The 15 m.y. geomagnetic reversal periodicity: a quantitative test. *Earth Planet Sci. Lett.*, **107**, 689.

Merrill, R. T. and McFadden, P. L. (1990). Paleomagnetism and the nature of the geodynamo. *Science*, **248**, 345.

Negi, J. G. and Tirwari, R. K. (1983). Matching long term periodicities of geomagnetic reversals and galactic motions of the solar system. *Geophys. Res. Lett.*, **10**, 713.

Nevanlinna, H. and Pesonen, L. J. (1983). Late Precambrian Keweenawan asymmetric polarities as analysed by axial offset dipole geomagnetic models. *J. Geophys. Res.*, **88**, 645.

Olson, P. (1983). Geomagnetic polarity reversals in a turbulent core. *Phys. Earth Planet. Int.*, **33**, 260.

Opdyke, N. D. (1969). The Jaramillo event as detailed in oceanic cores. In *The Application of Modern Physics to the Earth and Planetary Interiors*, ed. S. K. Runcorn, p. 549. New York: Wiley.

Opdyke, N. D., Glass, B., Hays, J. D. and Foster, J. (1966). Palaeomagnetic study of Antarctic deep-sea cores. *Science*, **154**, 349.

Opdyke N. D., Kent D. V. and Lowrie W. (1973). Details of magnetic polarity transitions recorded in a high deposition rate deep-sea core. *Earth Planet. Sci. Lett.*, **20**, 315.

Pal, P. C. and Creer, K. M. (1986). Geomagnetic reversal spurts and episodes of extraterrestrial catastrophism. *Nature*, **320**, 148.

Palmer, H. C. (1970). Palaeomagnetism and correlation of some Middle Keweenawan rocks, Lake Superior. *Can. J. Earth Sci.*, **7**, 1410.

Pesonen, L. J. and Halls, H. C. (1983). Geomagnetic field intensity and reversal asymmetry in late Precambrian Keweenawan rocks. *Geophys. J. Roy. Astr. Soc.*, **73**, 241.

Petrova, G. N. and Rassanova, G. V. (1976). Features of reversals. In *Principal Geomagnetic Field and Problems of Palaeomagnetism*, Part II. Moscow: Nauka.

Prévot, M., Mankinen, E.-A., Grommé, C. S. and Coe, R. S. (1985). The Steens Mountain (Oregon) geomagnetic polarity transition. 2. Field intensity variations and discussions of reversal models. *J. Geophys. Res.*, **90**, 10, 417.

Raisbeck, G. M., Yiou, F. Bourles, D. and Kent, D. V. (1985). Evidence for an increase in cosmogenic ^{10}Be during a geomagnetic reversal. *Nature*, **315**, 315.

Raup, D. M. (1985). Magnetic reversals and mass extinctions. *Nature*, **314**, 341.

Roberts, N. and Shaw, J. (1990). Secondary magnetisations cast doubt on exceptional palaeointensity values from the Lousetown Creek lavas of Nevada, USA. *Geophys. J. Int.*, **101**, 251.

Robertson, W. A. (1973). Pole positions from the Mamainse Point lavas and their bearing on a Keweenawan pole and polarity sequence. *Can. J. Earth Sci.*, **10**, 1541.

Shaw, J. (1975). Strong geomagnetic fields during a single Icelandic polarity transition. *Geophys. J. Roy. Astr. Soc.*, **40**, 345.

Shaw, J. (1977). Further evidence for a strong intermediate state of the palaeomagnetic field. *Geophys. J. Roy. Astr. Soc.*, **48**, 263.

Shaw, J., Dagley, P. and Mussett, A. (1982). The magnitude of the palaeomagnetic field in Iceland between 2 and 6 Myr ago. *Geophys. J. Roy. Astr. Soc.*, **68**, 211.

Shaw, J. and Wilson, R. L. (1977). The magnitude of the palaeomagnetic field during a polarity transition: a new technique and its application. *Phys. Earth Planet. Int.*, 13, 339.

Skiles, D. D. (1970). A method of inferring the direction of drift of the geomagnetic field from palaeomagnetic data. *J. Geomag. Geoelect.*, 22, 441.

Steinhauser, P. and Vincenz, S. A. (1973). Equatorial palaeopoles and behaviour of the dipole field during polarity transitions. *Earth Planet. Sci. Lett.*, 19, 113.

Stigler, S. M. (1987). A periodicity of magnetic reversals? *Nature*, 330, 26.

Stothers, R. B. (1986). Periodicity of the Earth's magnetic reversals. *Nature*, 332, 444.

Vadkovskii, V. N., Gurarii, G. Z. and Mannikon'van, M. R. (1980). Analysis of the process of geomagnetic field reversal. *Izv. Earth Phys.*, 16, 506.

Valet, J.-P., Tauxe, L. and Clark, D. R. (1988). The Matuyama–Brunhes transition recorded from Lake Tecopa sediments (California). *Earth Planet. Sci. Lett.*, 87, 463.

Van Zijl, J. S. V., Graham, K. W. T. and Hales, A. L. (1962). The palaeomagnetism of the Stormberg lavas. II, The behaviour of the magnetic field during a reversal. *Geophys. J. Roy. Astr. Soc.*, 7, 169.

Vine, F. J. (1966). Spreading of the ocean floor: new evidence. *Science*, 154, 1405.

Vitorello, I. and Van der Voo, R. (1977). Magnetic stratigraphy of Lake Michigan sediments obtained from cores of lacustrine clay. *Quatern. Res.*, 7, 398.

Wilson, R. L., Dagley, P. and McCormack, A. G. (1972). Palaeomagnetic evidence about the source of the geomagnetic field. *Geophys. J. Roy. Astr. Soc.*, 28, 213.

Chapter 4

Abrahamsen, N. (1982). Magnetostratigraphy. In *The Pleistocene/Holocene Boundary in South-western Sweden*, ed. O. Olausson. Uppsala: Sver. Geol. Undersok.

Abrahamsen, N. and Knudsen, K. L. (1979). Indication of a geomagnetic excursion in supposed Middle Weichselian interstadial marine clay at Rubjerg, Denmark. *Phys. Earth Planet. Int.*, 18, 238.

Abrahamsen, N. and Readman, P. W. (1980). Geomagnetic variations recorded in older (≥2300 BP) and younger Yolida Clay (≈14000 BP) at Norre Lyngly, Denmark. *Geophys. J. Roy. Astr. Soc.*, 62, 329.

Aksu, A. E. (1983). Short period geomagnetic excursions recorded in Pleistocene sediments of Baffin Bay and Davis Strait. *Geology*, 11, 537.

Banerjee, S. K., Lund, S. P. and Levi, S. (1979). Geomagnetic record in Minnesota Lake sediments – absence of the Gothenburg and Eriau excursions. *Geology*, 7, 588.

Barbetti, M. and Flude, K. (1979). Palaeomagnetic field strengths from sediments baked by lava flows of the Chaîne des Puys, France. *Nature*, 278, 153.

Barbetti, M. F. and McElhinny, M. W. (1972). Evidence of a geomagnetic excursion 30,000 yr BP. *Nature*, 239, 327.

Barbetti, M. F. and McElhinney, M. W. (1976). The Lake Mungo geomagnetic excursion. *Phil. Trans. Roy. Soc. A*, 281, 515.

Barbetti, M., Taborin, Y., Schmider, B. and Flude, K. (1980). Archaeomagnetic results from late Pleistocene hearths at Étiolles and Marsangy, France. *Archaeometry*, 22, 25.

Bard, E., Hamelin, B., Fairbanks, R. G. and Zinder, A. (1990a). Calibration of the ¹⁴C timescale over the past 30,000 years using mass-spectrometric U–Th ages from Barbados cords. *Nature*, 345, 405.

Bard, E., Hamelin, B., Fairbanks, R. G., Zinder, A., Mathieu, G. and Arnold, M. (1990b). U–Th and ¹⁴C ages of corals from Barbados and their use for calibrating the ¹⁴C timesale beyond 9000 years B.P. *Nucl. Instr. Methods Phys. Res.*, 52, 461.

Barton, C. E. and Polach, H. A. (1971). ¹⁴C ages and magnetic stratigraphy in three Australian maars. *Radiocarbon*, 22, 728.

Bingham, D. K. and Evans, M. E. (1975). Precambrian geomagnetic field reversal. *Nature*, 253, 332.

Bingham, D. K. and Stone, D. B. (1976). Evidence for geomagnetic field excursions and secular variation from the Wrangell Volcanoes of Alaska. *Can. J. Earth Sci.*, 13, 547.

Blow, R. A. and Hamilton, N. (1975). Palaeomagnetic evidence from DSDP cores of northward drift of India. *Nature*, 257, 570.

Blow, R. A. and Hamilton, N. (1978). Effect of compaction on the acquisition of a detrital remanent magnetisation in fine-grained sediments. *Geophys. J. Roy. Astr. Soc.*, 52, 13.

Bonhommet, N. and Babkine, J. (1967). Sur la présence d'aimantations inversées dans la Chaine des Puys. *C. R. Acad. Sci. Paris*, 264, 92.

Bonhommet, N. and Zahringer, J. (1969). Palaeomagnetism and potassium–argon age determination of the Laschamp geomagnetic polarity event. *Earth Planet. Sci. Lett.*, 6, 43.

Bucha, V. (1983). In *Magnetic Field and the Processes in the Earth's Interior*. Prague: Academia.

Champion, D. E., Dalrymple, G. B. and Kuntz, M. A. (1981). Radiometric and palaeomagnetic evidence for the Emperor reversed polarity event at 0.46 ± 0.005 Myr in basalt lava flows from the Eastern Snake River plain, Idaho. *Geophys. Res. Lett.*, 8, 1055.

Champion, D. E., Lanphere, M. A. and Kuntz, M. A. (1988). Evidence for a new geomagnetic reversal from lava flows in Idaho: Discussion of short polarity reversals in the Brunhes and late Matuyama polarity chrons. *J. Geophys. Res.*, 93, 11667.

Chauvin, A., Duncan, R. A., Bonhommet, N. and Levi, S. (1989). Palaeointensity of the Earth's magnetic field and K–Ar dating of the Louchardière volcanic flow, central France: New evidence for the Laschamp excursion. *Geophys. Res. Lett.*, 16, 1189.

Clark, H. C. and Kennett, J. P. (1973). Palaeomagnetic excursion recorded in latest Pleistocene deep-sea sediments, Gulf of Mexico. *Earth Planet. Sci. Lett.*, 19, 267.

Clement, B. M. and Kent, D. V. (1986/7). Short polarity intervals within the Matuyama transitional field records from hydraulic piston cored sediments from the North Atlantic. *Earth Planet. Sci. Lett.*, 81, 253.

Clement, B. M. and Martinson, D. G. (1992). A quantitative comparison of two palaeomagnetic records of the Cobb Mountain subchron from North Atlantic deep-sea sediments. *J. Geophys. Res.*, 97, 1735.

Coe, R. S. (1977). Source models to account for Lake Mungo palaeomagnetic excursion and their implications. *Nature*, 269, 49.

Condomines, M. (1978). Age of Olby–Laschamp geomagnetic polarity event. *Nature*, 276, 257.

Cox, A. (1968). Length of geomagnetic polarity intervals. *J. Geophys. Res.*, 73, 3247.

Creer, K. M., Anderson, T. W. and Lewis, C. F. M. (1976a). Late Quaternary geomagnetic stratigraphy recorded in Lake Erie sediments. *Earth Planet. Sci. Lett.*, 31, 37.

Creer, K. M., Gross, D. L. and Lineback, J. A. (1976b). Origin of regional geomagnetic variations recorded by Wisconsinan and Holocene sediments from Lake Michigan, USA and Lake Windermere, England. *Geol. Soc. Am. Bull.*, 87, 531.

Creer, K. M., Readman, P. W. and Jacobs, A. M. (1980). Palaeomagnetic and palaeontological dating of a section at Gioia Tauro, Italy: identification of the Blake event. *Earth Planet. Sci. Lett.*, 50, 289.

Creer, K. M., Thouveny, N. and Blunk, I. (1990). Climatic and geomagnetic influences on the Lac du Bouchet palaeomagnetic SV record through the last 110,000 years. *Phys. Earth Planet. Int.*, 64, 314.

Denham, C. R. (1974). Counter-clockwise motion of palaeomagnetic directions 24,000 years ago at Mono Lake, California. *J. Geomag. Geoelect.*, 26, 487.

Denham, C. R. (1976). Blake polarity episode in two cores from the greater Antilles outer ridge. *Earth Planet. Sci. Lett.*, **29**, 422.

Denham, C., Anderson, R. F. and Bacon, M. P. (1977). Palaeomagnetism and radio-chemical age estimates for the Brunhes polarity episodes. *Earth Planet. Sci. Lett.*, **35**, 384.

Denham, C. R. and Cox, A. (1971). Evidence that the Laschamp polarity event did not occur 13,300–30,400 years ago. *Earth Planet. Sci. Lett.*, **13**, 1981.

Dodson, R. E., Fuller, M. D. and Kean, W. F. (1977). Palaeomagnetic records of secular variation from Lake Michigan sediment cores. *Earth Planet. Sci. Lett.*, **34**, 387.

Doh, S. J. and Steele, W. K. (1981). The late Pleistocene geomagnetic field as recorded by sediments from Fargher Lake, Washington. *Trans. Am. Geophys. Union*, **62**, 851.

Doh, S. J. and Steele, W. K. (1983). The late Pleistocene geomagnetic field as recorded by sediments from Fargher Lake, Washington, USA. *Earth Planet. Sci. Lett.*, **63**, 385.

Eardley, A. J., Shuey, R., Gvosdetsky, V., Nash, W., Picard Dane, M., Grey, D. and Kukla, G. (1973). Lake cycles in the Bonneville Basin, Utah. *Geol. Soc. Am. Bull.*, **84**, 211.

Freed, W. K. and Healy, N. (1974). Excursions of the Pleistocene geomagnetic field recorded in Gulf of Mexico sediments. *Earth Planet. Sci. Lett.*, **24**, 99.

Geissman, J. W., Brown, L., Turrin, B. D., McFadden, L. D. and Harlan, S. S. (1990). Brunhes chron excursion/polarity episode recorded during the late Pleistocene, Albuquerque volcanoes, New Mexico, U.S.A. *Geophys. J. Int.*, **102**, 73.

Gillot, P. Y., Labeyrie, J., Laj, C., Valladas, G., Guérin, G., Poupeau, G. and Delibrias, G. (1979). Age of the Laschamp polarity palaeomagnetic excursion revisited. *Earth Planet. Sci. Lett.*, **42**, 444.

Guérin, G. and Valladas, G. (1980). Thermoluminescence dating of volcanic plagioclases. *Nature*, **286**, 697.

Hall, C. M. and York, D. (1978). K–Ar and $^{40}Ar/^{39}Ar$ age of the Laschamp geomagnetic polarity reversal. *Nature*, **274**, 462.

Hall, C. M., York, D. and Bonhommet, M. (1979). $^{40}Ar/^{39}Ar$ dating of the Laschamp event and associated volcanism in the Chaîne des Puys. *Trans. Am. Geophys. Union*, **60**, 244.

Hamilton, T. S. and Evans, M. E. (1983). A magnetostratigraphic and secular variation study of Level Mountain, northern British Columbia. *Geophys. J. Roy. Astr. Soc.*, **73**, 39.

Hanna, R. L. and Verosub, K. L. (1989). A review of lacustrine palaeomagnetic records from western North America, 0–40,000 years B.P. *Phys. Earth Planet. Int.*, **56**, 76.

Harrison, C. G. A. (1974). The palaeomagnetic record from deep-sea sediment cores. *Earth Sci. Rev.*, **10**, 1.

Harrison, C. G. A. and Ramirez, E. (1975). Areal coverage of spurious reversals of the Earth's magnetic field. *J. Geomag. Geoelect.*, **27**, 139.

Hayashida, A. (1980). Confirmation of a magnetic polarity episode in the Brunhes normal epoch; a preliminary report. *Rock Magn. Palaeogeophys.*, **7**, 85.

Hayashida, A. (1984). Reversed polarity episode in the late Brunhes epoch as confirmed in parallel sections of Pleistocene deposits in the Kinki District, Japan. *J. Geomag. Geoelect.*, **36**, 585.

Heller, F. (1980). Self-reversal of natural remanent magnetisation in the Olby–Laschamp lavas. *Nature*, **284**, 334.

Heller, F. and Liu, T.-S. (1982). Magnetostratigraphical dating of loess deposits in China. *Nature*, **300**, 431.

Heller, F. and Petersen, N. (1982). The Laschamp excursion. *Phil. Trans. Roy. Soc. A*, **306**, 169.

Herrero-Bervera, E., Helsley, C. E., Hammond, S. R. and Chitwood, L. A. (1989). A possible lacustrine palaeomagnetic record of the Blake episode from Pringle Falls, Oregon, U.S.A. *Phys. Earth Planet. Int.*, **56**, 112.

Heusser, C. J. and Heusser, L. E. (1980). Sequence of pumiceous tephra layers and the late Quaternary environmental record near Mount St Helens. *Science*, 210, 1007.

Hirooka, K. (1976). Some notes on the characteristics of geomagnetic excursion in late Pleistocene in Japan. *Palaeolimnology of Lake Biwa and the Japanese Pleistocene*, vol. 4, ed. S. Horie, p. 153. Otsu: Kyoto University.

Hirooka, K., Tobita, C., Yokoyama, T. and Nakaya, S. (1977). On the excursion of the latest Pleistocene recorded in Ontake Tephra, Ina, central Japan. *Rock Magn. Palaeogeophys.*, 4, 81.

Hsu, V., Merrill, D. L. and Shibuya, H. (1990). Palaeomagnetic transition records of the Cobb Mountain event from sediments of the Celebes and Sulu seas. *Geophys. Res. Lett.*, 17, 2069.

Huxtable, J. and Aitken, M. J. (1977). Thermoluminescent dating of Lake Mungo geomagnetic polarity excursion. *Nature*, 265, 40.

Huxtable, J., Aitken, M. J. and Bonhommet, N. (1978). Thermoluminescence dating of sediment baked by lava flows of the Chaîne des Puys. *Nature*, 275, 207.

Hyodo, M., Sunata, W. and Susanto, E. E. (1992). A long term geomagnetic excursion from Plio-Pleistocene sediments in Java. *J. Geophys. Res.*, 97, 9323.

Johnson, H. G., Kinoshita, H. and Merrill, R. T. (1975). Rock magnetism and palaeo-magnetism of some north Pacific deep-sea sediment cores. *Geol. Soc. Am. Bull.*, 86, 412.

Kawai, N. (1984). Palaeomagnetic study of the Lake Biwa sediments. In *Lake Biwa*, ed. S. Horie. Dordrecht: W. Junk.

Kawai, N., Sato, T., Sueishi, T. and Kobayashi, K. (1977). Palaeomagnetic study of deep-sea sediments from the Melanesian Basin. *J. Geomag. Geoelect.*, 29, 211.

Kawai, N., Yaskawa, K., Nakajima, T., Torrii, M. and Horie, S. (1972). Oscillating geomagnetic field with a recurring reversal discovered from Lake Biwa. *Proc. Japan Acad.*, 48, 186.

Kochegura, V. V. and Zubakov, V. A. (1978). Palaeomagnetic timescale of the Ponto-Caspian Plio-Pleistocene deposits. *Palaeogeog. Palaeoclimatol., Palaeoecol.*, 23, 151.

Kristjansson, L. and Gudmundsson, A. (1980). Geomagnetic excursion in late-glacial basalt outcrops in south-western Iceland. *Geophys. Res. Lett.*, 7, 337.

Kukla, G. J. and Koci, A. (1972). End of the last interglacial in the loess record. *Quatern. Res.*, 2, 374.

Kulikova, L. S. and Pospelova, G. A. (1976). Geomagnetic field secular variation in late Pleistocene upon alluvial sediments of Objriver. In *Palaeomagnetism of Mesozoic and Cenozoic of Siberia and the Far East*, pp. 95–112. Novosibirsk.

Lajoie, K. R. and Liddicoat, J. C. (1980). Refinement of the chronology and palaeomag-netic record at Mono Lake, California. *Trans. Am. Geophys. Union*, 61, 215.

Larson, E. E., Watson, D. E. and Jennings, W. (1971). Regional comparison of a Miocene geomagnetic transition in Oregon and Nevada. *Earth Planet. Sci. Lett.*, 11, 391.

Levi, S., Audunsson, H., Duncan, R. A., Kristjansson, L., Gillot, P.-Y. and Jakobsson, L. (1990). Late Pleistocene geomagnetic excursion in Icelandic lavas: confirmation of the Laschamp excursion. *Earth Planet. Sci. Lett.*, 96, 443.

Levi, S. and Karlin, R. (1989). A sixty thousand year palaeomagnetic record from Gulf of California sediments: secular variation, late Quaternary excursions and geomagnetic implications. *Earth Planet. Sci. Lett.*, 92, 219.

Liddicoat, J. C. (1992). Mono Lake excursion in Mono Basin, California, and at Carson Sink and Pyramid Lake, Nevada. *Geophys. J. Int.*, 108, 442.

Liddicoat, J. C. and Coe, R. S. (1979). Mono Lake geomagnetic excursion. *J. Geophys. Res.*, 84, 261.

Liddicoat, J. C., Lajoie, K. R. and Sarna-Wojcicki, A. M. (1982). Detection and dating of the Mono Lake excursion in the Lake Lahontan Sehoo Formation, Carson Sink, Nevada. *Trans. Am. Geophys. Union*, 63, 920.

Løvlie, R. and Holtedahl, H. (1980). Apparent palaeomagnetic low-inclination excursion on a preconsolidated continental shelf sediment. *Phys. Earth Planet Int.*, 22, 137.

Løvlie, R., Markussen, B., Sejrup, H. P. and Thiede, J. (1986). Magnetostratigraphy in three Arctic Ocean sediment cores: arguments for geomagnetic excursions within oxygen-isotope stage 2–3. *Phys. Earth Planet. Int.*, 43, 173.

Løvlie, R. and Sandnes, A. (1987). Palaeomagnetic excursions recorded in mid-Weichselian cave sediments from Skjonghelleren, Valderoy, Norway. *Phys. Earth Planet. Int.*, 45, 337.

Lund, S. P. and Banerjee, S. K. (1985). Late Quaternary palaeomagnetic field secular variation from two Minnesota lakes. *J. Geophys. Res.*, 90, 803.

Lund, S. P., Liddicoat, J. C., Lajoie, K. R., Henyey, J. L. and Robinson, S. W. (1988). Palaeomagnetic evidence for long term (10^4 year) memory and periodic behaviour in the Earth's core dynamo process. *Geophys. Res. Lett.*, 15, 1101.

MacDougal, I. and Chamalaun, F. H. (1966). Geomagnetic polarity scale of time. *Nature*, 212, 1415.

McFadden, P. L. and McElhinny, M. W. (1982). Variations in the geomagnetic dipole 2: statistical analysis of VDMs for the past 5 million years. *J. Geomag. Geoelect.*, 34, 163.

Manabe, K. I. (1977). Reversed magneto-zone in the late Pleistocene sediments from the Pacific coast of Odaka, northeast Japan. *Quatern. Res.*, 7, 372.

Maenaka, K. (1983). Magnetostratigraphic study on the Osaka group with special reference to the existence of pre- and post-Jaramillo episodes in the late Matuyama Polarity epoch. *Mem. Hanozono Univ.*, 14, 1.

Maenaka, K., Yokoyama, T. and Ishida, S. (1977). Palaemagnetic stratigraphy and biostratigraphy of the Plio-Pleistocene in the Kinki district, Japan. *Quatern. Res.*, 7, 341.

Mankinen, E. A. and Dalrymple, G. B. (1979). Revised geomagnetic polarity timescale for the interval 0–5 Myr BP. *J. Geophys. Res.*, 84, 615.

Mankinen, E. A., Donnelly, J. M. and Grommé, C. S. (1978). Geomagnetic polarity event recorded at 1.1 Myr BP on Cobb Mountain, Clear Lake volcanic field, California. *Geology*, 6, 653.

Mankinen, E. A., Donnelly-Nolan, J. M., Grommé, C. S. and Hearn, B. C. Jr. (1981). Palaeomagnetism of the Clear Lake Volcanics and new limits on the age of the Jaramillo Normal-Polarity event. *U.S. Geol. Surv. Prof. Pap.*, 1141, 67.

Mankinen, E. A. and Grommé, C. S. (1982). Palaeomagnetic data from the Coso Range, California and current status of the Cobb Mountain, normal geomagnetic polarity event. *Geophys. Res. Lett.*, 9, 1279.

Mankinen, E. A., Grommé, C. S., Dalrymple, G. B., Lamphere, M. A. and Bailey, R. (1986). Palaeomagnetism and K–Ar ages of volcanic rocks from Long Valley Caldera, California. *J. Geophys. Res.*, 91, 633.

Marino, R. J. and Ellwood, B. B. (1978). Anomalous magnetic fabric in sediments which record an apparent geomagnetic field excursion. *Nature*, 274, 581.

Marshall, M. A., Chauvin, A. and Bonhommet, N. (1988). Preliminary palaeointensity measurements and detailed magnetic analyses of basalts from the Skalamaelifell excursion, southwest Iceland. *J. Geophys. Res.*, 93, 11, 681.

Mazaud, A., Laj, C., Bard, E., Arnold, M. and Tric, E. (1991). Geomagnetic field control of ^{14}C production over the last 80 ky: implications for the radiocarbon timescale. *Geophys. Res. Lett.*, 18, 1885.

Mörner, N. A. (1976). Palaeomagnetism in deep-sea core A179–15: a reply. *Earth Planet. Sci. Lett.*, 29, 240.

Mörner, N. A. (1977). The Gothenburg magnetic excursion. *Quatern. Res.*, 7, 413.

Mörner, N. A. (1986). Geomagnetic excursions in late Brunhes time, European long core data. *Phys. Earth Planet. Int.*, 44, 47.

Mörner, N. A. and Lanser, J. P. (1974). Gothenburg magnetic 'flip'. Nature, 251, 408.

Mörner, N. A. and Lanser, J. P. (1975). Palaeomagnetism in deep sea core A179–15. Earth Planet. Sci. Lett., 26, 121.

Mörner, N. A., Lanser, J. P. and Hospers, J. (1971). Late Weichselian palaeomagnetic reversal. Nature Phys. Sci., 234, 173.

Nakajima, T., Yaskawa, K., Natsuhara, N., Kawai, N. and Horie, S. (1973). Very short period geomagnetic excursion 18 000 yr BP. Nature Phys. Sci., 244, 8.

Negrini, R. M., Verosub, K. L. and Davis, J. O. (1987). Long term nongeocentric axial dipole directions and a geomagnetic excursion from the Middle Pleistocene sediments of the Humboldt River Canyon, Pershing County, Nevada. J. Geophys. Res., 92, 10, 617.

Negrini, R. M., Verosub, K. L. and Davis, J. O. (1988). The middle to late Pleistocene geomagnetic field recorded in fine-grained sediments from Summer Lake, Oregon and Double Hotsprints, Nevada, USA. Earth Planet. Sci. Lett., 87, 173.

Ninkovich, D., Opdyke, N. D., Heezen, B. C. and Foster, J. H. (1966). Palaeomagnetic stratigraphy, rates of deposition and tephrachronology in North Pacific deep-sea sediments. Earth Planet. Sci. Lett., 1, 476.

Noel, M. and Tarling, D. H. (1975). The Laschamp 'event'. Nature, 253, 705.

Noltimier, H. C. and Colinvaux, P. A. (1976). Geomagnetic excursion from Imuruk Lake, Alaska. Nature, 259, 197.

Oberg, C. J. and Evans, M. E. (1977). Spectral analysis of Quaternary palaeomagnetic data from British Columbia and its bearing on geomagnetic secular variation. Geophys. J. Roy. Astr. Soc., 51, 691.

Opdyke, N. D. (1976). Discussion of paper by Mörner and Lanser concerning the palaeomagnetism of deep sea core A179–15. Earth Planet. Sci. Lett., 29, 238.

Opdyke, N. D., Shackleton, N. J. and Hays, J. D. (1974). The details of a magnetic excursion as seen in a piston core from the Southern Indian Ocean. Trans. Am. Geophys. Union, 55, 237.

Palmer, D. F., Henyey, T. L. and Dodson, R. E. (1979). Palaeomagnetic and sedimentological studies at Lake Tahoe, California, Nevada. Earth Planet. Sci. Lett., 46, 125.

Peirce, J. W. and Clark, M. J. (1978). Evidence from Iceland on geomagnetic reversal during the Wisconsin Ice Age. Nature, 273, 456.

Petrova, G. N. and Pospelova, G. A. (1990). Excursions of the magnetic field during the Brunhes chron. Phys. Earth Planet. Int., 63, 135.

Rais, A., Surmont, J., Gillot, P.-Y., Blanchard, E. and Laj, C. (1991). A volcanic record of the Blake event: preliminary results from Réunion Island (France). EOS Abstracts, Am. Geophys. Union, Fall Meeting, p. 139.

Rampino, M. R. (1981). Revised age estimates of Brunhes palaeomagnetic events: support for a link between geomagnetism and eccentricity. Geophys. Res. Lett., 8, 1047.

Ransom, C. J. (1973). Magnetism and archaeology. Nature, 242, 518.

Rea, D. K. and Blakely, R. J. (1975). Short wave length magnetic anomalies in a region of rapid sea-floor spreading. Nature, 255, 126.

Rieck, H. J., Sarna-Wojcicki, A., Meyer, C. E. and Adam, D. P. (1992). Magnetostratigraphy and teprhochronology of an upper Pliocene to Holocene record in lake sediments at Tulelake, northern California. Geol. Soc. Am. Bull., 104, 409.

Rolph, T. C., Shaw, J., Derbyshire, E. and Wang, J. (1989). A detailed geomagnetic record from Chinese loess. Phys. Earth Planet. Int., 56, 151.

Roperch, P., Bonhommet, N. and Levi, S. (1988). Palaeointensity of the Earth's magnetic field during the Laschamp excursion and its geomagnetic implications. Earth Planet. Sci. Lett., 88, 209.

Ryan, W. B. F. (1972). Stratigraphy of late Quaternary sediments in the eastern Mediterrean. In *The Mediterranean Sea: A Natural Sedimentation Laboratory*, ed D. J. Stanley. Stroudsburg: Dowden, Hutchinson & Ross.

Sasajima, S., Nishimura, S. and Hirooka, K. (1980). Studies on the Blake episode with special emphasis to East Asian results obtained. *Rock Magn. Palaeomagn.*, 7, 90.

Sandgren, P. (1986). Late Weichselian palaeomagnetic secular variation from the Torreberge Basin, South Sweden. *Phys. Earth Planet. Int.*, 43, 160.

Sarna-Wojcicki, A. M., Lajoie, K. R., Meyer, C. E., Adam, D. P. and Rieck, H. J. (1988). Tephrochronological correlation of upper Neogene sediments along the Pacific margin, conterminous United States. In *Non-glacial Quaternary of the United States, Decade N, Am. Geol.*, ed. R. Morrison. Boulder: Geol. Soc. Am.

Shibuya, H., Cassidy, J., Smith, I. E. M. and Itaya, T. (1992). A geomagnetic excursion in the Brunhes epoch recorded in New Zealand basalts. *Earth Planet. Sci. Lett.*, 111, 41.

Smith, D. J. and Foster, J. H. (1969). Geomagnetic reversal in Brunhes normal polarity epoch. *Science*, 163, 565.

Steiner, M. B. (1983). Geomagnetic excursion in the late Cretaceous. *Geophys. J. Roy. Astr. Soc.*, 73, 17.

Sueishi, T., Sato, T., Kawai, N. and Kobayashi, K. (1979). Short geomagnetic episodes in the Matuyama epoch. *Phys. Earth Planet. Int.*, 19, 1.

Sukroo, J. C., Christofel, D. A., Vella, P. and Topping, W. W. (1978). Rejecting evidence of Gothenburg geomagnetic reversal in New Zealand. *Nature*, 271, 650.

Thompson, R. and Berglund, B. (1976). Late Weichselian geomagnetic 'reversal' as a possible example of the reinforcement syndrome. *Nature*, 263, 490.

Thouveny, N. and Creer, K. M. (1992). Geomagnetic excursions in the past 600 ka: ephemeral secular variation features. *Geology*, 20, 399.

Torii, M., Yoshikawa, S. and Itihara, M. (1974). Paleomagnetism in the water laid volcanic ash layers in the Osaka Group, Sennan and Senpoku hills, southwest Japan. *Rock Magn. Paleogeophys.*, 2, 34.

Tric, E., Laj, C., Valet, J. P., Tucholka, P., Paterne, M. and Guichard, F (1991). The Blake geomagnetic event: transition geometry, dynamical characteristics and geomagnetic significance. *Earth Planet. Sci. Lett.*, 102, 1.

Tric, E., Valet, J. P., Tucholka, P., Paterne, M., Labeyrie, L., Guichard, F., Tauxe, L. and Fontugne, M. (1992). Paleointensity of the geomagnetic field during the last 80,000 yrs. *J. Geophys. Res.*, 97, 9337.

Tucholka, P. (1977). Magnetic polarity events in Polish loess profiles. *Biul. Inst. Geol., Warsaw*, 305, 117.

Tucholka, P., Fontugne, M., Gulchard, F. and Paterne, M. (1987). The Blake magnetic polarity episode in cores from the Mediterranean Sea. *Earth Planet. Sci. Lett.*, 86, 320.

Turner, G. M., Evans, M. E. and Hussin, I. B. (1982). A geomagnetic secular variation study (31 000–19 000 BP) in Western Canada. *Geophys. J. Roy. Astr. Soc.*, 71, 159.

Turner, P. and Vaughan, D. J. (1977). Evidence of rapid changes in the Permian geomagnetic field during the Zechstein marine transgression. *Nature*, 270, 593.

Valet, J.-P., Laj, C. and Tucholka, P. (1986). High resolution sedimentary record of a geomagnetic reversal. *Nature*, 322, 27.

Valladas, G., Gillot, P. Y., Poupeau, G. and Reyss, J. L. (1977). 5th Eur. Conf. Geochronology, Pisa.

van Hoof, A. A. M. and Langereis, C. G. (1992). The upper Kaena sedimentary geomagnetic reversal from southern Sicily. *J. Geophys. Res.*, 97, 6941.

Verosub, K. L. (1975). Palaeomagnetic excursions as magnetostragraphic horizons: a cautionary note. *Science*, 190, 48.

Verosub, K. L. (1977a). The absence of the Mono Lake geomagnetic excursion from the palaeomagnetic record of Clear Lake, California. *Earth Planet. Sci. Lett.*, 36, 219.

Verosub, K. L. (1977b). Depositional and post-depositional processes in the magnetisation of sediments. *Rev. Geophys. Space Phys.*, **15**, 129.

Verosub, K. L. and Banerjee, S. K. (1977). Geomagnetic excursions and their palaeomagnetic record. *Rev. Geophys. Space Phys.*, **15**, 145.

Verosub, K. L., Davis, J. O. and Valastro, S. Jr. (1980). A palaeomagnetic record from Pyramid Lake, Nevada, and its implication for proposed geomagnetic excursions. *Earth Planet. Sci. Lett.*, **49**, 141.

Wang Jingtai, Derbyshire, E. and Shaw, J. (1986). Preliminary magnetostratigraphy of Dabusan Lake, and Qaidam basin, Central Asia. *Phys. Earth Planet. Int.*, **44**, 41.

Watkins, N. D. (1968). Short period geomagnetic polarity events in deep-sea sedimentary cores. *Earth Planet. Sci. Lett.*, **4**, 341.

Watkins, N. D. (1972). Review of the development of the geomagnetic polarity timescale and discussion of prospects for its finer definition. *Geol. Soc. Am. Bull.*, **83**, 551.

Westgate, J. (1988). Isothermal, plateau fission track age of the late Pleistocene old Crow Tephra Alaska. *Geophys. Res. Lett.*, **15**, 376.

Westgate, J. A., Walter, R. C., Pearce, G. W. and Gorton, M. P. (1985). Distribution, stratigraphy, petrochemistry, and palaeomagnetism of the late Pleistocene, Old Crow Tephra in Alaska and the Yukon. *Can. J. Earth Sci.*, **22**, 893.

Wilson, R. L., Dagley, P. and McCormack, A. G. (1972). Palaeomagnetic evidence about the source of the geomagnetic field. *Geophys. J. Roy. Astr. Soc.*, **28**, 213.

Wilson, D. S. and Hey, R. N. (1981). The Galapagos axial magnetic anomaly: evidence for the Emperor reversal within the Brunhes and for a two-layer magnetic source. *Geophys. Res. Lett.*, **8**, 1051.

Wintle, A. G. (1973). Anomalous fading of thermoluminescence in mineral samples. *Nature*, **245**, 143.

Wintle, A. G. (1977). Thermoluminescence dating of minerals – traps for the unwary. *J. Electrostatics*, **3**, 281.

Wollin, G., Ericson, D. B., Ryan, W. B. F. and Foster, J. M. (1971). Magnetism of the Earth and climate changes. *Earth Planet. Sci. Lett.*, **12**, 175.

Yaskawa, K. (1974). Reversals, excursions and secular variations of the geomagnetic field in the Brunhes normal polarity epoch. In *Palaeolimnology of Lake Biwa and the Japanese Pleistocene*, ed. S. Horie, Vol. 2, p. 77. Otsu: Kyoto University.

Yaskawa, K., Nakajima, T., Kawai, N., Torii, M., Notsuhara, N. and Horie, S. (1973). Palaeomagnetism of a core from Lake Biwa (I). *J. Geomag. Geoelect.*, **25**, 447.

Zhu, R. X., Zhou, L. P., Laj, C., Mazaud, A. and Ding, Z. L. (1994). The Blake geomagnetic polarity episode recorded in Chinese loess. *Geophys. Res. Lett.* **21**, 697.

Chapter 5

Aldridge, K. D. and Jacobs, J. A. (1974). Mortality curves for normal and reversed polarity intervals of the Earth's magnetic field. *J. Geophys. Res.*, **79**, 4944.

Allan, D. W. (1958). Reversals of the Earth's magnetic field. *Nature*, **182**, 469.

Allan, D. W. (1962). On the behaviour of systems of coupled dynamos. *Proc. Camb. Phil. Soc.*, **58**, 671.

Blakely, R. J. (1974). Geomagnetic reversals and crustal spreading rates during the Miocene. *J. Geophys. Res.*, **79**, 2979.

Blakely, R. J. and Cox, A. (1972). Evidence for short geomagnetic polarity intervals in the early Cenozoic. *J. Geophys. Res.*, **77**, 7065.

Bloxham, J. and Gubbins, D. (1985). The secular variation of the Earth's magnetic field. *Nature*, **317**, 777.

Bloxham, J. and Gubbins, D. (1987). Thermal core–mantle interactions. *Nature*, **325**, 54.

Bochev, A. (1969). Two and three dipoles approximating the Earth's main magnetic field. *Pure Appl. Geophys.*, **74**, 29.

Bogue, S. W. and Coe, R. S. (1982). Successive palaeomagnetic reversal records from Kauai. *Nature*, **295**, 399.

Bogue, S. W. and Coe, R. S. (1984). Transitional palaeointensities from Kauai, Hawaii, and geomagnetic reversal models. *J. Geophys. Res.*, **89**, 10341.

Brock, A. (1971). An experimental study of palaeosecular variation. *Geophys. J. Roy. Astr. Soc.*, **24**, 303.

Bullard, E. C. (1955). The stability of a homopolar dynamo. *Proc. Camb. Phil. Soc.*, **51**, 744.

Bullard, E. C. (1978). The disk dynamo. In *Topics in Non-linear Dynamics*, ed. S. Jorna. AIP Conf. Proc. No 46.

Bullard, E. C. and Gellman, H. (1954). Homogeneous dynamos and terrestrial magnetism. *Phil. Trans. Roy. Soc. A*, **247**, 213.

Busse, F. H. (1975). A model of the geodynamo. *Geophys. J. Roy. Astr. Soc.*, **42**, 437.

Cande, S. C. and Kent, D. V. (1992). Ultrahigh resolution marine magnetic anomaly profiles: a record of continuous palaeointensity variations? *J. Geophys. Res.*, **97**, 15, 075.

Cande, S. C. and LaBrecque, J. L. (1974). Behaviour of the Earth's palaeomagnetic field from small scale marine magnetic anomalies. *Nature*, **247**, 26.

Chandrasekhar, S. (1961). *Hydrodynamic and Hydromagnetic Stability*, Ch. v. Oxford: Clarendon Press.

Chauvin, A., Roperch, P. and Duncan, R. A. (1990). Records of geomagnetic reversals from volcanic islands of French Polynesia. 2. Paleomagnetic study of a flow sequence (1.2–0.6 Ma) from the island of Tahiti and discussion of reversal models. *J. Geophys. Res.*, **95**, 2727.

Cook, A. E. and Roberts, P. H. (1970). The Rikitake two disc dynamo system. *Proc. Camb. Phil. Soc.*, **68**, 547.

Clement, B. M. (1992). Evidence for dipolar fields during the Cobb Mountain geomagnetic polarity reversals. *Nature*, **358**, 405.

Clement, B. M. (1987). Palaeomagnetic evidence of reversals resulting from helicity fluctuations in a turbulent core. *J. Geophys. Res.*, **92**, 10629.

Clement, B. M. and Kent, D. V. (1984). A detailed record of the lower Jaramillo polarity transition from a southern hemisphere, deep-sea sediment core. *J. Geophys. Res.*, **89**, 1049.

Constable, C. (1990). A simple statistical model for geomagnetic reversals. *J. Geophys. Res.*, **95**, 4587.

Constable, C. G. and Parker, R. L. (1988). Statistics of the geomagnetic secular variation for the past 5 m.y. *J. Geophys. Res.*, **93**, 11, 569.

Courtillot, V. and Besse, J. (1987). Magnetic field reversals, polar wander and core-mantle coupling. *Science*, **237**, 1140.

Cox, A. (1968). Length of geomagnetic polarity intervals. *J. Geophys. Res.*, **73**, 3247.

Cox, A. (1969). Geomagnetic reversals. *Science*, **163**, 237.

Cox, A. (1970). Reconciliation of statistical models for reversals. *J. Geophys. Res.*, **75**, 7501.

Cox, A. (1975a). The frequency of geomagnetic reversals and the symmetry of the non-dipole field. *Rev. Geophys. Space Phys.*, **13**, 35.

Cox, A. (1975b). Symmetric and asymmetric geomagnetic reversals as a renewal process. In *Proc. Takesi Nagata Conference: Magnetic Fields, Past and Present*. Pittsburgh: University of Pittsburgh.

Cox, A. (1981). A stochastic approach towards understanding the frequency and polarity bias of geomagnetic reversals. *Phys. Earth Planet. Int.*, **24**, 178.

Crossley, D., Jensen, O. and Jacobs, J. (1986). The stochastic excitation of reversals in simple dynamos. *Phys. Earth Planet. Int.*, **42**, 143.

Creer, K. M. and Ispir, Y. (1970). An interpretation of the behaviour of the geomagnetic field during polarity transitions. *Phys. Earth Planet. Int.*, **2**, 283.

Dagley, P. and Lawley, E. (1974). Palaeomagnetic evidence for the transitional behaviour of the geomagnetic field. *Geophys. J. Roy. Astr. Soc.*, **36**, 577.

Dodson, R., Dunn, J. R., Fuller, M., Williams, I., Ito, H., Schmidt, V. A. and Wu Yu, M. (1978). Paleomagnetic record of a late Tertiary field reversal. *Geophys. J. Roy. Astr. Soc.*, **53**, 373.

Fuller, M., Williams, K. and Hoffman, K. A. (1979). Paleomagnetic records of geomagnetic field reversals and the morphology of the transitional fields. *Rev. Geophys. Space Phys.*, **17**, 179.

Gaffin, S. (1989). Analysis of scaling in the geomagnetic polarity reversal record. *Phys. Earth Planet. Int.*, **57**, 284.

Gubbins, D. (1975). Numerical solution of the hydromagnetic dynamo problem. *Geophys. J. Roy. Astr. Soc.*, **42**, 295.

Gubbins, D. (1987). Mechanism for geomagnetic polarity reversals. *Nature*, **326**, 167.

Gubbins, D. and Bloxham, J. (1987). Morphology of the geomagnetic field and implications for the geodynamo. *Nature*, **325**, 509.

Gubbins, D. and Masters, T. G. (1979). Driving mechanisms for the Earth's dynamo. *Adv. Geophys.*, **21**, 1.

Gubbins, D. and Zhang, K. (1993). Symmetry properties of the dynamo equations for palaomagnetism and geomagnetism. *Phys. Earth Planet. Int.*, **75**, 225.

Harland, W. B., Cox, A., Llewellyn, P. G., Pichton, C. A. S., Smith, A. G. and Walters, R. (1983). *A Geologic Time Scale*. Cambridge: Cambridge University Press.

Harrison, C. G. A. (1969). What is the true rate of reversals of the Earth's magnetic field? *Earth Planet. Sci. Lett.*, **6**, 186.

Heirtzler, J. R., Dickson, G. O., Herron, E. M., Pitman, W. C. and Le Pichon, X. (1968). Marine magnetic anomalies, geomagnetic field reversals and motions of the ocean floor and continents. *J. Geophys. Res.*, **73**, 2119.

Herrero-Bervera, E. and Theyer, F. (1986). Non axisymmetric behaviour of Olduvai and Jaramillo polarity transitions recorded in north-central Pacific deep-sea sediments. *Nature*, **322**, 159.

Herzenberg, A. (1958). Geomagnetic dynamos. *Phil. Trans. Roy. Soc. A*, **250**, 543.

Hillhouse, J. and Cox, A. (1976). Brunhes–Matuyama polarity transition. *Earth Planet. Sci. Lett.*, **29**, 51.

Hoffman, K. A. (1977). Polarity transition records and the geomagnetic dynamo. *Science*, **196**, 1329.

Hoffman, K. A. (1979). Behaviour of the geodynamo during reversal: a phenomenological model. *Earth Planet. Sci. Lett.*, **44**, 7.

Hoffman, K. A. (1981). Quantitative description of the geomagnetic field during the Matuyama–Brunhes polarity transition. *Phys. Earth Planet. Int.*, **24**, 229.

Hoffman, K. A. (1982). The testing of geomagnetic reversal models: recent development. *Phil. Trans. Roy. Soc. A*, **306**, 147.

Hoffman, K. A. (1986). Transitional field behaviour from southern hemisphere lavas: evidence for two-stage reversals of the geodynamo. *Nature*, **320**, 228.

Hoffman, K. A. (1991). Long-lived transitional states of the geomagnetic field and the two dynamo families. *Nature*, **354**, 273.

Hoffman, K. A. and Fuller, M. (1978). Transitional field configuration and geomagnetic reversal. *Nature*, **273**, 715.

Honkura, Y. and Matsushima, M. (1988). Fluctuation of the non-dipole magnetic field and its implication for the process of geomagnetic polarity reversal in the Cox model. *J. Geophys. Res.*, **93**, 11,631.

Hoshi, M. and Kono, M. (1988). Rikitake two-disc dynamo system: statistical properties and growth of instability. *J. Geophys. Res.*, **93**, 11,643.

Huppert, H. E. and Moore, D. R. (1976). Non-linear double diffusive convection. *J. Fluid Mech.*, **78**, 821.

Irving, E. and Pullaiah, G. (1976). Reversals of the geomagnetic field, magnetostratigraphy and relative magnitude of palaeosecular variation in the Phanerozoic. *Earth Sci. Rev.*, **12**, 35.

Ito, K. (1980). Chaos in the Rikitake two-disc dynamo system. *Earth Planet. Sci. Lett.*, **51**, 451.

Ito, H. M. (1988). Stochastic disc dynamo as a model of reversals of the Earth's magnetic field. *J. Statist. Phys.*, **53**, 19.

Kerridge, D. J. and Wilkinson, I. (1983). Spectral analyses of the reversal waveforms of the magnetic field produced by an experimental homogeneous dynamo. *Geophys. J. Roy. Astr. Soc.*, **72**, 310.

Kono, M. (1972). Mathematical models of the Earth's magnetic field. *Phys. Earth Planet. Int.*, **5**, 140.

Kono, M. (1987). Rikitake two disc dynamo and palaeomagnetism. *Geophys. Res. Lett.*, **14**, 21.

Kristjansson, L. and McDougall, I. (1982). Some aspects of the late Tertiary geomagnetic field in Iceland. *Geophys. J. Roy. Astr. Soc.*, **68**, 273.

Kropachev, E. P. (1971a). Numerical solutions of the equations of the dynamo theory of terrestrial magnetism. 1. Method. *Geomagn. Aeron.*, **11**, 585.

Kropachev, E. P. (1971b). Numerical solutions of the equations of the dynamo theory of terrestrial magnetism. 2: Results. *Geomag. Aeron.*, **11**, 737.

LaBrecque, J. L., Kent, D. V. and Cande, S. C. (1977). Revised magnetic polarity timescale for late Cretaceous and Cenozoic time. *Geology*, **5**, 330.

Laj, C., Guitton, S., Kissel, C. and Mazaud, A. (1988). Complex behaviour of the geomagnetic field during three successive polarity reversals, 11–12 m.y. B. P. *J. Geophys. Res.*, **93**, 11,655.

Laj, C., Mazaud, A., Weeks, R., Fuller, M. and Herrero-Bervera, E. (1991). Geomagnetic reversal paths. *Nature*, **351**, 447.

Laj, C., Nordemann, D. and Pomeau, Y. (1979). Correlation function analysis of geomagnetic field reversals. *J. Geophys. Res.*, **84**, 4511.

Lebovitz, N. R. (1969). The equilibrium stability of a system of disc dynamos. *Proc. Camb. Phil. Soc.*, **56**, 154.

Lee, S. and Lilley, F. E. M. (1986). On palaeomagnetic data and dynamo theory. *J. Geomag. Geoelect.*, **38**, 797.

Levy, E. H. (1972a). Effectiveness of cyclonic convection for producing the geomagnetic field. *Astrophys. J.*, **171**, 621.

Levy, E. H. (1972b). Kinematic reversal schemes for the geomagnetic dipole. *Astrophys. J.*, **171**, 635.

Levy, E. H. (1972c). On the state of the geomagnetic field and its reversals. *Astrophys. J.*, **175**, 573.

Lilley, F. E. M. (1970a). On kinematic dynamos. *Proc. Roy. Soc. A*, **316**, 153.

Lilley, F. E. M. (1970b). Geomagnetic reversals and the position of the North magnetic pole. *Nature*, **227**, 1336.

Loper, D. (1978). The gravitationally powered dynamo. *Geophys. J. Roy. Astr. Soc.*, **54**, 389.

Lorenz, E. N. (1963). Deterministic non-periodic flow. *J. Atmos. Sci.*, **20**, 130.

Lowes, F. J. and Wilkinson, I. (1963). Geomagnetic dynamo: a laboratory model. *Nature*, **198**, 1158.

Lowrie, W. and Kent, D. V. (1983). Geomagnetic reversal frequency since the late Cretaceous. *Earth Planet. Sci. Lett.*, **62**, 305.

Lutz, T. M. and Watson, G. S. (1988). Effects of long-term variation on the frequency spectrum of the geomagnetic reversal record. *Nature*, **334**, 240.

McFadden, P. L. (1984). Statistical tools for the analysis of geomagnetic reversal sequences. *J. Geophys. Res.*, **89**, 3363.

McFadden, P. L. and McElhinny, M. W. (1982). Variations in the geomagnetic dipole. 2. Statistical analysis of VDMs for the past 5 million years. *J. Geomag. Geoelect.*, **34**, 163.

McFadden, P. L. and Merrill, R. T. (1984). Lower mantle convection and geomagnetism. *J. Geophys. Res.*, **89**, 3354.

McFadden, P. L. and Merrill, R. T. (1986). Geodynamo energy source constraints from palaeomagnetic data. *Phys. Earth Planet. Int.*, **43**, 22.

McFadden, P. L. and Merrill, R. T. (1993). Inhibition and geomagnetic field reversals. *J. Geophys. Res.*, **98**, 6189.

McFadden, P. L., Merrill, R. T., Lowrie, W. and Kent, D. V. (1987). The relative stabilities of the reverse and normal polarity states of the Earth's magnetic field. *Earth Planet. Sci. Lett.*, **82**, 373.

McFadden, P. L., Merrill, R. T. and McElhinny, M. W. (1985). Non-linear processes in the geodynamo: palaeomagnetic evidence. *Geophys. J. Roy. Astr. Soc.*, **83**, 111.

McFadden, P. L., Merrill, R. T., McElhinny, M. W. and Lee, S. (1991). Reversals of the Earth's magnetic field and temporal variations of the dynamo families. *J. Geophys. Res.*, **96**, 3923.

Malkus, W. V. R. (1972). Reversing Bullard's dynamo. *Trans. Am. Geophys. Union*, **53**, 617.

Mandelbrot, B. B. (1983). *The Fractal Geometry of Nature.* San Francisco: Freeman.

Mankinen, E. A., Prévot, M., Grommé, C. S. and Coe, R. S. (1985). The Steens Mountain (Oregon) geomagnetic polarity transition. 1. Directional history, duration of episodes, and rock magnetism. *J. Geophys. Res.*, **90**, 10,393.

Marzocchi, W. and Mulargia, F. (1990). Statistical analysis of the geomagnetic reversal sequences. *Phys. Earth Planet. Int.*, **61**, 149.

Mathews, J. H. and Gardner, W. K. (1963). Field reversals of 'Palaeomagnetic' type in coupled disk dynamos. *US Naval Res. Lab. Rep.* **5886**.

Mazaud, A. and Laj, C. (1989). Simulation of geomagnetic polarity reversals by a model of interacting dipole sources. *Earth Planet. Sci. Lett.*, **92**, 299.

Merrill, R. T., McElhinny, M. W. and Stevenson, D. J. (1979). Evidence for long term asymmetries in the Earth's magnetic field and possible implications for dynamo theories. *Phys. Earth Planet. Int.*, **20**, 75.

Merrill, R. T. and McFadden, P. L. (1988). Secular variation and the origin of geomagnetic field reversals. *J. Geophys. Res.*, **93**, 11,589.

Merrill, R. T. and McFadden, P. L. (1990). Paleomagnetism and the nature of the geodynamo. *Science*, **248**, 345.

Merrill, R. T., McFadden, P. L. and McElhinny, M. W. (1990). Paleomagnetic tomography of the core mantle boundary. *Phys. Earth Planet. Int.*, **64**, 87.

Muller, R. A. and Morris, D. E. (1986). Geomagnetic reversals from impacts on the Earth. *Geophys. Res. Lett.*, **13**, 1177.

Nagata, T. (1969). Length of geomagnetic polarity intervals. *J. Geomag. Geoelect.*, **21**, 701.

Naidu, P. S. (1971). Statistical structure of geomagnetic field reversals. *J. Geophys. Res.*, **76**, 2649.

Naidu, P. S. (1974). Are geomagnetic field reversals independent? *J. Geomag. Geoelect.*, **26**, 101.

Naidu, P. S. (1975). Second-order statistical structure of geomagnetic field reversals. *J. Geophys. Res.*, **80**, 803.

Naidu, P. S. (1976). Comment on 'Second-order statistical structure of geomagnetic field reversals' by P. S. Naidu; by T. J. Ulrych and R. W. Clayton: Reply. *J. Geophys. Res.*, **81**, 1034.

Nozières, P. (1978). Reversals of the Earth's magnetic field: an attempt at a relaxation model. *Phys. Earth Planet. Int.*, **17**, 55.

Olson, P. (1983). Geomagnetic polarity reversals in a turbulent core. *Phys. Earth Planet. Int.*, **33**, 260.

Olson, P. and Hagee, V. L. (1990). Geomagnetic polarity reversals, transition field structure and convection in the outer core. *J. Geophys. Res.*, **95**, 4609.

Opdyke, N. D., Kent, D. V. and Lowrie, W. (1973). Details of magnetic polarity transitions recorded in a high-deposition rate deep-sea core. *Earth Planet. Sci. Lett.*, **20**, 315.

Pal, P. C. and Roberts, P. H. (1988). Long-term polarity stability and strength of the geomagnetic dipole. *Nature*, **331**, 702.

Parker, E. N. (1969). The occasional reversal of the geomagnetic field. *Astrophys. J.*, **158**, 815.

Phillips, J. D. (1977). Time variation and asymmetry in the statistics of geomagnetic reversal sequences. *J. Geophys. Res.*, **82**, 835.

Phillips, J. D., Blakely, R. J. and Cox, A. (1975). Independence of geomagnetic polarity intervals. *Geophys. J. Roy. Astr. Soc.*, **43**, 747.

Phillips, J. D. and Cox, A. (1976). Spectral analysis of geomagnetic reversal timescales. *Geophys. J. Roy. Astr. Soc.*, **45**, 19.

Prévot, M., Derder, M. E., McWilliams, M. and Thompson, J. (1990). Intensity of the Earth's magnetic field: evidence for a Mesozoic dipole low. *Earth Planet. Sci. Lett.*, **97**, 129.

Rikitake, T. (1958). Oscillations of a system of disk dynamo. *Proc. Camb. Phil. Soc.*, **54**, 89.

Robbins, K. G. (1977). A new approach to sub-critical instability and turbulent transitions in a simple dynamo. *Math. Proc. Camb. Phil. Soc.*, **82**, 309.

Roberts, P. H. (1978). Diffusive instabilities in magnetohydrodynamic convection. In *Les instabilités hydrodynamiques en convection libré, forces et mixté* (*Lecture Notes in Physics*, Vol. 72), ed. J. C. Legras and J. K. Platten.

Roberts, P. and Stix, M. (1972). α-effect dynamos, by the Bullard–Gellman formalism. *Astron. Astrophys.*, **18**, 453.

Schloessin, H. H. and Jacobs, J. A. (1980). Dynamics of a fluid core with inward growing boundaries. *Can. J. Earth Sci.*, **17**, 72.

Schneider, D. A. and Kent, D. V. (1988). The palaeomagnetic field from equatorial deep-sea sediments: axial symmetry and polarity asymmetry. *Science*, **242**, 252.

Schouten, H. and Denham, C. R. (1979). Modelling the oceanic magnetic source layer. In *Deep Drilling in the Atlantic Ocean: Ocean Crust*, ed. H. Talwani, C. G. Harrison and D. E. Hayes, p. 151. AGU Maurice Ewing Series, Vol. 2.

Seki, M. and Ito, K. (1993). A phase transition model for geomagnetic polarity reversals. *J. Geomag. Geoelect.*, **45**, 79.

Sherwood, G. J. and Shaw, J. (1991). The relationship between magnetic field strength and reversal frequency: preliminary palaeointensity results from the Cretaceous Quiet Zone. *Studia Geophys. Geod.*, **35**, 325.

Sherwood, G. J., Shaw, J., Baer, G. and Mallik, S. B. (1993). The strength of the geomagnetic field during the Cretaceous Quiet Zone: Palaeointensity results from Israeli and Indian lavas. *J. Geomag. Geoelect.*, **45**, 339.

Shimizu, M. and Honkura, Y. (1985). Statistical nature of polarity reversals of the magnetic field in coupled-disc dynamo models. *J. Geomag. Geoelect.*, **37**, 455.

Sommerville, R. C. J. (1967). In *Woods Hole Oceanographic Inst. Rep.* No 67–54, Vol. 2, p. 132.

Sparrow, C. (1982). The Lorentz equations: Bifurcations, chaos and strange attractors. In *Applied Mathematical Sciences*, Vol. 41. Berlin: Springer-Verlag.

Stevenson, A. F. and Wolfson, S. J. (1966). Calculations of the dynamo problem of the Earth's magnetic field. *J. Geophys. Res.*, **71**, 4446.

Tacier, J.-D., Switzer, P. and Cox, A. (1975). A model relating undetected geomagnetic polarity intervals to the observed rate of reversals. *J. Geophys. Res.*, **880**, 4446.

Theyer, F., Herrero-Bervero, E., Hsu, V. and Hammond, S. R. (1985). The zonal harmonic model of polarity transitions: a test using successive reversals. *J. Geophys. Res.*, **90**, 1963.

Tritton, D. J. (1989). Deterministic chaos, geomagnetic reversals, and the spherical pendulum. In *Geomagnetism and Palaeomagnetism*, ed. F. J. Lowes *et al.* Dordrecht: Kluwer.

Ulrych, T. J. and Clayton, R. W. (1976). Comment on 'Second-order statistical structure of geomagnetic field reversals' by P. S. Naidu. *J. Geophys. Res.*, **81**, 1033.

Valet, J.-P. and Laj, C. (1981). Palaeomagnetic record of the successive Miocene geomagnetic reversals in western Crete. *Earth Planet. Sci. Lett.*, **54**, 53.

Valet, J.-P. and Laj, C. (1984). Invariant and changing transitional field configurations in a sequence of geomagnetic reversals. *Nature*, **311**, 552.

Verosub, K. L. (1975a). Alternative to the geomagnetic self-reversing dynamo. *Nature*, **253**, 707.

Verosub, K. L. (1975b). A method of determining a generalised representation of geomagnetic field sources. *Geophys. J. Roy. Astr. Soc.*, **41**, 127.

Watanabe, H. (1981). Non-steady state of a hydromagnetic $\alpha\omega$-dynamo and its application to the geomagnetic reversals. *J. Geomag. Geoelect.*, **33**, 531.

Williams, I. S. and Fuller, M. (1981a). A far-sided R→N VGP path from a reversal recorded in the Agno batholith. *Trans. Am. Geophys. Union*, **62**, 853.

Williams, I. S. and Fuller, M. (1981b). Zonal harmonic models of reversal transition fields. *J. Geophys. Res.*, **86**, 11, 657.

Williams, I. S. and Fuller, M. (1982). A Miocene polarity transition (R→N) from the Agno batholith, Luzon. *J. Geophys. Res.*, **87**, 9408.

Williams, I., Weeks, R. and Fuller, M. (1988). A model for transition fields during geomagnetic reversals. *Nature*, **332**, 719.

Wilson, R. L. (1972). Palaeomagnetic differences between normal and reversed field sources, and the problem of far-sided and right-handed pole positions. *Geophys. J. Roy. Astr. Soc.*, **28**, 295.

Zhang, K. K. and Busse, F. H. (1988). Finite amplitude convection and magnetic field generation in a rotating spherical shell. *Geophys. Astrophys. Fluid Dyn.*, **44**, 33.

Zhang, K. K. and Busse, F. H. (1989). Convection driven magnetohydrodynamic dynamos in rotating spherical shells. *Geophys. Astrophys. Fluid Dyn.*, **49**, 97.

Chapter 6

Athanassopoulos, J., Fuller, M., Weeks, R. and Williams, I. S. (1993). Atlas of reversal records. *Trans. Am. Geophys. Union* EθS, **74**(16), 109.

Bloxham, J. and Gubbins, D. (1985). The secular variation of the Earth's magnetic field. *Nature*, **317**, 777.

Bloxham, J. and Jackson, A. (1991). Fluid flow near the surface of Earth's outer core. *Rev. Geophys.*, **29**, 97.

Bogue, S. W. and Coe, R. S. (1982). Successive palaeomagnetic reversal records from Kauai. *Nature*, **295**, 399.

Bogue, S. W. and Coe, R. S. (1984). Transitional palaeointensities from Kauai Hawaii and geomagnetic reversal models. *J. Geophys. Res.*, **89**, 10341.

Braginsky, S. J. (1964). Kinematic models of the Earth's hydromagnetic dynamo. *Geomag. Aeron.*, **4**, 572.

Chauvin, A., Roperch, P. and Duncan, R. A. (1990). Records of geomagnetic reversals from volcanic islands of French Polynesia. 2. Palaeomagnetic study of a flow sequence (1.2–0.6 Ma) from the island of Tahiti and discussion of reversal models. *J. Geophys. Res.*, **95**, 2727.

Clement, B. M. (1991). Geographical distribution of transitional VGPs: evidence for non-zonal equatorial symmetry during the Matuyama–Brunhes geomagnetic reversal. *Earth Planet. Sci. Lett.*, **104**, 48.

Clement, B. M. (1992). Evidence for dipolar fields during the Cobb Mountain geomagnetic polarity reversals. *Nature*, **358**, 405.

Clement, B. M. and Kent, D. V. (1985). A comparison of two sequential geomagnetic polarity transitions (upper Olduvai and Lower Jaramillo) from the Southern Hemisphere. *Phys. Earth Planet. Int.*, **39**, 301.

Clement, B. M. and Kent, D. V. (1986). Geomagnetic polarity transition records from five hydraulic piston core sites in the North Atlantic. Init. Rept. DSDP, Vol. 94. US Govt Printing Office, Washington, D.C., p. 831.

Clement, B. M. and Kent, D. V. (1987). Short polarity intervals within the Matuyama: Transitional field records from hydraulic piston core sites in the North Atlantic. *Earth Planet. Sci. Lett.*, **81**, 253.

Clement, B. M. and Kent, D. V. (1991). A southern hemispheric record of the Matuyama–Brunhes polarity reversal. *Geophys. Res. Lett.*, **18**, 81.

Clement, B. M. and Martinson, D. G. (1992). A quantitative comparison of two palaeomagnetic records of the Cobb Mountain subchron from North Atlantic deep-sea sediments. *J. Geophys. Res.*, **97**, 1735.

Coe, R. S., Hsu, V. and Theyer, F. (1985) *Trans. Am. Geophys. Union E0S*, **66**, 872 (Abstract).

Coe, R. S. and Prévot, M. (1989). Evidence suggesting extremely rapid field variation during a geomagnetic reversal. *Earth Planet. Sci. Lett.*, **92**, 292.

Constable, C. G. (1990). Simple statistical model for geomagnetic reversals. *J. Geophys. Res.*, **95**, 4587.

Constable C. (1992). Link between geomagnetic reversal paths and secular variation of the field over the past 5 Myr. *Nature*, **358**, 230.

Courtillot, V., Le Mouel, J.-L. and Ducruix, J. (1984). On Backus' mantle filter theory and the 1969 geomagnetic impulse. *Geophys. J. Roy. Astr. Soc.*, **78**, 619.

Courtillot, V., Valet, J.-P., Hulot, G. and Le Mouel, J.-L. (1992). The Earth's magnetic field: which geometry? *Trans. Am. Geophys. Union E0S* **73**, 337.

Creer, K. M. and Tucholka, P. (1982). Secular variation as recorded in lake sediments: a discussion of North American and European results. *Phil. Trans. Roy. Soc. A*, **306**, 87.

Cuong, P. G. and Busse, F. H. (1981). Generation of magnetic fields by convection in a rotating sphere I. *Phys. Earth Planet. Int.*, **24**, 272.

Dodson, R., Dunn, J. R., Fuller, M., Williams, I., Ito, H., Schmidt, V. A. and Wu Yu, M. (1978). Palaeomagnetic record of a late Tertiary field reversal. *Geophys. J. Roy. Astr. Soc.*, **53**, 373.

Dziewonski, A. M. and Woodhouse, J. H. (1987). Global images of the Earth's interior. *Science*, **236**, 37.

Egbert, G. (1992). Sampling bias in VGP longitudes. *Geophys. Res. Lett.*, **19**, 2353.

Grommé, C. S., Mankinen, E. A., Prévot, M. and Coe, R. S. (1985). Steens Mountain geomagnetic polarity transition is a single phenomenon. *Nature*, **318**, 487.

Gubbins, D. (1987). Mechanism for geomagnetic polarity reversals. *Nature*, **326**, 167.

Gubbins, D. and Bloxham, J. (1987). Morphology of the geomagnetic field and implications for the geodynamo. *Nature*, **325**, 509.

Gubbins, D. and Coe, R. S. (1993). Longitudinally confined geomagnetic reversal paths from non-dipolar transition fields. *Nature*, **362**, 51.

Gubbins, D. and Roberts, N. (1983). Use of the frozen flux approximation in the interpretation of archaeomagnetic and palaeomagnetic data. *Geophys. J. Roy. Astr. Soc.*, **73**, 675.

Helsley, C. E., Herrero-Bervera, E., Keating, B., Fuller, M. and Laj, C. (1989). Reliability of deep-sea polarity transition records. *Trans. Am. Geophys. Union* EθS, **70**, 1072.

Herrero-Bervera, E., Helsley, C. E., Hammond, S. R. and Chitwood, L. A. (1989). A possible palaeomagnetic record of the Blake episode from Pringle Falls, Oregon, USA. *Phys. Earth Planet. Int.*, **56**, 112.

Herrero-Bervera, E. and Khan, M. A. (1992). Olduvai termination: detailed palaeomagnetic analysis of a north central Pacific core. *Geophys. J. Int.*, **108**, 535.

Herrero-Bervera, E. and Theyer, F. (1986). Non-axisymmetric behaviour of Olduvai and Jaramillo polarity transitions recorded in north-central Pacific deep-sea sediments. *Nature*, **322**, 159.

Herrero-Bervera, E., Theyer, F. and Helsley, C. E. (1987). Olduvai onset polarity transition: two detailed palaeomagnetic records from North Central Pacific sediments. *Phys. Earth Planet. Int.*, **49**, 325.

Hide, R. (1982). On the role of rotation in the generation of magnetic fields by fluid motions. *Phil. Trans. Roy. Soc. A*, **306**, 223.

Hillhouse, J. W. and Cox, A. (1976). Brunhes–Matuyama polarity transition. *Earth Planet. Sci. Lett.*, **29**, 51.

Hoffman, K. A. (1977). Polarity transition records and the geomagnetic dynamo. *Science*, **196**, 1329.

Hoffman, K. A. (1979). Behaviour of the geodynamo during reversals: a phenomenological model. *Earth Planet. Sci. Lett.*, **44**, 7.

Hoffman, K. A. (1981). Quantitative description of the geomagnetic field during the Matuyama–Brunhes transition. *Phys. Earth Planet. Int.*, **24**, 229.

Hoffman, K. A. (1982). The testing of geomagnetic reversal models: recent developments. *Phil. Trans. Roy. Soc. A*, **306**, 147.

Hoffman, K. A. (1986). Transitional field behaviour from southern hemisphere lavas: evidence for two-stage reversals of the geodynamo. *Nature*, **320**, 228.

Hoffman, K. A. (1991). Long-lived transitional states of the geomagnetic field and the two dynamo families. *Nature*, **354**, 273.

Hoffman, K. A. (1992). Dipolar reversal states of the geomagnetic field and core–mantle dynamics. *Nature*, **359**, 789.

Karlin, R., Lyle, M. and Heath, G. R. (1987). Authigenic magnetite formation in suboxic marine sediments. *Nature*, **326**, 490.

Keating, B. H. and Ishimatsu, J. (1992). Brunhes–Matuyama and Jaramillo polarity transitions in high latitude southern hemisphere sediments. *Suppl. EθS, Am. Geophys. Union, Fall Meeting*, p. 148.

Laj, C., Guitton, S. and Kissel, C. (1987). Rapid changes and near stationarity of the geomagnetic field during a polarity reversal. *Nature*, **330**, 145.

Laj, C., Guitton, S., Kissel, C. and Mazaud, A. (1988). Complex behaviour of the geomagnetic field during three successive polarity reversals 11–12 my. B.P. *J. Geophys. Res.*, **93**, 11, 655.

Laj, C., Mazaud, A., Weeks, R., Fuller, M. and Herrero-Bervera, E. (1991). Geomagnetic reversal paths. *Nature*, **351**, 447.

Laj, C., Mazaud, A., Weeks, R., Fuller, M. and Herrero-Bervera, E. (1992a). Geomagnetic reversal paths. *Nature*, **359**, 111.

Laj, C., Mazaud, A., Weeks, R., Fuller, M. and Herrero-Bervera, E. (1992b). Statistical assessment of the preferred longitudinal bands for recent geomagnetic reversal records. *Geophys. Res. Lett.*, **19**, 2003.

Langereis, C. G., van Hoof, A. A. M. and Rochette, P. (1992). Longitudinal confinement of geomagnetic reversal paths as a possible sedimentary artefact. *Nature*, **358**, 226.

Larson, E. E., Watson, D. E. and Jennings, W. (1971). Regional comparison of a

Miocene geomagnetic transition in Oregon and Nevada. *Earth Planet. Sci. Lett.*, 11, 391.

Le Mouel, J.-L. (1984). Outer core geostrophic flow and secular variation of Earth's geomagnetic field. *Nature*, 311, 734.

Liddicoat, J. C. (1982). Gauss/Matuyama polarity transition. *Phil. Trans. Roy. Soc. A*, 306, 121.

Linssen, J. H. (1988). Preliminary results of a study of four successive sedimentary geomagnetic reversals from the Mediterranean (upper Thvera, Lower and upper Sidufjall and Lower Nunivak). *Phys. Earth Planet. Int.*, 52, 207.

Linssen, J. H. (1991). Properties of Pliocene Sedimentary Geomagnetic Reversal Records from the Mediterranean. Ph.D. thesis, University of Utrecht.

Liu, H., An, J. and Wang, J. (1993). Preliminary study of the Olduvai termination recorded in the Red Loam in southeast Shanxi Province, China. *J. Geomag. Geoelect.*, 45, 331.

McFadden, P. L., Barton, C. E. and Merrill, R. T. (1993). Do virtual geomagnetic poles follow preferred paths during geomagnetic reversals? *Nature*, 361, 342.

Mankinen, E. A., Prévot, M., Grommé, C. S. and Coe, R. S. (1985). The Steens Mountain (Oregon) geomagnetic polarity transition. 1. Directional history, duration of episodes, and rock magnetism. *J. Geophys. Res.*, 90, 10.

Martinson, D. G., Manke, W. and Stoffa, P. (1982). An inverse approach to signal correlation. *J. Geophys. Res.*, 87, 4807.

Merrill, R. T. and McFadden, P. L. (1988). Secular variation and the origin of geomagnetic reversals. *J. Geophys. Res.*, 93, 11, 589.

Niitsuma, N. (1971). Detailed study of the sediments recording the Matuyama–Brunhes geomagnetic reversal. *Tohoku Univ. Sci. Rep. 2nd Ser. (Geol.)*, 43, 1.

Okada, M. and Niitsuma, N. (1989). Detailed palaeomagnetic records during the Brunhes–Matuyama geomagnetic reversal, and a direct determination of depth lag for magnetization in marine sediments. *Phys. Earth Planet. Int.*, 56, 133.

Olson, P., Silver, P. G. and Carlson, R. W. (1990). The large scale structure of convection in the Earth's mantle. *Nature*, 344, 209.

Prévot, M. and Camps, P. (1993). Absence of preferred longitude sectors for poles from volcanic records of geomagnetic reversals. *Nature*, 366, 53.

Prévot, M., Mankinen, E. A., Coe, R. S. and Grommé, C. S. (1985a). The Steens Mountain (Oregon) geomagnetic polarity transition. 2. Field intensity and discussion of reversal models. *J. Geophys. Res.*, 90, 10, 417.

Prévot, M., Maninken, E. A., Coe, R. S. and Grommé, C. S. (1985b). How the geomagnetic field vector reverses polarity. *Nature*, 316, 230.

Quidelleur, X. and Valet, J.-P. (1994). Palaeomagnetic records of excursions and reversals: possible biases caused by magnetization artefacts. *Phys. Earth Planet. Int.*, 82, 27.

Roberts, N. and Fuller, M. (1990). Similarity of new palaeomagnetic data from the Santa Rosa Mountains with those from Steens Mountain gives wide regional evidence for a two-stage process of geomagnetic field reversal. *Geophys. J. Int.*, 100, 521.

Rochette, P. (1990). Rationale of geomagnetic reversals versus remanence recording processes in rocks: a critical review. *Earth Planet. Sci. Lett.*, 98, 33.

Rolph, T. C. (1993). The Matuyama–Jaramillo R→N transition recorded in a loess section near Lanzhou, P. R. China. *J. Geomag. Geoelect.*, 45, 301.

Roperch, P. and Chauvin, A. (1987). Transitional geomagnetic field behaviour: volcanic records from French Polynesia. *Geophys. Res. Lett.*, 14, 151.

Roperch, P. and Duncan, R. A. (1990). Records of geomagnetic reversals from volcanic islands of French Polynesia. 1. Palaeomagnetic study of a polarity transition in a lava sequence from the Island of Huahine. *J. Geophys. Res.*, 95, 2713.

Runcorn, S. K. (1956). The magnetism of the Earth's body. In *Handbach der Physik, Band XLVII, Geophysik 1*, ed. J. Bartels. Berlin: Springer-Verlag.

Runcorn, S. K. (1992). Polar path in geomagnetic reversals. *Nature*, **356**, 654.

Shaw, J. (1975). Strong geomagnetic fields during a single Icelandic polarity transition. *Geophys. J. Roy. Astr. Soc.*, **40**, 345.

Sun, D., Shaw, J., An, Z. and Rolph, T. (1993). Matuyama/Brunhes (M/B) transition recorded in Chinese loess. *J. Geomag. Geoelect.*, **45**, 319.

Theyer, F., Herrero-Bervera, E., Hsu, V. and Hammond, S. R. (1985). The zonal harmonic model of polarity transitions: a test using successive reversals. *J. Geophys. Res.*, **90**, 1963.

Tric, E., Laj, C., Jehanno, C., Valet, J.-P., Kissel, C., Mazaud, A. and Iaccarino, S. (1991a). High resolution record of the upper Olduvai transition from Po Valley (Italy) sediments: support for dipolar transition geometry? *Phys. Earth Planet. Int.*, **65**, 319.

Tric, E., Laj, C., Valet, J.-P., Tucholka, P., Paterne, M. and Guichard, F. (1991b). The Blake geomagnetic event: transition geometry, dynamical characteristics and geomagnetic significance. *Earth Planet. Sci. Lett.*, **102**, 1.

Valet, J.-P. and Laj, C. (1981). Palaeomagnetic record of two successive Miocene geomagnetic reversals in western Crete. *Earth Planet. Sci. Lett.*, **54**, 53.

Valet, J.-P. and Laj, C. (1984). Invariant and changing transitional field configurations in a sequence of geomagnetic reversals. *Nature*, **311**, 552.

Valet, J.-P., Laj, C. and Langereis, C. G. (1983). Two different R–N geomagnetic reversals with identical VGP paths recorded at the same site. *Nature*, **304**, 330.

Valet, J.-P., Laj, C. and Langereis, C. G. (1988a). Sequential geomagnetic reversals recorded in upper Tortonian marine clays in western Crete (Greece). *J. Geophys. Res.*, **93**, 1131.

Valet, J.-P., Laj, C. and Tucholka, P. (1985a). Volcanic record of reversal. *Nature*, **316**, 217.

Valet, J.-P., Laj, C. and Tucholka, P. (1985b). Steens Mountain geomagnetic polarity transition is a single phenomenon: Reply. *Nature*, **318**, 487.

Valet, J.-P., Laj, C. and Tucholka, P. (1986). High resolution sedimentary record of a geomagnetic reversal. *Nature*, **322**, 27.

Valet, J.-P., and Meynadier, L. (1993). Geomagnetic field intensity and reversals during the last four million years. *Nature*, **366**, 234.

Valet, J.-P., Tauxe, L. and Clark, D. R. (1988b). The Matuyama–Brunhes transition recorded from Lake Tecopa sediments (California). *Earth Planet. Sci. Lett.*, **87**, 463.

Valet, J.-P., Tauxe, L. and Clement, B. M. (1989). Equatorial and mid-latitude records of the last geomagnetic reversal from the Atlantic Ocean. *Earth Planet. Sci. Lett.*, **94**, 371.

Valet, J.-P., Tucholka, P., Courtillot, V. and Meynadier, L. (1992). Palaeomagnetic constraints on the geometry of the geomagnetic field during reversals. *Nature*, **356**, 400.

van Hoof, A. A. M. (1993). The Gilbert/Gauss sedimentary geomagnetic reversal record from southern Sicily. *Geophys. Res. Lett.*, **20**, 835.

van Hoof, A. A. M. and Langereis, C. G. (1991). Reversal records in marine marls and delayed acquisition of remanent magnetization. *Nature*, **351**, 223.

van Hoof, A. A. M. and Langereis, C. G. (1992a). The upper Kaena sedimentary geomagnetic reversal record from southern Sicily. *J. Geophys. Res.*, **97**, 6941.

van Hoof, A. A. M. and Langereis, C. G. (1992b). The upper and Lower Thvera sedimentary geomagnetic reversal records from southern Sicily. *Earth Planet. Sci. Lett.*, **114**, 59.

van Hoof, A. A. M., van Os, B. J. H. and Langereis, C. G. (1993a). The upper and lower Nunivak sedimentary transitional records from southern Sicily. *Phys. Earth Planet. Int.*, **77**, 297.

van Hoof, A. A. M., van Os, B. J. H., Rademakers, J. G., Langereis, C. G. and de Lange, C. J. (1993b). A palaeomagnetic and geochemical record of the upper Cochiti reversal and two subsequent precessional cycles from southern Sicily (Italy). *Earth Planet. Sci. Lett.*, 117, 235.

Watkins, N. D. (1965a). Palaeomagnetism of the Columbia plateaus. *J. Geophys. Res.*, 70, 1379.

Watkins, N. D. (1965b). Frequency of extrusion of some Miocene lavas in Oregon during an apparent transition of the polarity of the geomagnetic field. *Nature*, 206, 801.

Watkins, N. D. (1969). Non-dipole behaviour during an upper Miocene geomagnetic polarity transition in Oregon. *Geophys. J. Roy. Astr. Soc.*, 17, 121.

Watkins, N. D. and Nougier, J. (1973). Excursions and secular variation of the Brunhes epoch geomagnetic field in the Indian Ocean region. *J. Geophys. Res.*, 78, 6060.

Weeks, R., Fuller, M., Laj, C., Mazaud, A. and Herrero-Bervera, E. (1992). Sedimentary records of reversal transitions – magnetization smoothing artefact or geomagnetic field behaviour? *Geophys. Res. Lett.*, 19, 2007.

Weeks, R. J., Fuller, M. and Williams, I. (1988). A model for transitional field geometries involving low-order zonals and drifting non-dipole harmonics. *J. Geophys. Res.*, 93, 11613.

Whaler, K. A. (1982). Geomagnetic secular variation and fluid motion at the core surface. *Phil. Trans. Roy. Soc. A*, 306, 235.

Williams, I. and Fuller, M. (1981). Zonal harmonic models of reversal transition fields. *J. Geophys. Res.*, 86, 11, 657.

Williams, I., Weeks, R. and Fuller, M. (1988). A model for transition fields during geomagnetic reversals. *Nature*, 322, 719.

Yoshimura, H. (1980). Non-linear astrophysical dynamos: autonomous and sporadic field reversals of steady dynamos and the polarity transition phenomenon of the geodynamo. *Astrophys. J.*, 235, 625.

Zhang, K. K. and Busse, F. H. (1988). Finite amplitude convection and magnetic field generation in a rotating spherical shell. *Geophys. Astrophys. Fluid Dyn.*, 44, 33.

Zhang, K. K. and Busse, F. H. (1989). Convection driven magnetohydrodynamic dynamos in rotating spherical shells. *Geophys. Astrophys. Fluid Dyn.*, 49, 97.

Zhu, R., Ding, Z., Wu, H., Huang, B. and Jiang, Li (1993). Details of magnetic polarity transition recorded in Chinese loess. *J. Geomag. Geoelect.*, 45, 289.

Chapter 7

Alvarez, W., Arthur, M. A., Fischer, A. G., Lowrie, W., Napoleone, G., Premoli Silva I. and Roggenthen, W. M. (1977). Upper Cretaceous–Palaeocene magnetic stratigraphy at Gubbio, Italy. V. Type section for the late Cretaceous-Palaeocene geomagnetic reversal timescale. *Geol. Soc. Am. Bull.*, 88, 383.

Alvarez, W. and Lowrie, W. (1978). Upper Cretaceous palaeomagnetic stratigraphy at Moria (Umbrian Apennines, Italy): verification of the Gubbio section. *Geophys. J. Roy. Astr. Soc.*, 55, 1.

Anon. (1979). Magnetostratigraphic polarity units. A supplementary chapter of the International Subcommission on Stratigraphic Classification. *International Stratigraphic Guide Geology*, 7, 578.

Arthur, M. A. and Fischer, A. G. (1977). Upper Cretaceous-Palaeocene magnetic stratigraphy at Gubbio, Italy. I. Lithostratigraphy and sedimentology. *Geol. Soc. Am. Bull.*, 88, 367.

Baksi, A. K., Hsu, V., McWilliams, M. O. and Farrar, E. (1992). [40] Ar/[39]Ar dating of the Brunhes–Matuyama geomagnetic field reversal. *Science*, 256, 356.

Berger, A. and Loutre, M. F. (1988). New insolation values for the climate of the last 10 million years. *Sci. Inst. Astr. Geophys.*, George Lemaitre Univ. Cath. Louvain-la Neuve.

Berggren, W. A., Kent, D. V. and Flynn, J. J. (1985a). Jurassic to Paleogene: Part 2. Paleogene geochronology and chronostratigraphy. In *The Chronology of the Geological Record*, ed. N. J. Snelling, p. 141. Geological Society Memoir No. 10.

Berggren, W. A., Kent, D. V., Obradovich, J. D. and Swisher III, C. C. (1992). Toward a revised Palaeogene geochronology. In *Eocene–Oligocene Climatic and Biotic Evolution*, ed. D. R. Prothero and W. A. Berggren. Princeton University Press.

Berggren, W. A., Kent, D. V. and Van Couvering, J. A. (1985b). The Neogene: Part 2. Neogene geochronology and chronostratigraphy. In *The Chronology of the Geological Record*, ed. N. J. Snelling p. 211. Geological Society Memoir No. 10.

Berggren, W. A., Kent, D. V. and Van Couvering, J. A. (1985c). Cenozoic geochronology. *Geol. Soc. Am. Bull.*, 96, 1407.

Blakely, R. J. (1974). Geomagnetic reversals and crustal spreading rates during the Miocene. *J. Geophys. Res.*, 79, 2979.

Bralower, T. J. (1987). Valanginian to Aptian calcareous nannofossil stratigraphy and correlation with the upper M-sequence magnetic anomalies. *Marine Micropaleontol.*, 11, 293.

Bralower, T. J., Thierstein, H. R. and Monechi, S. (1989). Calcareous nanno-fossil zonation of the Jurassic–Cretaceous boundary interval and correlation with the geomagnetic polarity timescale. *Marine Micropaleontol.*, 14, 143.

Briden, J. C., Rex, D. C., Faller, A. M. and Tomblin, J. F. (1979). K–Ar geochronology and palaeomagnetism of volcanic rocks in the Lesser Antilles island arc. *Phil. Trans. Roy. Soc. A*, 291, 485.

Bryan, G. M., Markl, R. G. and Sheridan, R. E. (1980). IPOD site surveys in the Blake–Bahama basin. *Marine Geol.*, 35, 43.

Bryan, N. B. and Duncan, R. A. (1983). Age and provenance of clastic horizons from hole 516F. Init. Rept DSDP, Vol. 72. US Govt Printing Office, Washington, D.C., p. 475.

Cande, S. C. and Kent, D. V. (1992). A new geomagnetic polarity timescale for the late Cretaceous and Cenozoic. *J. Geophys. Res.*, 97, 13917.

Cande, S. E. and LaBrecque, J. L. (1974). Behaviour of the Earth's palaeomagnetic field from small scale marine magnetic anomalies. *Nature*, 247, 26.

Cande, S. C., LaBrecque, J. L., Larson, R. L., Pitman W. C. III, Golovchenko, X. and Haxby, W. F. (1989). Magnetic lineations of the world's ocean basins, map with text. Am. Assoc. Petrol. Geol., Tulsa, Okla.

Cande, S. C., Larson, R. L. and LaBrecque, J. L. (1978). Magnetic lineations in the Pacific Jurassic quiet zone. *Earth Planet. Sic. Lett.*, 41, 434.

Cao, J. X., Xu, Q. Z., Zhang, Y. T. and Chen, F. H. (1985). A study on the loess/palaesol sequence and the environmental evolution at Jiuzhoutai, Lanzhou. *Mon. J. Lanzhou Univ.*, 24, 118.

Champion, D. E., Lamphere, M. A. and Kuntz, M. A. (1988). Evidence for a new geomagnetic reversal from lava flows in Idaho: discussion of short polarity intervals in the Brunhes and late Matuyama polarity chrons. *J. Geophys. Res.*, 93, 11667.

Channell, J. E. T., Bralower, T. J. and Grandesso, P. (1987). Biostratigraphic correlation of Mesozoic polarity chrons CM1 to CM23 at Capriolo and Xausa (Southern Alps, Italy). *Earth Planet. Sci. Lett.*, 85, 203.

Channell, J. E. T. and Erba, E. (1992). Early Cretaceous polarity chrons CM0 to CM11 recorded in northern Italian land sections near Bressica. *Earth Planet. Sci. Lett.*, 108, 161.

Channell, J. E. T., Lowrie, W. and Medizza, F. (1979). Middle and early Cretaceous magnetic stratigraphy from the Cismon section, northern Italy. *Earth Planet. Sci. Lett.*, 42, 153.

Channell, J. E. T., Ogg, J. G. and Lowrie, W. (1982). Geomagnetic polarity in the early Cretaceous and Jurassic. *Phil. Trans. Roy. Soc. A*, 306, 137.

Clement, B. M., Kent, D. V. and Opdyke, N. D. (1982). Brunhes–Matuyama polarity in three deep-sea sediment cores. *Phil. Trans. Roy. Soc. A*, **306**, 113.

Cox, A. and Dalrymple, G. B. (1967). Geomagnetic polarity epochs – Nunivak Island, Alaska. *Earth Planet. Sci. Lett.*, **3**, 173.

Cox, A., Doell, R. R. and Dalrymple, G. B. (1964). Reversals of the Earth's magnetic field. *Science*, **44**, 1537.

Divenere, V. J. and Opdyke, N. D. (1990). Palaeomagnetism of the Maringouin and Shepody formation, New Brunswick: a Namurian magnetic stratigraphy. *Can. J. Earth Sci.*, **27**, 803.

Divenere, V. J. and Opdyke, N. D. (1991). Magnetic polarity stratigraphy in the uppermost Mississippian Mauch Chunk Formation, Pottsville, Pennsylvania. *Geology*, **19**, 127.

Doell, R. R. and Dalrymple, G. B. (1973). Potassium–argon ages and palaeomagnetism of the Waianae and Koolau volcanic series, Oahu, Hawaii. *Bull. Geol. Soc. Am.*, **84**, 1217.

Evans, A. L. (1970). Geomagnetic polarity reversals in a late Tertiary lava sequence from the Akaroa volcano, New Zealand. *Geophys. J. Roy. Astr. Soc.*, **21**, 163.

Flynn, J. J. (1986). Correlation and geochronology of Middle Eocene strata from the western United States. *Palaeogeog. Palaeochron. Palaeoecol.*, **55**, 335.

Foster, J. H. and Opdyke, N. D. (1970). Upper Miocene to Recent magnetic stratigraphy in deep-sea sediments. *J. Geophys. Res.*, **75**, 4465.

Haag, M. and Heller, F. (1991). Late Permian to Early Triassic magnetostratigraphy. *Earth Planet. Sci. Lett.*, **107**, 42.

Hailwood, E. A. (1989). *Magnetostratigraphy*. Geological Society Special Report No. 19.

Hailwood, E. A., Bock, W., Costa, L., Dupeuble, P. A., Mueller, C. and Schnitkev, D. (1979). *Chronology and biostratigraphy of northeast Atlantic sediments, DSDP Leg 48*. US Govt Printing Office, Washington, D.C., p. 1119.

Hardenbol, J. and Berggren, W. A. (1978). A new Palaeocene numerical timescale. *Contributions to the Geologic TimeScale, Studies of Geology*, 6. Am. Assoc. Petrol. Geol., 213.

Harland, W. B., Armstrong, R. L., Cox, A. V., Craig, L. E., Smith, A. G. and Smith, D. G. (1990). *A Geologic TimeScale, 1989*. Cambridge: Cambridge University Press.

Harland, W. B., Cox, A. V., Llewellyn, P. G., Picton, C. A. G., Smith, A. G. and Walters, R. (1982). *A Geologic Time Scale*. Cambridge: Cambridge University Press.

Harrison, C. G. A. (1966). The palaeomagnetism of deep-sea sediments. *J. Geophys. Res.*, **71**, 3033.

Harrison, C. G. A. and Funnell, B. M. (1964). Relationship of palaeomagnetic reversals and micropalaeontology in two late Cenozoic cores from the Pacific Ocean. *Nature*, **204**, 566.

Hays, J. D. and Opdyke, N. D. (1967). Antarctic radiolaria, magnetic reversals and climatic change. *Science*, **158**, 1001.

Hays, J. D., Saito, T., Opdyke, N. D. and Burckle, L. H. (1969). Pliocene–Pleistocene sediments of the equatorial Pacific; their palaeomagnetic, biostratigraphic, and climatic record. *Geol. Soc. Am. Bull.*, **80**, 1481.

Heirtzler, J. R., Dickson, G. O., Herron, E. M., Pitman, W. C. III and Le Pichon, X. (1968). Marine magnetic anomalies, geomagnetic field reversals, and motions of the ocean floor and continents. *J. Geophys. Res.*, **73**, 2119.

Heller, F., Lowrie, W., Li, H. and Wang, J. (1988). Magnetostratigraphy of the Permo-Triassic boundary section at Shangsi (Guangyuan, Sichuan Province, China). *Earth Planet. Sci. Lett.*, **88**, 348.

Hilgen, F. J. (1991a). Astronomical calibration of Gauss to Matuyama sapropels in the Mediterranean and implication for the geomagnetic polarity timescale. *Earth Planet. Sci. Lett.*, **104**, 226.

Hilgen, F. J. (1991b). Extension of the astronomically calibrated (polarity) timescale to the Miocene/Pliocene boundary. *Earth Planet. Sci. Lett.*, **107**, 349.

Hilgen, F. J. and Langereis, C. G. (1989). Periodicities of carbonate cycles in the Pliocene of Sicily: discrepancies with the quasi-periodicities of the Earth's orbital cycles. *Terra Nova*, **1**, 409.

Imbrie, J. and Imbrie, J. Z. (1980). Modelling the climatic response to orbital variations. *Science*, **207**, 943.

Irving, E. and Parry, L. G. (1963). The magnetism of some Permian rocks from New South Wales. *Geophys. J. Roy. Astr. Soc.*, **7**, 395.

Johnson, R. G. (1982). Brunhes–Matuyama magnetic reversal dated at 790,000 yr BP by marine astronomical correlations. *Quatern. Res.*, **17**, 135.

Johnson, N. M. and McGee, V. E. (1983). Magnetic polarity stratigraphy: stochastic properties of data, sampling problems, and the evaluation of interpretations. *J. Geophys. Res.*, **88**, 1213.

Keating, B. H. and Helsley, C. E. (1978). Magnetostratigraphy of Cretaceous-age sediments from sites 361, 363, and 364. Init. Rept DSDP, Vol. 40, US Govt Printing Office, Washington, D.C., p. 459.

Kennett, J. P. (Ed.) (1980). *Magnetic Stratigraphy of Sediments*. Dowden: Hutchingson & Ross.

Kennett, J. P., Watkins, N. D. and Vella, P. (1971). Palaeomagnetic chronology of Pliocene–early Pleistocene climates and the Plio–Pleistocene boundary in New Zealand. *Science*, **171**, 276.

Kent, D. V. and Gradstein, F. M. (1985). A Cretaceous and Jurassic chronology. *Geol. Soc. Am. Bull.*, **96**, 1419.

Khramov, A. N. (1957). Palaeomagnetism: the basis of a new method of correlation and subdivision of sedimentary strata. *Dokl. Akad. Nauk Earth Sci. Sect. Proc.*, **112**, 129.

Kristjansson, L., Fridleifsson, I. B. and Watkins, N. D. (1978). Palaeomagnetism of the Esja area, SW Iceland. *Trans. Am. Geophys. Union*, **59**, 270.

LaBrecque, J. L. and Hsu, K. J. (1983). DSDP Leg 73: contributions to Palaeogene stratigraphy in nomenclature, chronology and sedimentation rates. *Palaeogeog., Palaeoclimatol. Palaeoecol.*, **42**, 91.

LaBrecque, J. L., Kent, D. V. and Cande, S. C. (1977). Revised magnetic polarity timescale for Late Cretaceous and Cenozoic time. *Geology*, **5**, 330.

Larson, R. L. and Hilde, T. W. C. (1975). A revised timescale of magnetic reversals for the early Cretaceous and late Jurassic. *J. Geophys. Res.*, **80**, 2586.

Larson, R. L. and Pitman, W. C. III (1972). Worldwide correlation of Mesozoic magnetic anomalies and its implications. *Geol. Soc. Am. Bull.*, **83**, 3645.

Le Pichon, X. and Heirtzler, J. R. (1968). Magnetic anomalies in the Indian Ocean and sea-floor spreading. *J. Geophys. Res.*, **73**, 2101.

Liddicoat, J. C. (1993). Matuyama/Brunhes polarity transition near Bishop, California. *Geophys. J. Int.*, **112**, 497.

Liddicoat, J. C., Opdyke, N. D. and Smith, G. I. (1980). Palaeomagnetic polarity in a 930 m core from Searles Valley, California. *Nature*, **286**, 22.

Liu, X. M., Liu, T. S., Xu, T. G. and Chen, M. Y. (1988). The Chinese loess in Xifeng. I. the primary study on magnetostratigraphy of a loess profile in Xifeng area, Gansu province. *Geophys. J. Roy. Astr. Soc.*, **92**, 345.

Lowrie, W. and Alvarez, W. (1977). Upper Cretaceous–Palaeocene magnetic stratigraphy at Gubbio, Italy. III. Upper Cretaceous magnetic stratigraphy. *Geol. Soc. Am. Bull.*, **88**, 374.

Lowrie, W. and Alvarez, W. (1981). One hundred million years of geomagnetic polarity history. *Geology*, **9**, 392.

Lowrie, W., Alvarez, W., Napoleone, G., Perch-Nielsen, K., Premoli Silva, I. and

Tormarkine, M. (1982). Palaeogene magnetic stratigraphy in Umbrian pelagic carbonate rocks: the Contessa sections. *Geol. Soc. Am. Bull.*, **93**, 414.

Lowrie, W., Alvarez, W., Premoli Silva, I. and Monechi, S. (1980a). Lower Cretaceous magnetic stratigraphy in Umbian pelagic carbonate rocks. *Geophys. J. Roy. Astr. Soc.*, **60**, 283.

Lowrie, W., Channell, J. E. T. and Alvarez, W. (1980b). A review of magnetic stratigraphy investigations in Cretaceous pelagic carbonate rocks. *J. Geophys. Res.*, **85**, 3597.

Luterbacher, H. P. and Premoli Silva, I. (1964). Biostratigrafia del limite Cretaceo–Terziario nelli Appennino Centrale. *Riv. Ital. Palaeontol. Studigrafia*, **70**, 67.

McDougall, I. (1979). The present status of the geomagnetic polarity timescale. In *The Earth: its Origin, Structure and Evolution*, ed. M. W. McElhinny. New York: Academic Press.

McDougall, I., Brown, F. H., Cerling, T. E. and Hillhouse, J. W. (1992). A reappraisal of the geomagnetic polarity timescale to 4 Ma using data from the Turkana basin, East Africa. *Geophys. Res. Lett.*, **19**, 2349.

McDougall, I., Kristjansson, L. and Saemundsson, K. (1984). Magnetostratigraphy and geochronology of northwest Iceland. *J. Geophys. Res.*, **89**, 7029.

McDougall, I., Saemundsson, K., Johannesson, H., Watkins, N. D. and Kristjansson, L. (1977). Extension of the geomagnetic polarity timescale to 6.5 M yr: K–Ar dating, geological and palaeomagnetic study of a 3500 m lava succession in western Iceland. *Geol. Soc. Am. Bull.*, **88**, 1.

McDougall, I., Watkins, N. D., Walker, G. P. L. and Kristjansson, L. (1976). Potassium–argon and palaeomagnetic analysis of Icelandic lava flows: limits on the age of anomaly 5. *J. Geophys. Res.*, **81**, 1505.

McDowell, F. W., Wilson, J. A. and Clark, J. (1973). K–Ar dates for biotite from two palaeontologically significant localities: Duchnesre Formation, Utah and Chadron Formation, South Dakota. *Isochron/West*, **7**, 11.

McIntosh, W. C., Geissman, J. W., Chapin, C. E., Kunk, M. J. and Henry, C. D. (1992). Calibration of the latest Eocene–Oligocene geomagnetic polarity timescale using $^{40}Ar/^{39}Ar$ dated ignimbrites. *Geology*, **20**, 459.

Mankinen, E. A. and Dalrymple, G. B. (1979). Revised geomagnetic polarity timescale for the interval 0–5 Myr BP. *J. Geophys. Res.*, **84**, 615.

Mankinen, E. A., Donnelly-Nolan, J. M., Grommé, C. S. and Hearn, B. C., Jr. (1981). Palaeomagnetism of the Clear Lake volcanics and new limits on the age of the Jaramillo normal polarity event. *U.S. Geol. Surv. Prof. Pap.*, **1141**, 67.

Miller, K. G., Kahn, M. J., Aubry, M.-P., Berggren, W. A., Kent, D. V. and Melillo, A. (1985). Oligocene to Miocene bio-, magneto-, and isotope stratigraphy of the western North Atlantic. *Geology*, **13**, 257.

Montanari, A., Drake, R., Bice, D. M., Alvarez, W., Curtis, G. H., Turrin, B. D. and De Paolo, D. J. (1985). Radiometric timescale for the upper Eocene and Oligocene based on K/Ar and Rb/Sr dating of volcanic biotites from the pelagic sequences of Gubbio, Italy. *Geology*, **13**, 596.

Napoleone, G., Premoli Silva, I., Heller, F., Cheli, P., Corezzi, S. and Fischer, A. G. (1983). Eocene magnetic stratigraphy at Gubbio, Italy and its implications for Palaeocene geochronology. *Geol. Soc. Am. Bull.*, **94**, 181.

Ness, G., Levi, S. and Couch, R. (1980). Marine magnetic anomaly timescales for the Cenozoic and Late Cretaceous; a précis, critique and synthesis. *Rev. Geophys. Space Phys.*, **18**, 753.

Ninkovich, D., Opdyke, N. D., Heezen, B. C. and Foster, J. H. (1966). Palaeomagnetic stratigraphy, rates of deposition and tephra-chronology in North Pacific deep-sea sediments. *Earth Planet. Sci. Lett.*, **1**, 476.

Nocchi, M., Parasi, G., Monaco, P., Monechi, S., Mandile, M., Napoleone, G., Ripepe,

M., Orlando, M., Premoli Silva, I. and Brice, D. M. (1986). The Eocene–Oligocene boundary in the Umbrian pelagic regression. In *Terminal Eocene Events*, Developments in Palaeontology and Stratigraphy, Vol. 9, ed. C. Pomerol and I. Premoli Silva, pp. 25–40. Amsterdam: Elsevier.

Obradovich, J. D. and Cobban, W. A. (1975). A timescale for the late Cretaceous of the western interior of North America. *Geol. Assoc. Can. Spec. Pap.*, 13, 31.

Obradovich, J. D., Sutter, J. F. and Kunk, M. J. (1986). Magnetic polarity chron tie-points for the Cretaceous and early Tertiary. *Terra Cognita*, 6, 140.

Odin, G. S., Montanari, A., Deino, A., Drake, R., Guise, P. G., Kreuzer, H. and Rex, D. C. (1991). Reliability of volcano-sedimentary biotite ages across the Eocene–Oligocene boundary (Apennines, Italy). *Chem. Geol. Isotope Geosci. Sect.*, 86, 203.

Ogg, J. G. and Steiner, M. B. (1984). Jurassic magneto polarity timescale: current status and compilation. In: *International Symposium on Jurassic Stratigraphy, Erlingen, Sept. 1984*, ed. O. Michelsen and A. Zeiss, Vol. 3, p. 777. Copenhagen: Geological Survey of Denmark.

Okada, M. and Niitsuma, N. (1989). Detailed palaeomagnetic records during Brunhes–Matuyama geomagnetic reversal, and direct determination of depth lag for magnetization in marine sediments. *Phys. Earth Planet. Int.*, 56, 133.

Opdyke, N. D. (1972). Palaeomagnetism of deep-sea cores. *Rev. Geophys. Space Phys.*, 10, 213.

Opdyke, N. D., Burckle, L. H. and Todd, A. (1974). The extension of magnetic timescale in sediments of the Central Pacific Ocean. *Earth Planet. Sci. Lett.*, 22, 300.

Opdyke, N. D., Glass, B., Hays, J. D. and Foster, J. H. (1966). Palaeomagnetic study of Antarctic deep-sea cores. *Science*, 154, 349.

Pitman, W. C. III, Herron, E. M. and Heirtzler, J. R. (1968). Magnetic anomalies in the Pacific and sea-floor spreading. *J. Geophys. Res.*, 73, 2069.

Preisinger, A., Zobetz, E., Gratz, A. J., Lahodynsky, R., Becke, M., Mauritsch, H. J., Eider, G., Grass, F., Rogl, F., Stradner, H. and Surenian, R. (1986). The Cretaceous–Tertiary boundary in the Gosan Basin, Austria. *Nature*, 322, 794.

Premoli Silva, I. (1977). Upper Cretaceous–Palaeocene magnetic stratigraphy at Gubbio, Italy. II. Biostratigraphy. *Geol. Soc. Am. Bull.*, 88, 371.

Prothero, D. R., Denham, C. R. and Farmer, H. G. (1982). Oligocene calibration of the magnetic polarity timescale. *Geology*, 10, 650.

Renne, P. R., Walter, R. C., Verosub, K. L., Sweitzer, M. and Aronson, J. L. (1993). New data from Hadar (Ethiopia) support orbitally tuned timescale to 3.3 Ma. *Geophys. Res. Lett.*, 20, 1067.

Roggenthen, W. M. and Napoleone, G. (1977). Upper Cretaceous–Palaeocene magnetic stratigraphy at Gubbio, Italy. IV. Upper Maastrichtian–Palaeocene magnetic stratigraphy. *Geol. Soc. Am. Bull.*, 88, 378.

Ruddiman, W. F., Raymo, M. E., Martinson, D., Clement, B. and Backman, J. (1989). Pleistocene evolution: northern hemisphere ice sheets and north Atlantic ocean. *Paleoceanography*, 4, 353.

Rutten, M. C. (1959). Palaeomagnetic reconnaissance of mid-Italian volcanoes. *Geol. Mijnbouw.*, 21, 373.

Saemundsson, K., Kristjansson, L., McDougall, I. and Watkins, N. D. (1980). K–Ar dating, geological and palaeomagnetic study of a 5-km lava succession in northern Iceland. *J. Geophys. Res.*, 85, 3268.

Shackleton, N. J., Berger, A. and Peltier, W. R. (1990). An alternative astronomical calibration of the lower Pleistocene timescale based on ODP site 677. *Trans. Roy. Soc. Edinb., Earth Sci.*, 81, 251.

Shackleton, N. J. and Opdyke, N. D. (1973). Oxygen isotope and palaeomagnetic stratigraphy of equatorial Pacific core V28–238: oxygen isotope temperatures and ice volumes on a 10^5 and 10^6 year scale. *Quatern. Res.*, 3, 39.

Spell, T. L. and McDougall, I. (1992). Revisions to the age of the Brunhes–Matuyama boundary and the Pleistocene geomagnetic polarity timescale. *Geophys. Res. Lett.*, 19, 1181.

Steiger, R. H. and Jäger, E. (1977). Subcommission on geochronology: convention on the use of decay constants in geo- and cosmochronology. *Earth Planet. Sci. Lett.*, 36, 359.

Steiner, M. B. and Ogg, J. G. (1984). Early and Middle Jurassic magnetic polarity timescale. In International Symposium on *Jurassic Stratigraphy, Erlingen, Sept. 1984*, ed. O. Michelsen and A. Zeiss, Vol. 3. Copenhagen: Geological Survey of Denmark.

Steiner, M., Ogg, J., Zhang, Z. and Sun, S. (1989). The late Permian/early Triassic magnetic polarity timescale and plate motions of south China. *J. Geophys. Res.*, 94, 7343.

Swisher, C. C. and Knox, R. W. O'B. (1991). The age of the Palaeocene/Eocene boundary: $^{40}Ar/^{39}Ar$ dating of the lower part of NP10, North Sea basin and Denmark. Int. Geol. Corr. Project 308, Brussels meeting, Dec.

Talwani, M., Windisch, C. C. and Langseth, M. G., Jr (1971). Reykyanes ridge crest: a detailed geophysical study. *J. Geophys. Res.* 76, 473.

Tarduno, J. A. (1990). Brief reversed polarity interval during the Cretaceous normal polarity superchron. *Geology*, 18, 683.

Tarduno, J. A., Lowrie, W., Sliter, W. V., Bralower, T. J. and Heller, F. (1992). Reversed polarity characteristic magnetizations in the Albian Contessa section, Umbrian Apennines, Italy: implications for the existence of a mid-Cretaceous mixed polarity interval. *J. Geophys. Res.*, 97, 241.

Tarduno, J. A., Sliter, W. V., Bralower, T. J., McWilliams, M., Premoli Silva, I. and Ogg, J. (1989). M sequence reversals recorded in DSDP sediment cores from the western mid-Pacific mountains and Magellan Rise. *Geol. Soc. Am. Bull.*, 101, 1306.

Tauxe, L., Deino, A. D., Behrensmeyer, A. R. and Potts, R. (1992). Pinning down the Brunhes/Matuyama and upper Jaramillo boundaries: a reconciliation of orbital and isotopic timescales. *Earth Planet. Sci. Lett.*, 109, 561.

Theyer, F. and Hammond, S. R. (1974). Cenozoic magnetic timescale in deep-sea cores: completion of the Neogene. *Geology*, 2, 487.

Vandenberg, J., Klootwijk, C. T. and Wonders, A. A. H. (1978). The late Mesozoic and Cenozoic movements of the Umbrian peninsular: further paleomagnetic data from the Umbrian sequence. *Geol. Soc. Am. Bull.*, 89, 133.

Vandenberg, J. and Wonders, A. A. H. (1980). Palaeomagnetism of late Mesozoic pelagic limestones from the southern Alps. *J. Geophys. Res.*, 85, 3623.

van Hinte, J. E. (1976a). A Cretaceous timescale. *Am. Assoc. Petrol. Geol. Bull.*, 60, 498.

van Hinte, J. E. (1976b). A Jurassic timescale. *Am. Assoc. Petrol. Geol. Bull.*, 60, 489.

Walter, R. C. (1994). Age of Lucy and the First Family: single-crystal $^{40}Ar/^{39}Ar$ dating of the Denen Dora and lower Kada Hadar members of the Hadar Formation, Ethiopia. *Geology*, 22, 6.

Watkins, N. D. (1976). Polarity subcommission sets up some guidelines. *Geotimes*, 21, 18.

Zheng, H., An, Z. and Shaw, J. (1992). New contributions to Chinese Plio–Pleistocene magnetostratigraphy. *Phys. Earth Planet. Int.*, 70, 146.

Zijderveld, J. D. A., Hilgen, F. J., Langereis, C. G., Verhallen, P. J. J. M. and Zachariasse, W. J. (1991). Integrated magnetostratigraphy and biostratigraphy of the upper Pliocene–lower Pleistocene from the Monte Singa and Crotone areas in Calabria, Italy. *Earth Planet. Sci. Lett.*, 107, 697.

Chapter 8

Ahrens, T. J. and Hager, B. H. (1987). Heat transport in D″; Problems and paradoxes. *EOS, Trans. Am. Geophys. Union*, 68, 1493.

Alvarez, L. W., Alvarez, W., Asaro, F. and Michel, H. V. (1980). Extraterrestrial cause for the Cretaceous/Tertiary extinction. *Science*, 208, 1095.

Alvarez, W. and Muller, R. A. (1984). Evidence from crater ages for periodic impacts on the Earth. *Nature*, 308, 718.

Arthur, M. A., Dean, W. A. and Schlanger, S. O. (1985). Variation in the global carbon cycle during the Cretaceous related to climate, volcanism and changes in atmospheric CO_2. In *The Carbon Cycle and Atmospheric CO_2: Natural Variations Archean to Present*, ed. E. T. Sundquist and W. S. Broecker, American Geophysics Union Monograph 32, p. 504.

Baksi, A. K., Hsu, V., McWilliams, M. O. and Farrar, E. (1992). $^{40}Ar/^{39}Ar$ dating of the Brunhes–Matuyama geomagnetic field reversal. *Science*, 256, 356.

Berger, A. (1976). Obliquity and precession for the last 5,000,000 years. *Astron. Astrophys.*, 51, 127.

Berger, A. (1977a). Support for the astronomical theory of climatic change. *Nature*, 269, 44.

Berger, A. (1977b). Long term variations of the Earth's orbital elements. *Celestial Mech.*, 15, 53.

Berger, A., Loutre, M. F. and Laskar, J. (1992). Stability of the astronomical frequencies over the Earth's history for paleoclimate studies. *Science*, 255, 560.

Black, D. I. (1967). Cosmic-ray effects and faunal extinctions at geomagnetic field reversals. *Earth Planet. Sci. Lett.*, 3, 225.

Blakemore, R. (1975). Magnetotactic bacteria. *Science*, 190, 377.

Blakemore, R. P., Frankel, R. B. and Kalmijn, A. J. (1980). South-seeking magnetotactic bacteria in the southern hemisphere. *Nature*, 286, 384.

Bloxham, J. and Gubbins, D. (1987). Thermal core–mantle interactions. *Nature*, 325, 511.

Bonatti, E. and Gartner, S. (1973). Caribbean climate during Pleistocene ice ages. *Nature*, 244, 563.

Broecker, W. S. (1992). Climate cycles. Upset for Milankovitch theory. *Nature*, 359, 779.

Bullard, E. C. (1968). Reversals of the Earth's magnetic field. *Phil. Trans. Roy. Soc. A*, 263, 481.

Chappel, J. (1975). On possible relationships between Quaternary glaciation, geomagnetism and vulcanism. *Earth Planet. Sci. Lett.*, 26, 370.

Chave, A. D. and Denham, C. R. (1979). Climatic changes, magnetic intensity variations and fluctuations of the eccentricity of the Earth's orbit during the past 2,000,000 years and a mechanism which may be responsible for the relationship – a discussion. *Earth Planet. Sci. Lett.*, 44, 150.

Claeys, P., Casier, J.-G. and Margolis, S. V. (1992). Microtektites and mass extinctions: evidence for a late Devonian asteroid impact. *Science*, 257, 1102.

Clube, S. V. M. and Napier, W. M. (1984). Terrestrial catastrophism – nemesis or galaxy? *Nature*, 311, 635.

Courtillot, V. and Besse, J. (1987). Magnetic field reversals, polar wander, and core–mantle coupling. *Science*, 237, 1140.

Cox, K. G. (1991). A superplume in the mantle. *Nature*, 352, 564.

Crain, I. K. and Crain, P. L. (1970). New stochastic model for geomagnetic reversals. *Nature*, 228, 39.

Crain, I. K., Crain, P. L. and Plaut, M. G. (1969). Long period Fourier spectrum of geomagnetic reversals. *Nature*, 223, 283.

Creer, K. M., Readman, P. W. and Jacobs, A. M. (1980). Palaeomagnetic and palaeontological dating of a section at Gioia Tauro, Italy: identification of the Blake event. *Earth Planet. Sci. Lett.*, 50, 289.

Crutzen, P. J., Isaksen, I. S. A. and Reid, G. C. (1975). Solar proton events: stratospheric sources of nitric oxide. *Science*, 189, 457.

Davis, M., Hut, P. and Muller, R. A. (1984). Extinction of species by periodic comet showers. *Nature*, **308**, 715.

Denham, C. R. (1976). Blake polarity episode in two cores from the Greater Antilles outer ridge. *Earth Planet. Sci. Lett.*, **29**, 422.

Doake, C. S. M. (1977). A possible effect of ice ages on the Earth's magnetic field. *Nature*, **267**, 415.

Doake, C. S. M. (1978). Climatic change and geomagnetic field reversals: a statistical correlation. *Earth Planet. Sci. Lett.*, **38**, 313.

Durrani, S. A. and Khan, H. A. (1971). Ivory Coast microtektites: fission track age and geomagnetic reversals. *Nature*, **232**, 320.

Edwards, R. L. and Gallup, C. D. (1993). Dating of the Devil's Hole calcite vein. *Science*, **259**, 1626.

Frankel, R. B., Blakemore, R. P., Torres de Araujo, F. F., Esquivel, M. S. and Danon, J. (1981). Magnetotactic bacteria at the geomagnetic equator. *Science*, **212**, 1269.

Fuller, M. and Weeks, R. (1992). Superplumes and superchrons. *Nature*, **356**, 16.

Funaki, M., Sakai, H. and Matsunaga, T. (1989). Identification of the magnetic poles on strong magnetic grains from meteorites using magnetotactic bacteria. *J. Geomag. Geoelect.*, **41**, 77.

Funaki, M., Sakai, H., Matsunaga, T. and Hirose, S. (1992). The S pole distribution on magnetic grains in pyroxenite determined by magnetotactic bacteria. *Phys. Earth Planet. Int.*, **70**, 253.

Gaffin, S. (1987). Phase difference between sea level and magnetic reversal rate. *Nature*, **329**, 816.

Gallee, H., van Ypersele, J. P., Fichefet, T., Marsiat, I., Tricot, C. and Berger, A. (1992). Simulation of the last glacial cycle by a coupled, sectorially averaged climate ice-sheet model. 2. Response to insolation and CO_2 variation. *J. Geophys. Res.*, **97**, 15, 713.

Glass, B. P. and Heezen, B. C. (1967). Tektites and geomagnetic reversals. *Sci. Am.*, **217**(7), 32.

Glass, B. P. and Zwart, P. A. (1979). The Ivory Coast microtektite strewn field: new data. *Earth Planet. Sci. Lett.*, **43**, 336.

Grieve, R. A. F., Sharpton, V. L., Goodacre, A. K. and Garvin, J. B. (1985/86). A perspective on the evidence for periodic cometay impacts on Earth. *Earth Planet. Sci. Lett.*, **76**, 1.

Gubbins, D. (1983). The influence of extrinsic pressure changes on the Earth's dynamo. *Phys. Earth Planet. Int.*, **33**, 255.

Gubbins, D. (1987). Mechanism for geomagnetic polarity reversals. *Nature*, **326**, 167.

Gubbins, D. and Richards, M. (1986). Coupling of the core dynamo and mantle: thermal or topographic? *Geophys. Res. Lett.*, **13**, 1521.

Haq, B. U., Hardenbol, J. and Vail, P. R. (1988). Mesozoic and Cenozoic chronostratigraphy and cycles of sea-level change. In *Sea-level Changes: An Integrated Approach*, ed. C. K. Wilgus *et al.*, Soc. Econ. Palaeont. Mineral. Spec. Publ. 42, p. 71.

Harland, W. B., Armstrong, R. L., Cox, A. V., Craig, L. E., Smith, A. G. and Smith, D. G. (1990). *A Geologic Time Scale 1989*. Cambridge: Cambridge University Press.

Harland, W. R., Cox, A. V., Llwellyn, P. G., Pickton, C. A. G., Smith, A. G. and Walters, R. (1982). *A Geologic Time Scale*. Cambridge: Cambridge University Press.

Harrison, C. G. A. (1968). Evolutionary processes and reversals of the Earth's magnetic field. *Nature*, **217**, 46.

Harrison, C. G. A. (1974). The palaeomagnetic record from deep-sea sediment cores. *Earth Sci. Rev.*, **10**, 1.

Harrison, C. G. A. and Funnel, B. M. (1964). Relationship of palaeomagnetic reversals and micropalaeontology in two late Cenozoic cores from the Pacific Ocean. *Nature*, **204**, 566.

Hays, J. D., Imbrie, J. and Shackleton, N. J. (1976). Variations in the Earth's orbit: Pacemaker of the Ice ages? *Science*, **194**, 1121.

Hays, J. D. and Opdyke, N. D. (1967). Antarctic radiolaria, magnetic reversals and climatic change. *Science*, **158**, 1001.

Hays, J. D., Saito, T., Opdyke, N. D. and Burckle, L. (1969). Pliocene/Pleistocene sediments of the Equatorial Pacific; their palaeomagnetic, biostratigraphic and climatic record. *Geol. Soc. Am. Bull.*, **80**, 1481.

Heirtzler, J. R. (1970). The palaeomagnetic field as inferred from marine magnetic studies. *J. Geomag. Geoelect.*, **22**, 197.

Helsley, C. E. and Steiner, M. B. (1974). Palaeomagnetism of the lower Triassic Moenkopi Formation. *Geol. Soc. Am. Bull.*, **85**, 457.

Herbert, T. D. (1992). Paleomagnetic calibration of Milankovitch cyclicity in lower Cretaceous sediments. *Earth Planet. Sci. Lett.*, **112**, 15.

Herbert, T. D. and D'Hondt, S. L. (1990). Precessional climate cyclicity in Late-Cretaceous – early Tertiary marine sediments: a high resolution chronometer of Cretaceous–Tertiary boundary events. *Earth Planet. Sci. Lett.*, **99**, 263.

Hide, R. (1967). Motion of the Earth's core and mantle and variations of the main geomagnetic field. *Science*, **157**, 55.

Hilgen, F. J. (1991a). Astronomical calibration of Gauss to Matuyama sapropels in the Mediterranean and implication for the geomagnetic polarity timescale. *Earth Planet. Sci. Lett.*, **104**, 226.

Hilgen, F. J. (1991b). Extension of the astronomically calibrated (polarity) timescale to the Miocene/Pliocene boundary. *Earth Planet. Sci. Lett.*, **107**, 349.

Hoffman, A. (1985). Patterns of family extinction: dependence on definitions and geologic timescale. *Nature*, **315**, 659.

Honda, S., Yuen, D. A., Balachandar, S. and Reuteler, D. (1993). Three dimensional instabilities of mantle convection with multiple phase transitions. *Science*, **259**, 1308.

Imbrie, J., Hays, J. D., Martinson, D. G., McIntyre, A., Mix, A. C., Morley, J. J., Pisias, N. G., Prell, W. L. and Shackleton, N. J. (1984). The orbital theory of Pleistocene climate: support from a revised chronology of the marine $\delta^{18}O$ record. In *Milankovitch and Climate*, Part I, ed. A. L. Berger *et al.* Hingham, Mass.: D. Reidel.

Imbrie, J. and Imbrie, J. Z. (1980). Modelling the climatic response to orbital variations. *Science*, **207**, 943.

Imbrie, J., Mix, A. C. and Martinson, D. G. (1993). Milankovitch theory viewed from Devil's Hole. *Nature*, **363**, 531.

Irving, E. (1966). Palaeomagnetism of some carboniferous rocks of New South Wales and its relation to geological events. *J. Geophys. Res.*, **71**, 6025.

Irving, E., North, F. K. and Couillard, R. (1974). Oil, climate and tectonics. *Can. J. Earth Sci.*, **11**, 1.

Irving, E. and Pullaiah, C. (1976). Reversals of the geomagnetic field, magnetostratigraphy and relative magnitude of palaeosecular variation in the Phanerozoic. *Earth Sci. Rev.*, **12**, 35.

Jacobs, J. A. (1981). Heat flow and reversals of the Earth's magnetic field. *J. Geomag. Geoelect.*, **33**, 527.

Jenkyns, H. C. (1980). Cretaceous anoxic events: from continents to oceans. *J. Geol. Soc.*, **137**, 171.

Johnson, R. G. (1978). Initial glacial eustatic sea-level fall in the early Wisconsin calculated from temperature corrected isotope ratios in cores. *Geol. Soc. Am. Abstr.*, **10**, 429.

Johnson, R. G. (1982). Brunhes–Matuyama magnetic reversal dated at 790,000 yr BP by marine-astronomical correlations. *Quatern. Res.*, **17**, 135.

Jones, G. M. (1977). Thermal interaction of the core and the mantle and long term behaviour of the geomagnetic field. *J. Geophys. Res.*, **82**, 1703.

Kawai, N., Yaskawa, K., Nakajima, T., Torii, M. and Horie, S. (1972). Oscillating geomagnetic field with a recurring reversal discovered from Lake Biwa. *Proc. Japan Acad.*, **48**, 186.

Kawai, N., Yaskawa, K., Nakajima, T., Torii, M. and Natsuhara, N. (1975). Voices of geomagnetism from Lake Biwa. In *Palaeolimnology of Lake Biwa and the Japanese Pleistocene*, ed. S. Horie, Vol. 3, p. 143. Otsu: Kyota University.

Kellogg, L. H. and King, S. D. (1993). Effect of mantle plumes on the growth of D″ by reaction between the core and mantle. *Geophys. Res. Lett.*, **20**, 379.

Kennett, J. P. and Watkins, N. D. (1970). Geomagnetic polarity change, volcanic maxima and faunal extinction in the South Pacific. *Nature*, **227**, 930.

Kent, D. V. (1977). An estimate of the duration of the faunal change at the Cretaceous –Tertiary Boundary. *Geology*, **5**, 769.

Kent, D. V. (1982). Apparent correlation of palaeomagnetic intensity and climatic records in deep-sea sediments. *Nature*, **299**, 538.

Kitchell, J. A. and Pena, D. (1984). Periodicity of extinction in the geologic past: deterministic versus stochastic explanations. *Science*, **226**, 689.

Knittle, E. and Jeanloz, R. (1989). Simulating the core–mantle boundary: an experimental study of high-pressure reactions between silicates and liquid iron. *Geophys. Res. Lett.*, **16**, 609.

Knittle, E. and Jeanloz, R. (1991). Earth's core-mantle boundary: results of experiments at high pressures and temperatures. *Science*, **251**, 1438.

Kominz, M. A. (1984). Oceanic ridge volumes and sea-level change – an error analysis. In *Interregional Unconformities and Hydrocarbon Accumulation*, Mem. 36, ed. J. S. Schlee. Tulsa, Okla.: Amer. Assoc. Petrol. Geol.

Kominz, M. A., Heath, G. R., Ku, T.-L. and Pisias, N.-G. (1979). Brunhes timescales and the interpretation of climatic change. *Earth Planet. Sci. Lett.*, **45**, 394.

Kominz, M. A. and Pisias, N. G. (1979). Pleistocene climate: Deterministic or stochastic. *Science*, **204**, 171.

Kukla, G., Berger, A., Lotti, R. and Brown, J. (1981). Orbital signature of interglacials. *Nature*, **290**, 295.

Larson, R. L. (1991a). Latest pulse of Earth: Evidence for a mid-Cretaceous superplume. *Geology*, **19**, 547.

Larson, R. L. (1991b). Geological consequences of superplumes. *Geology*, **19**, 963.

Larson, R. L. and Olson, P. (1991). Mantle plumes control magnetic reversal frequency. *Earth Planet. Sci. Lett.*, **107**, 437.

Lay, T. and Helmberger, D. V. (1983). A lower mantle S wave triplication and the shear velocity structure of D″. *Geophys. J. Roy. Astr. Soc.*, **75**, 799.

Liu, H.-S. (1992). Frequency variations of the Earth's obliquity and the 100-k yr ice-age cycles. *Nature*, **358**, 397.

Loper, D. E. (1992). On the correlation between mantle plume flux and the frequency of reversals of the geomagnetic field. *Geophys. Res. Lett.*, **19**, 25.

Loper, D. E. and Eltayeb, I. A. (1986). On the stability of the D″ layer. *Geophys. Astrophys. Fluid Dyn.*, **36**, 229.

Loper, D. E. and McCartney, K. (1986). Mantle plumes and the periodicity of magnetic field reversals. *Geophys. Res. Lett.*, **13**, 1525.

Loper, D. E., McCartney, K. and Buzyna, G. (1988). A model of correlated episodicity in magnetic field reversals, climate and mass extinctions. *J. Geol.*, **96**, 1.

Loper, D. E. and Stacey, F. D. (1983). The dynamical and thermal structure of deep mantle plumes. *Phys. Earth Planet. Int.*, **3**, 304.

Ludwig, K. R., Simmons, K. R., Szabo, B. J., Winograd, I. J., Landwehr, J. M., Riggs, A. G. and Hoffman, R. J. (1992). Mass-spectrometric ^{230}Th–^{234}U–^{238}U dating of the Devil's Hole calcite vein. *Science*, **258**, 284.

Ludwig, K. R., Simmons, K. R., Winograd, I. J., Szabo, B. J., Landwehr, J. M. and Riggs, A. G. (1993a). Last interglacial in Devil's Hole, Reply. *Nature*, **362**, 596.

Ludwig, K. R., Simmons, K. R., Winograd, I. J., Szabo, B. J., Landwehr, J. M. and Riggs, A. G. (1993b). Dating of the Devil's Hole calcite vein, Response. *Science*, 259, 1626.

Lutz, T. M. (1987). Limitations to the statistical analysis of episodic and periodic models of geologic time series. *Geology*, 15, 1115.

Maasch, K. A. and Saltzman, B. (1990). A low order dynamical model of global climatic variability over the full Pleistocene. *J. Geophys. Res.*, 95, 1955.

McElhinny, M. W. (1971). Geomagnetic reversals during the Phanerozoic. *Science*, 172, 157.

McElhinny, M. W. (1973). *Palaeomagnetism and Plate Tectonics*. Cambridge: Cambridge University Press.

McFadden, P. L. and Merrill, R. T. (1984). Lower mantle convection and geomagnetism. *J. Geophys. Res.*, 89, 3354.

McFadden, P. L. and Merrill, R. T. (1986). Geodynamo energy source constraints from paleomagnetic data. *Phys. Earth Planet. Int.*, 43, 22.

Margolis, S. V. and Herman, Y. (1980). Northern hemisphere sea-ice and glacial development in the late Cenozoic. *Nature*, 286, 145.

Marzocchi, W., Mulargia, F. and Paruolo, P. (1992). The correlation of geomagnetic reversals and mean sea level in the last 150 m.y. *Earth Planet. Sci. Lett.*, 111, 383.

Matsuda, T., Endo, J., Osakabe, N. and Tonomura, A. (1983). Morphology and structure of biogenic magnetite particles. *Nature*, 302, 411.

Matthews, R. K. (1972). Dynamics of the ocean cryosphere system: Barbados data. *Quatern. Res.*, 2, 368.

de Menocal, P. B., Ruddiman, W. F. and Kent, D. V. (1990). Depth of post-depositional acquisition in deep-sea sediments: a case study of the Brunhes–Matuyama reversal and oxygen isotope Stage 19.1. *Earth Planet. Sci. Lett.*, 99, 1.

Milankovitch, M. (1941). *K. Serb. Akad. Beogr. Spec. Publ.*, 132 (translated by Israel Program for Scientific Translation, Jerusalem, 1969).

Moore, T. C., Pisias, N. G. and Dunn, D. A. (1982). Carbonate time series of the Quaternary and late Miocene sediments in the Pacific ocean; a spectral comparison. *Marine Geol.*, 46, 217.

Morelli, A. and Dziewonski, A. M. (1987). Topography of the core–mantle boundary and lateral homogeneity of the liquid core. *Nature*, 325, 678.

Morgan, W. J. (1971). Convection plumes in the lower mantle. *Nature*, 230, 42.

Morgan, W. J. (1972). Plate motions and deep mantle convection. *Geol. Soc. Am. Mem.*, 132, 7.

Muller, R. A. and Morris, D. E. (1986). Geomagnetic reversals from impacts on the Earth. *Geophys. Res. Lett.*, 13, 1177.

Negi, J. G., Agrawal, P. K., Pandey, O. P. and Singh, A. P. (1993). A possible K–T boundary bolide impact site offshore near Bombay and triggering of rapid Deccan volcanism. *Phys. Earth Planet. Int.*, 76, 189.

Negi, J. G. and Tiwari, R. K. (1983). Matching long term periodicities of geomagnetic reversals and galactic motions of the solar system. *Geophys. Res. Lett.*, 10, 713.

Newell, N. D. (1963). Crises in the history of life. *Sci. Am.*, 208 (2), 76.

Ninkovich, D., Opdyke, N., Heezen, B. C. and Foster, J. H. (1966). Palaeomagnetic stratigraphy and tephrochronology in North Pacific deep-sea sediments. *Earth Planet. Sci. Lett.*, 1, 476.

Oerlemans, J. (1980). Model experiments on the 100,000 yr glacial cycle. *Nature*, 287, 430.

Olson, P. and Hagee, V. L. (1990). Geomagnetic polarity reversals, transition field structure and convection in the outer core. *J. Geophys. Res.*, 95, 4609.

Olson, P., Schubert, G. and Anderson, C. (1987). Plume formations in the D″ layer and the roughness of the core–mantle boundary. *Nature*, 327, 409.

Pal, P. C. and Roberts, P. H. (1988). Long term polarity stability and strength of the geomagnetic dipole. *Nature*, 331, 702.

Pandey, O. P. and Negi, G. (1987). Global volcanism, biological mass extinctions and the galactic vertical motion of the solar system. *Geophys. J. Roy. Astr. Soc.*, 89, 857.

Panella, G. (1972). Paleontological evidence on the Earth's rotational history since early Precambrian. *Astrophys. Space Sci.*, 16, 212.

Rampino, M. R. (1979). Possible relationships between changes in global volume, geomagnetic excursions and the eccentricity of the Earth's orbit. *Geology*, 7, 584.

Rampino, M. R. (1981). Revised age estimates of Brunhes palaeomagnetic events: support for a link between geomagnetism and eccentricity. *Geophys. Res. Lett.*, 8, 1047.

Rampino, M. R. and Caldeira, K. (1993). Major episodes of geologic change: correlations, time structure and possible causes. *Earth Planet. Sci. Lett.*, 114, 215.

Rampino, M. R. and Stothers, R. B. (1984). Terrestrial mass extinctions, cometary impacts and the sun's motion perpendicular to the galactic plane. *Nature*, 308, 709.

Raup, D. M. and Sepkoski, J. J. (1984). Periodicity of extinctions in the geologic past. *Proc. Natl Acad. Sci. USA*, 81, 801.

Reid, G. C., Isaksen, I. S. A., Holzer, T. E. and Crutzen, P. J. (1976). Influence of ancient solar-proton events on the evolution of life. *Nature*, 259, 177.

Ruddiman, W. F. and McIntyre, A. (1976). Northeast Atlantic palaeoclimatic changes over the past 600,000 years. *Geol. Soc. Am. Mem.*, 145, 111.

Savin, S. M. (1977). The history of the Earth's surface temperature during the past 100 million years. *Am. Rev. Earth Planet. Sci.*, 5, 319.

Schneider, D. A. and Kent, D. V. (1990). Ivory Coast microtektites and geomagnetic reversals. *Geophys. Res. Lett.*, 17, 163.

Schneider, D. A., Kent, D. V. and Mello, G. A. (1992). A detailed chronology of the Australasian impact event, the Brunhes–Matuyama geomagnetic polarity reversal, and global climate change. *Earth Planet. Sci. Lett.*, 111, 395.

Schwartz, R. D. and James, P. B. (1984). Periodic mass extinctions and the sun's oscillation about the galactic plane. *Nature*, 308, 712.

Sepkoski, J. J. (1989). Periodicity in extinction and the problem of catastrophism in the history of life. *J. Geol. Soc.*, 146, 7.

Shackleton, N. J. (1976). Oxygen-isotope evidence relating to the end of the last interglacial at the substage 5e to 5d transition about 115,000 years ago. *Geol. Soc. Am. Abstr.*, 8, 1099.

Shackleton, N. J. (1977). Carbon 13 in Uvigerina: Tropical rainforest history and the Equatorial Pacific carbonate dissolution cycles. In *The Fate of Fossil Fuel CO_2 in the Oceans*, ed. N. R. Anderson and A. Malahoff, p. 401. New York: Plenum Press.

Shackleton, N. J. (1993). Last interglacial in Devil's Hole. *Nature*, 362, 596.

Shackleton, N. J., Berger, A. and Peltier, W. R. (1990). An alternative astronomical calibration of the Lower Pleistocene timescale based on ODP site 677. *Trans. Roy. Soc. Edinb., Earth Sci.*, 81, 251.

Shackleton, N. J. and Opdyke, N. D. (1976). Oxygen isotope and palaeomagnetic stratigraphy of Pacific core V28–239, Late Pliocene to Latest Pleistocene. *Geol. Soc. Am. Mem.*, 145, 449.

Sharpton, V. L., Dalrymple, G. B., Marín, L. E., Ryder, G., Schuraytz, B. C. and Urrutia-Fucugauchi, J. (1992). New links between the Chicxulub impact structure and the Cretaceous/Tertiary boundary. *Nature*, 359, 819.

Shaw, H. F. and Wasserburg, G. J. (1982). Age and provenance of the target materials for tektites and possible impactites as inferred from Sm–Nd and Rb–Sr systematics. *Earth Planet. Sci. Lett.*, 60, 155.

Sheridan, R. E. (1983). Phenomena of pulsation tectonics related to the breaking up of the eastern North American continental margin. *Tectonophysics*, 94, 169.

Shoemaker, E. M. (1984). Large body impacts through geologic time. In *Patterns of Change in Earth Evolution*, ed. H. D. Holland and A. F. Trendal, p 15. Berlin: Springer-Verlag.

Simpson, J. F. (1966). Evolutionary pulsation and geomagnetic polarity. *Geol. Soc. Am. Bull.*, 77, 197.

Sleep, N. H. (1988). Gradual entrainment of a chemical layer at the base of the mantle by overlying convection. *Geophys. J. Roy. Astr. Soc.*, 95, 437.

Smith, J. D. and Foster, J. H. (1969). Geomagnetic reversal in Brunhes normal polarity epoch. *Science*, 163, 565.

Stacey, F. D. (1975). Thermal regime of the Earth's interior. *Nature*, 255, 44.

Stacey, F. D. (1991). Effects on the core of structure within D″. *Geophys. Astrophys. Fluid Dyn.*, 60, 157.

Stacey, F. D. and Loper, D. E. (1983). The thermal boundary layer interpretation of D″ and its role as a plume source. *Phys. Earth Planet. Int.*, 33, 45.

Steinbach, V. and Yuen, D. A. (1992). The effects of multiple phase transitions on Venusian mantle convection. *Geophys. Res. Lett.*, 19, 2243.

Stothers, R. B. (1987). Do slow orbital periodicities appear in the record of Earth's magnetic reversals? *Geophys. Res. Lett.*, 14, 1087.

Swisher, C. C. III *et al.* (1992). Coeval ^{40}Ar/^{39}Ar ages of 65.0 million years ago from Chicxulub crater meltrock and Cretaceous–Tertiary boundary tektites. *Science*, 257, 954.

Tauxe, L., Deino, A. D., Behrensmeyer, A. K. and Potts, R. (1992). Pinning down the Brunhes/Matuyama and upper Jaramillo boundaries: a reconciliation of orbital and isotopic timescales. *Earth Planet. Sci. Lett.*, 109, 561.

Thaddeus, P. and Chanan, G. A. (1985). Cometary impacts, molecular clouds, and the motion of the sun perpendicular to the galactic plane. *Nature*, 314, 73.

Tissot, B. (1979). Effects on prolific petroleum source rocks and major rock deposits caused by sea-level changes. *Nature*, 277, 463.

Torbett, M. V. and Smoluchowski, R. (1984). Orbital stability of the unseen solar companion linked to periodic extinction events. *Nature*, 311, 641.

Uffen, R. J. (1963). Influence of the Earth's core on the origin and evolution of life. *Nature*, 198, 143.

Ulrych, T. J. (1972). Maximum entropy power spectrum of truncated sinusoids. *J. Geophys. Res.*, 77, 1396.

Verosub, K. L. and Banerjee, S. K. (1977). Geomagnetic excursions and their palaeomagnetic record. *Rev. Geophys. Space Phys.*, 15, 145.

Waddington, C. J. (1967). Palaeomagnetic field reversals and cosmic radiation. *Science*, 158, 913.

Watkins, N. D. and Goodell, H. G. (1967). Geomagnetic polarity change and faunal extinction in the Southern Ocean. *Science*, 156, 1083.

Weertman, J. (1961). Stability of ice age sheets. *J. Geophys. Res.*, 66, 3783.

Weinstein, S. A. (1993). Catastrophic overturn of the Earth's mantle driven by multiple phase changes and internal heat generation. *Geophys. Res. Lett.*, 20, 101.

Westermann, G. (1984). Gauging the duration of stages: a new appraisal for the Jurassic. *Episodes*, 7, 26.

Whitmire, D. P. and Jackson, IV, A. A. (1984). Are periodic mass extinctions driven by a distant solar companion? *Nature*, 308, 713.

Whitmire, D. P. and Matese, J. J. (1985). Periodic comet showers and planet X. *Nature*, 313, 36.

Wigley, T. M. L. (1976). Spectral analysis and the astronomical theory of climatic change. *Nature*, 264, 629.

Wilson, J. T. (1963). A possible origin of the Hawaiian islands. *Can. J. Phys.*, 41, 863.

Wilson, J. T. (1965). Convection currents and continental drift: evidence from ocean islands suggesting movement in the Earth. *Phil. Trans. Roy. Soc. A*, **258**, 145.

Winograd, I. J., Coplen, T. B., Landwehr, J. M., Riggs, A. C., Ludwig, K. R., Szabo, B. J., Kolesar, P. T. and Revesz, R. M. (1992). Continuous 500,000-year climate record from vein calcite in Devil's Hole, Nevada. *Science*, **258**, 255.

Wollin, G., Ericson, D. B. and Ryan, W. B. F. (1971a). Variations in magnetic intensity and climate changes. *Nature*, **232**, 549.

Wollin, G., Ericson, D. B., Ryan, W. B. F. and Foster, J. H. (1971b). Magnetism of the Earth and climatic changes. *Earth Planet. Sci. Lett.*, **12**, 175.

Wollin, G., Ryan, W. B. F. and Ericson, D. B. (1978). Climatic changes, magnetic intensity variations and fluctuations of the eccentricity of the Earth's orbit during the past 2,000,000 years and a mechanism which may be responsible for the relationship. *Earth Planet. Sci. Lett.*, **41**, 395.

Wollin, G., Ryan, W. B. F., Ericson, D. B. and Foster, J. H. (1977). Palaeoclimate, palaeomagnetism and the eccentricity of the Earth's orbit. *Geophys. Res. Lett.*, **4**, 267.

Yaskawa, K. (1974). Reversals, excursions and secular variations of the geomagnetic field in the Brunhes normal polarity epoch. In *Palaeolimnology of Lake Biwa and the Japanese Pleistocene*, ed. S. Horie, Vol. 2, p. 77. Otsu: Kyoto University.

Yuen, D. A. and Peltier, W. R. (1980). Mantle plumes and the thermal stability of the D″ layer. *Geophys. Res. Lett.*, **7**, 625.

Zhang, K. and Gubbins, D. (1992). On convection in the Earth's core driven by lateral temperature variations in the lower mantle. *Geophys. J. Int.*, **108**, 247.

Index

Printed in the United States
By Bookmasters